A Short Course in Differential Equations

A Short Course in
Differential
Equations
Sixth Edition

Earl D. Rainville

Late Professor of Mathematics
University of Michigan

Phillip E. Bedient

Professor of Mathematics
Franklin and Marshall College

Macmillan Publishing Co., Inc.
New York
Collier Macmillan Publishers
London

Copyright © 1981, Macmillan Publishing Co., Inc.

Printed in the United States of America

This book appears as the first 16 chapters of *Elementary
Differential Equations*, Sixth Edition © 1981 by Macmillan
Publishing Co., Inc. Earlier editions of *Elementary Differential
Equations* copyright 1949 and 1952, © 1958, and copyright
© 1964 and 1969 by Macmillan Publishing Co., Inc. Some
material is from *The Laplace Transform: An Introduction*,
© 1963 by Earl D. Rainville.

Macmillan Publishing Co., Inc.
866 Third Avenue, New York, New York 10022

Collier Macmillan Canada, Ltd.

Library of Congress Cataloging in Publication Data

Rainville, Earl David
  A short course in differential equations.

  Appears as the first 16 chapters of [the
authors'] Elementary differential equations, sixth
edition, c1981."
  Includes index.
  1. Differential equations. I. Bedient, Philip
Edward, 1922–   joint author. II. Title.
QA371.R32   1981   515.3'5   80-13416
ISBN 0-02-397760-4

Printing: 45678   Year: 45678

ISBN   0-02-397760-4

Preface

to the Sixth Edition

This new edition of Professor Rainville's book maintains the simple and direct style of earlier editions and makes some modest changes. The balance between developing techniques for solving equations and the theory necessary to support those techniques is essentially unchanged. However, the variety and number of applications has been increased and placed as early in the text as is feasible.

The omission of infinite series methods has made possible, even within the bounds of a small book, a thorough treatment of topics essential to a first course. A longer book, *Elementary Differential Equations*, Sixth Edition, contains an extensive treatment of both power series and Fourier series methods, and also an introduction to the application of the Laplace transform technique to boundary value problems involving partial differential equations.

The author wishes to thank those students and colleagues at Franklin and Marshall College whose suggestions and support have been most helpful.

Phillip E. Bedient

Lancaster, Pennsylvania

Contents

12   Inverse Transforms

13   Linear Systems of Equations

14   Electric Circuits and Networks

A Short Course in Differential Equations

Definitions, Elimination of Arbitrary Constants

1. Examples of differential equations

In physics, engineering, and chemistry and, increasingly, in such subjects as biology, physiology, and economics it is necessary to build a mathematical model to represent certain problems. It is often the case that these mathematical models involve the search for an unknown function that satisfies an equation in which derivatives of the unknown function play an important role. Such equations are called differential equations. As in equation (3) below, a derivative may be involved implicitly through the presence of differentials. Our aim is to find methods for solving differential equations; that is, to find the unknown function or functions that satisfy the differential equation.

The following are examples of differential equations:

$$\frac{dy}{dx} = \cos x, \tag{1}$$

$$\frac{d^2y}{dx^2} + k^2 y = 0, \tag{2}$$

$$(x^2 + y^2)\, dx - 2xy\, dy = 0, \tag{3}$$

$$\frac{\partial u}{\partial t} = h^2 \left(\frac{\partial^2 u}{\partial x^2} + \frac{\partial^2 u}{\partial y^2} \right), \tag{4}$$

$$L\frac{d^2i}{dt^2} + R\frac{di}{dt} + \frac{1}{C}i = E\omega \cos \omega t, \tag{5}$$

$$\frac{\partial^2 V}{\partial x^2} + \frac{\partial^2 V}{\partial y^2} = 0, \tag{6}$$

$$\left(\frac{d^2 w}{dx^2} \right)^3 - xy\frac{dw}{dx} + w = 0, \tag{7}$$

$$\frac{d^3 x}{dy^3} + x\frac{dx}{dy} - 4xy = 0, \tag{8}$$

$$\frac{d^2 y}{dx^2} + 7\left(\frac{dy}{dx} \right)^3 - 8y = 0, \tag{9}$$

$$\frac{d^2 y}{dt^2} + \frac{d^2 x}{dt^2} = x, \tag{10}$$

$$x\frac{\partial f}{\partial x} + y\frac{\partial f}{\partial y} = nf. \tag{11}$$

When an equation involves one or more derivatives with respect to a particular variable, that variable is called an *independent* variable. A variable is called *dependent* if a derivative of that variable occurs.

In the equation

$$L\frac{d^2i}{dt^2} + R\frac{di}{dt} + \frac{1}{C}i = E\omega \cos \omega t \tag{5}$$

i is the dependent variable, t the independent variable, and L, R, C, E, and ω are called parameters. The equation

$$\frac{\partial^2 V}{\partial x^2} + \frac{\partial^2 V}{\partial y^2} = 0 \tag{6}$$

has one dependent variable V and two independent variables.

Since the equation

$$(x^2 + y^2)\, dx - 2xy\, dy = 0 \tag{3}$$

may be written

$$x^2 + y^2 - 2xy\frac{dy}{dx} = 0$$

or

$$(x^2 + y^2)\frac{dx}{dy} - 2xy = 0,$$

we may consider either variable to be dependent, the other being the independent one.

Exercise

Identify the independent variables, the dependent variables, and the parameters in the equations given as examples in this section.

2. Definitions

The *order* of a differential equation is the order of the highest-ordered derivative appearing in the equation. For instance,

$$\frac{d^2 y}{dx^2} + 2b\left(\frac{dy}{dx}\right)^3 + y = 0 \tag{1}$$

is an equation of "order two." It is also referred to as a "second-order equation."

More generally, the equation

$$F(x, y, y', \ldots, y^{(n)}) = 0 \tag{2}$$

is called an "*n*th-order" ordinary differential equation. Under suitable restrictions on the function F, equation (2) can be solved explicitly for $y^{(n)}$ in terms of the other $n + 1$ variables $x, y, y', \ldots, y^{(n-1)}$, to obtain

$$y^{(n)} = f(x, y, y', \ldots, y^{(n-1)}). \tag{3}$$

For the purposes of this book we shall assume that this is always possible. Otherwise, an equation of the form of equation (2) may actually represent more than one equation of the form of equation (3).

For example, the equation

$$x(y')^2 + 4y' - 6x^2 = 0$$

actually represents the two different equations,

$$y' = \frac{-2 + \sqrt{4 + 6x^3}}{x} \quad \text{or} \quad y' = \frac{-2 - \sqrt{4 + 6x^3}}{x}.$$

A function ϕ, defined on an interval $a < x < b$, is called a solution of the differential equation (3), provided that the n derivatives of the function exist on the interval $a < x < b$ and

$$\phi^{(n)}(x) = f(x, \phi(x), \phi'(x), \dots, \phi^{(n-1)}(x)),$$

for every x in $a < x < b$.

For example, let us verify that

$$y = e^{2x}$$

is a solution of the equation

$$\frac{d^2y}{dx^2} + \frac{dy}{dx} - 6y = 0. \tag{4}$$

We substitute our tentative solution into the left member of equation (4) and find that

$$\frac{d^2y}{dx^2} + \frac{dy}{dx} - 6y = 4\,e^{2x} + 2\,e^{2x} - 6\,e^{2x} \equiv 0,$$

which completes the desired verification.

All of the equations we shall consider in Chapter 2 are of order one, and hence may be written

$$\frac{dy}{dx} = f(x, y).$$

For such equations it is sometimes convenient to use the definitions of elementary calculus to write the equation in the form

$$M(x, y)\,dx + N(x, y)\,dy = 0. \tag{5}$$

A very important concept in the study of differential equations is that of linearity. An ordinary differential equation of order n is called linear if it may be written in the form

$$b_0(x)\frac{d^ny}{dx^n} + b_1(x)\frac{d^{n-1}y}{dx^{n-1}} + \cdots + b_{n-1}(x)\frac{dy}{dx} + b_n(x)y = R(x).$$

For example, equation (1) above is nonlinear, and equation (4) is linear. The equation

$$x^2y'' + xy' + (x^2 - n^2)y = 4x^3$$

is also linear.

The notion of linearity may be extended to partial differential equations. For example, the equation

$$b_0(x, y)\frac{\partial w}{\partial x} + b_1(x, y)\frac{\partial w}{\partial y} = R(x, y)$$

is the general first-order linear partial differential equation with two independent variables and

$$b_0(x, y)\frac{\partial^2 w}{\partial x^2} + b_1(x, y)\frac{\partial^2 w}{\partial x\,\partial y} + b_2(x, y)\frac{\partial^2 w}{\partial y^2}$$

$$+ b_3(x, y)\frac{\partial w}{\partial x} + b_4(x, y)\frac{\partial w}{\partial y} + b_5(x, y)w = R(x, y)$$

is the general second-order linear partial differential equation with two independent variables.

Exercises

For each of the following, state whether the equation is ordinary or partial, linear or nonlinear, and give its order.

1. $\dfrac{d^2x}{dt^2} + k^2x = 0.$

2. $\dfrac{\partial^2 w}{\partial t^2} = a^2\dfrac{\partial^2 w}{\partial x^2}.$

3. $(x^2 + y^2)\,dx + 2xy\,dy = 0.$

4. $y' + P(x)y = Q(x).$

5. $y''' - 3y' + 2y = 0.$

6. $yy'' = x.$

7. $\dfrac{\partial^2 u}{\partial x^2} + \dfrac{\partial^2 u}{\partial y^2} + \dfrac{\partial^2 u}{\partial z^2} = 0.$

8. $\dfrac{d^4 y}{dx^4} = w(x).$

9. $x\dfrac{d^2 y}{dt^2} - y\dfrac{d^2 x}{dt^2} = c_1.$

10. $L\dfrac{di}{dt} + Ri = E.$

11. $(x + y)\,dx + (3x^2 - 1)\,dy = 0.$

12. $x(y'')^3 + (y')^4 - y = 0.$

13. $\left(\dfrac{d^3 w}{dx^3}\right)^2 - 2\left(\dfrac{dw}{dx}\right)^4 + yw = 0.$

14. $\dfrac{dy}{dx} = 1 - xy + y^2.$

15. $y'' + 2y' - 8y = x^2 + \cos x.$

16. $a\,da + b\,db = 0.$

3. The elimination of arbitrary constants

In practice, differential equations arise in many ways, some of which we shall encounter later. There is one way of arriving at a differential equation, however, that is useful in that it gives us a feeling for the kinds of solutions to be expected. In this section we shall start with a relation involving arbitrary constants and, by elimination of those arbitrary constants, come to a differential equation consistent with the original relation. In a sense we start with the answer and find the problem.

Methods for the elimination of arbitrary constants vary with the way in which the constants enter the given relation. A method that is efficient for one problem may be poor for another. One fact persists throughout. Because each differentiation yields a new relation, the number of derivatives that need be used is the same as the number of arbitrary constants to be eliminated. We shall in each case determine the differential equation that is

(a) Of order equal to the number of arbitrary constants in the given relation.
(b) Consistent with that relation.
(c) Free from arbitrary constants.

EXAMPLE (a): Eliminate* the arbitrary constants c_1 and c_2 from the relation

$$y = c_1 e^{-2x} + c_2 e^{3x}. \tag{1}$$

Since two constants are to be eliminated, obtain the two derivatives,

$$y' = -2c_1 e^{-2x} + 3c_2 e^{3x}, \tag{2}$$

$$y'' = 4c_1 e^{-2x} + 9c_2 e^{3x}. \tag{3}$$

The elimination of c_1 from equations (2) and (3) yields

$$y'' + 2y' = 15c_2 e^{3x};$$

the elimination of c_1 from equations (1) and (2) yields

$$y' + 2y = 5c_2 e^{3x}.$$

Hence

$$y'' + 2y' = 3(y' + 2y),$$

or

$$y'' - y' - 6y = 0.$$

Another method for obtaining the differential equation in this example proceeds as follows: We know from a theorem in elementary algebra that the three equations (1), (2), and (3), considered as equations in the two unknowns c_1 and c_2, can have solutions only if

$$\begin{vmatrix} -y & e^{-2x} & e^{3x} \\ -y' & -2e^{-2x} & 3e^{3x} \\ -y'' & 4e^{-2x} & 9e^{3x} \end{vmatrix} = 0. \tag{4}$$

Since e^{-2x} and e^{3x} cannot be zero, equation (4) may be rewritten, with the factors -1, e^{-2x}, and e^{3x} removed, as

---

* By differentiations and pertinent legitimate mathematical procedures. Elimination by erasure, for instance, is not permitted.

$$\begin{vmatrix} y & 1 & 1 \\ y' & -2 & 3 \\ y'' & 4 & 9 \end{vmatrix} = 0$$

from which the differential equation

$$y'' - y' - 6y = 0$$

follows immediately.

This latter method has the advantage of making it easy to see that the elimination of the constants $c_1, c_2, \ldots, c_n$ from a relation of the form

$$y = c_1 e^{m_1 x} + c_2 e^{m_2 x} + \cdots + c_n e^{m_n x}$$

will always lead to a linear differential equation

$$a_0 \frac{d^n y}{dx^n} + a_1 \frac{d^{n-1} y}{dx^{n-1}} + \cdots + a_{n-1} \frac{dy}{dx} + a_n y = 0,$$

in which the coefficients $a_0, a_1, \ldots, a_n$ are constants. The study of such differential equations will receive much of our attention.

EXAMPLE (b): Eliminate the constant a from the equation

$$(x - a)^2 + y^2 = a^2.$$

Direct differentiation of the relation yields

$$2(x - a) + 2yy' = 0,$$

from which

$$a = x + yy'.$$

Therefore, using the original equation, we find that

$$(yy')^2 + y^2 = (x + yy')^2,$$

or

$$y^2 = x^2 + 2xyy',$$

which may be written in the form

$$(x^2 - y^2) \, dx + 2xy \, dy = 0.$$

Another method will be used in this example as an illustration of a device that is often helpful. The method is based upon the isolation of an arbitrary constant.

The equation

$$(x - a)^2 + y^2 = a^2$$

may be put in the form

$$x^2 + y^2 - 2ax = 0,$$

or

$$\frac{x^2 + y^2}{x} = 2a.$$

Then differentiation of both members leads to

$$\frac{x(2x\,dx + 2y\,dy) - (x^2 + y^2)\,dx}{x^2} = 0,$$

or

$$(x^2 - y^2)\,dx + 2xy\,dy = 0,$$

as desired.

It is interesting to speculate here about the significance of $x = 0$ upon the argument just used. The student should draw a few of the members of this family of circles and observe what is peculiar about their behavior at $x = 0$.

EXAMPLE (c):   Eliminate B and α from the relation

$$x = B\cos(\omega t + \alpha), \qquad (5)$$

in which ω is a parameter (not to be eliminated).

First we obtain two derivatives of x with respect to t:

$$\frac{dx}{dt} = -\omega B\sin(\omega t + \alpha), \qquad (6)$$

$$\frac{d^2x}{dt^2} = -\omega^2 B\cos(\omega t + \alpha). \qquad (7)$$

Comparison of equations (5) and (7) shows at once that

$$\frac{d^2x}{dt^2} + \omega^2 x = 0.$$

EXAMPLE (d):   Eliminate c from the equation

$$cxy + c^2x + 4 = 0.$$

At once we get

$$c(y + xy') + c^2 = 0.$$

Since $c \neq 0$,

$$c = -(y + xy')$$

and substitution into the original equation leads us to the result

$$x^3(y')^2 + x^2yy' + 4 = 0.$$

Our examples suggest that in a certain sense the totality of solutions of an nth-order equation depends on n arbitrary constants. The sense in which this is true will be stated in Sections 5 and 12.

Exercises

In each of the following eliminate the arbitrary constants.

1. $x \sin y + x^2 y = c$.   ANS. $(\sin y + 2xy) dx + (x \cos y + x^2) dy = 0$.
2. $3x^2 - xy^2 = c$.   ANS. $(6x - y^2) dx - 2xy \, dy = 0$.
3. $xy^2 - 1 = cy$.   ANS. $y^3 \, dx + (xy^2 + 1) dy = 0$.
4. $cx^2 + x + y^2 = 0$.   ANS. $(x + 2y^2) dx - 2xy \, dy = 0$.

5. $x = A \sin(\omega t + \beta)$; ω a parameter, not to be eliminated.   ANS. $\dfrac{d^2 x}{dt^2} + \omega^2 x = 0$.

6. $x = c_1 \cos \omega t + c_2 \sin \omega t$; ω a parameter.   ANS. $\dfrac{d^2 x}{dt^2} + \omega^2 x = 0$.

7. $y = cx + c^2 + 1$.   ANS. $y = xy' + (y')^2 + 1$.

8. $y = mx + \dfrac{h}{m}$; h a parameter, m to be eliminated.   ANS. $y = xy' + \dfrac{h}{y'}$.

9. $y^2 = 4ax$.   ANS. $2x \, dy - y \, dx = 0$.
10. $y = ax^2 + bx + c$.   ANS. $y''' = 0$.
11. $y = c_1 + c_2 e^{3x}$.   ANS. $y'' - 3y' = 0$.
12. $y = 4 + c_1 e^{3x}$.   ANS. $y' - 3y = -12$.
13. $y = c_1 + c_2 e^{-4x}$.   ANS. $y'' + 4y' = 0$.
14. $y = c_1 e^x + c_2 e^{-x}$.   ANS. $y'' - y = 0$.
15. $y = x + c_1 e^x + c_2 e^{-x}$.   ANS. $y'' - y = -x$.
16. $y = c_1 e^{2x} + c_2 e^{3x}$.   ANS. $y'' - 5y' + 6y = 0$.
17. $y = x^2 + c_1 e^{2x} + c_2 e^{3x}$.   ANS. $y'' - 5y' + 6y = 6x^2 - 10x + 2$.
18. $y = c_1 e^x + c_2 x e^x$.   ANS. $y'' - 2y' + y = 0$.
19. $y = A e^{2x} + Bx e^{2x}$.   ANS. $y'' - 4y' + 4y = 0$.
20. $y = c_1 e^{2x} \cos 3x + c_2 e^{2x} \sin 3x$.   ANS. $y'' - 4y' + 13y = 0$.
21. $y = c_1 e^{ax} \cos bx + c_2 e^{ax} \sin bx$; a and b are parameters.
     ANS. $y'' - 2ay' + (a^2 + b^2)y = 0$.
22. $y = c_1 x + c_2 e^x$.   ANS. $(x - 1)y'' - xy' + y = 0$.
23. $y = c_1 x^2 + c_2 e^{-x}$.   ANS. $x(x + 2)y'' + (x^2 - 2)y' - 2(x + 1)y = 0$.
24. $y = x^2 + c_1 x + c_2 e^{-x}$.   ANS. $(x + 1)y'' + xy' - y = x^2 + 2x + 2$.
25. $y = c_1 x^2 + c_2 e^{2x}$.   ANS. $x(1 - x)y'' + (2x^2 - 1)y' - 2(2x - 1)y = 0$.

4. Families of curves

An equation involving a parameter, as well as one or both of the co-ordinates of a point in a plane, may represent a family of curves, one curve corresponding to each value of the parameter. For instance, the equation

$$(x - c)^2 + (y - c)^2 = 2c^2, \tag{1}$$

or

$$x^2 + y^2 - 2c(x + y) = 0, \tag{2}$$

may be interpreted as the equation of a family of circles, each having its center on the line $y = x$ and each passing through the origin. Figure 1 shows several elements, or members, of this family.

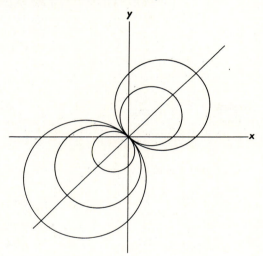

FIGURE 1

If the constant c in equation (1) or in equation (2) is treated as an arbitrary constant and eliminated as in the preceding section, the result is called the *differential equation of the family represented by equation* (1). In this example, the elimination of c is easily performed by isolating c, then differentiating throughout the equation with respect to x. Thus, from

$$\frac{x^2 + y^2}{x + y} = 2c$$

we find that

$$\frac{(x + y)(2x\,dx + 2y\,dy) - (x^2 + y^2)(dx + dy)}{(x + y)^2} = 0.$$

Therefore

$$(x^2 + 2xy - y^2)\,dx - (x^2 - 2xy - y^2)\,dy = 0 \tag{3}$$

is the differential equation of the family of circles represented by equation (1).

Note that equation (3) associates a definite slope with each point (x, y) in the plane

$$\frac{dy}{dx} = \frac{x^2 + 2xy - y^2}{x^2 - 2xy - y^2}, \tag{4}$$

except where the denominator on the right in equation (4) vanishes. When the denominator vanishes, the curve passing through that point must have a vertical tangent. From

$$x^2 - 2xy - y^2 = 0$$

we see that

$$y = (-1 + \sqrt{2})x \tag{5}$$

or

$$y = (-1 - \sqrt{2})x. \tag{6}$$

In Figure 2, the straight lines (5) and (6) appear along with the family (1). It is seen that the lines (5) and (6) cut the members of the family of circles in precisely those points of vertical tangency.

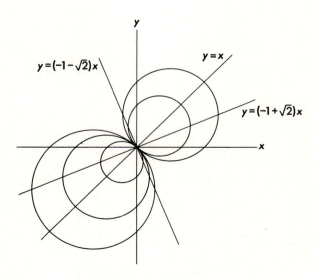

FIGURE 2

For a two-parameter family of curves, the differential equation will be of order two, and such a simple geometric interpretation is not available.

EXAMPLE (a):   Find the differential equation of the family of parabolas (Figure 3), having their vertices at the origin and their foci on the y-axis.

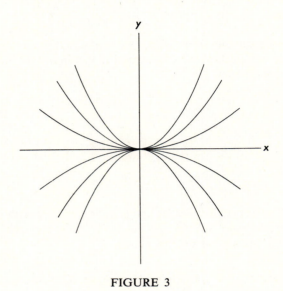

FIGURE 3

An equation of this family of parabolas is

$$y = ax^2, \tag{7}$$

so that

$$y' = 2ax. \tag{8}$$

It follows that

$$xy' - 2y = 0 \tag{9}$$

is the differential equation of the family. We note that (9) is a first-order linear differential equation.

EXAMPLE (b):   Find the differential equation of the family of circles (Figure 4) having their centers on the y-axis.

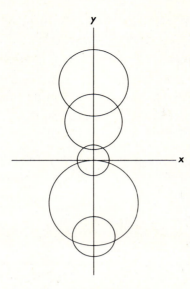

FIGURE 4

Because a member of the family of circles of this example may have its center anywhere on the y-axis and its radius of any magnitude, we are dealing with the two-parameter family

$$x^2 + (y - b)^2 = r^2. \tag{10}$$

We shall eliminate both b and r and obtain a second-order differential equation for the family (10).

At once

$$x + (y - b)y' = 0,$$

from which

$$\frac{x + yy'}{y'} = b.$$

Then

$$\frac{y'[1 + yy'' + (y')^2] - y''(x + yy')}{(y')^2} = 0,$$

so the desired differential equation is

$$xy'' - (y')^3 - y' = 0.$$

Exercises

In each exercise, obtain the differential equation of the family of plane curves described and sketch several representative members of the family.

1. Straight lines through the origin.   ANS. $y\,dx - x\,dy = 0.$

2. Straight lines through the fixed point (h, k); h and k not to be eliminated.

ANS. $(y - k)\,dx - (x - h)\,dy = 0.$

3. Straight lines with slope and y-intercept equal.   ANS. $y\,dx - (x + 1)\,dy = 0.$

4. Straight lines with slope and x-intercept equal.   ANS. $(y')^2 = xy' - y.$

5. Straight lines with algebraic sum of the intercepts fixed as k.

ANS. $(xy' - y)(y' - 1) + ky' = 0.$

6. Straight lines at a fixed distance p from the origin.

ANS. $(xy' - y)^2 = p^2[1 + (y')^2].$

7. Circles with center at the origin.   ANS. $x\,dx + y\,dy = 0.$

8. Circles with center on the x-axis.   ANS. $yy'' + (y')^2 + 1 = 0.$

9. Circles with fixed radius r and tangent to the x-axis.

ANS. $(y \pm r)^2(y')^2 + y^2 \pm 2ry = 0.$

10. Circles tangent to the x-axis.   ANS. $[1 + (y')^2]^3 = [yy'' + 1 + (y')^2]^2.$

11. Circles with center on the line $y = -x$, and passing through the origin.

ANS. $(x^2 - 2xy - y^2)\,dx + (x^2 + 2xy - y^2)\,dy = 0.$

12. Circles of radius unity. Use the fact that the radius of curvature is 1.

ANS. $(y'')^2 = [1 + (y')^2]^3.$

13. All circles. Use the curvature.   ANS. $y'''[1 + (y')^2] = 3y'(y'')^2.$

14. Parabolas with vertex on the x-axis, with axis parallel to the y-axis, and with distance from focus to vertex fixed as a.   ANS. $a(y')^2 = y.$

15. Parabolas with vertex on the y-axis, with axis parallel to the x-axis, and with distance from focus to vertex fixed as a.   ANS. $x(y')^2 = a.$

16. Parabolas with axis parallel to the y-axis and with distance from vertex to focus fixed as a.   ANS. $2ay'' = 1.$

17. Parabolas with axis parallel to the x-axis and with distance from vertex to focus fixed as a.   ANS. $2ay'' + (y')^3 = 0.$

18. Work exercise 17, using differentiation with respect to y.   ANS. $2a\dfrac{d^2x}{dy^2} = 1.$

19. Use the fact that

$$\frac{d^2x}{dy^2} = \frac{d}{dy}\left(\frac{dx}{dy}\right) = \frac{dx}{dy}\frac{d}{dx}\left(\frac{dx}{dy}\right) = \frac{dx}{dy}\frac{d}{dx}\left(\frac{dy}{dx}\right)^{-1} = \frac{-y''}{(y')^3}$$

to prove that the answers to exercises 17 and 18 are equivalent.

20. Parabolas with vertex and focus on the x-axis.   ANS. $yy'' + (y')^2 = 0.$
21. Parabolas with axis parallel to the x-axis.   ANS. $y'y''' - 3(y'')^2 = 0.$
22. Central conics with center at the origin and vertices on the coordinate axes.
$$\text{ANS.}\quad xyy'' + x(y')^2 - yy' = 0.$$

23. The confocal central conics

$$\frac{x^2}{a^2 + \lambda} + \frac{y^2}{b^2 + \lambda} = 1$$

with a and b held fixed.   ANS. $(xy' - y)(yy' + x) = (a^2 - b^2)y'.$
24. The cubics $cy^2 = x^2(x - a)$ with a held fixed.   ANS. $2x(x - a)y' = y(3x - 2a).$
25. The cubics of exercise 24 with c held fixed and a to be eliminated.
$$\text{ANS.}\quad 2cy(xy' - y) = x^3.$$
26. The quartics $c^2y^2 = x(x - a)^3$ with a held fixed.   ANS. $2x(x - a)y' = y(4x - a).$
27. The quartics of exercise 26 with c held fixed and a to be eliminated.
$$\text{ANS.}\quad c^2(2xy' - y)^3 = 27x^4y.$$

28. The strophoids $y^2 = \dfrac{x^2(a + x)}{a - x}.$ ANS. $(x^4 - 4x^2y^2 - y^4)\,dx + 4x^3y\,dy = 0.$

29. The cissoids $y^2 = \dfrac{x^3}{a - x}.$ ANS. $2x^3y' = y(y^2 + 3x^2).$

30. The trisectrices of Maclaurin $y^2(a + x) = x^2(3a - x).$
$$\text{ANS.}\quad (3x^4 - 6x^2y^2 - y^4)\,dx + 8x^3y\,dy = 0.$$
31. Circles through the intersections of the circle $x^2 + y^2 = 1$ and the line $y = x$. Use the "$u + kv$" form; that is, the equation

$$x^2 + y^2 - 1 + k(y - x) = 0.$$

$$\text{ANS.}\quad (x^2 - 2xy - y^2 + 1)\,dx + (x^2 + 2xy - y^2 - 1)\,dy = 0.$$
32. Circles through the fixed points $(a, 0)$ and $(-a, 0)$. Use the method of exercise 31.
$$\text{ANS.}\quad 2xy\,dx + (y^2 + a^2 - x^2)\,dy = 0.$$
33. The circles $r = 2a(\sin\theta - \cos\theta).$ ANS. $(\cos\theta - \sin\theta)\,dr + r(\cos\theta + \sin\theta)\,d\theta = 0.$
34. The cardioids $r = a(1 - \sin\theta).$ ANS. $(1 - \sin\theta)\,dr + r\cos\theta\,d\theta = 0.$
35. The cissoids $r = a\sin\theta\tan\theta.$ (See exercise 29.)
$$\text{ANS.}\quad \sin\theta\cos\theta\,dr - r(1 + \cos^2\theta)\,d\theta = 0.$$

36. The strophoids $r = a(\sec\theta + \tan\theta).$ ANS. $\dfrac{dr}{d\theta} = r\sec\theta.$

37. The trisectrices of Maclaurin $r = a(4\cos\theta - \sec\theta).$ (See exercise 30.)
$$\text{ANS.}\quad \cos\theta(4\cos^2\theta - 1)\,dr + r\sin\theta(4\cos^2\theta + 1)\,d\theta = 0.$$

Equations of Order One

5. The isoclines of an equation

In this chapter we shall study several elementary methods for solving first-order differential equations. Before we begin that study, we shall take a brief look at some of the basic geometrical ideas that are involved.

Consider the equation of order one

$$\frac{dy}{dx} = f(x, y). \tag{1}$$

We can think of equation (1) as a machine that assigns to each point (a, b) in the domain of f some direction with slope $f(a, b)$. We can thus speak of the direction field of the differential equation. In a real sense any solution of equation (1) must have a graph, which at each point has the direction equation (1) requires.

One way to visualize this basic idea is to draw a short mark at each point to indicate the direction associated with that point. This can be done rather systematically by first drawing curves called isoclines, that is, curves along which the direction indicated by equation (1) is fixed.

EXAMPLE (a): Consider the equation

$$\frac{dy}{dx} = y. \tag{2}$$

The isoclines are the straight lines $f(x, y) = y = c$. For each value of c we obtain a line in which, at each point, the direction dictated by the differential equation is that number c. For example, at each point along the line $y = 1$, equation (2) determines a direction of slope 1. In Figure 5 we have drawn several of these isoclines, indicating the direction associated with each isocline by short markers. If one starts at any point in the plane and moves along a curve whose direction is always in the direction of the direction marks, then a solution curve is obtained. Several solution curves have been drawn in Figure 5.

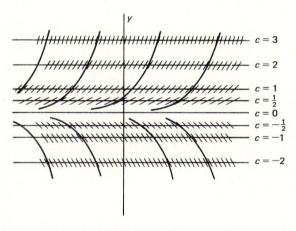

FIGURE 5

EXAMPLE (b): Use the method of isoclines to sketch some of the solution curves for the equation

$$\frac{dy}{dx} = x^2 + y^2. \tag{3}$$

Here the isoclines will be the circles $x^2 + y^2 = c$, with $c > 0$. When $c = \frac{1}{4}$, the isocline has radius $\frac{1}{2}$; for $c = 1$, radius 1; for $c = 4$, radius 2. In Figure 6 we have drawn these isoclines, marking each of them with the appropriate direction indicator, and finally sketching several curves that represent solutions of equation (3).

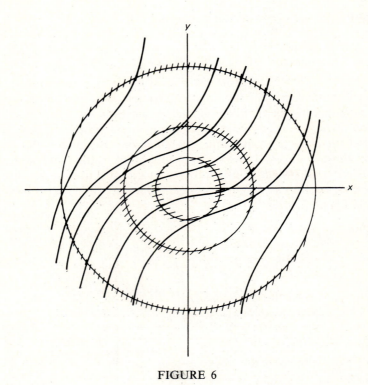

FIGURE 6

Exercises

For each of the following differential equations, draw several isoclines with appropriate direction markers and sketch several solution curves for the equation.

1. $\dfrac{dy}{dx} = x.$

2. $\dfrac{dy}{dx} = \dfrac{y}{x}.$

3. $\dfrac{dy}{dx} = \dfrac{2y}{x}.$

4. $\dfrac{dy}{dx} = y - x.$

5. $\dfrac{dy}{dx} = x + y + 1.$

6. $\dfrac{dy}{dx} = x - y - 1.$

7. $\dfrac{dy}{dx} = 2x - y.$

8. $\dfrac{dy}{dx} = y - x^2.$

6. An existence theorem

It should be clear even to the casual reader that drawing isoclines is not a practical tool for finding solutions to any differential equations other than those involving the simplest of functions. Before discussing some of the analytic techniques for finding solutions, we will state an important theorem concerning the existence and uniqueness of solutions, a theorem which will be discussed in detail in Chapter 15.

Consider the equation of order one

$$\frac{dy}{dx} = f(x, y). \tag{1}$$

Let T denote the rectangular region defined by

$$|x - x_0| \leqq a \qquad \text{and} \qquad |y - y_0| \leqq b,$$

a region with the point (x_0, y_0) at its center. Suppose that f and $\partial f / \partial y$ are continuous functions of x and y in T.

Under the conditions imposed on $f(x, y)$ above, an interval exists about $x_0, |x - x_0| \leqq h$, and a function $y(x)$ which has the properties:

(a) $y = y(x)$ is a solution of equation (1) on the interval $|x - x_0| \leqq h$;
(b) On the interval $|x - x_0| \leqq h$, $y(x)$ satisfies the inequality $|y(x) - y_0| \leqq b$;
(c) At $x = x_0$, $y = y(x_0) = y_0$;
(d) $y(x)$ is unique on the interval $|x - x_0| \leqq h$ in the sense that it is the only function that has all of the properties (a), (b), and (c).

The interval $|x - x_0| \leqq h$ may or may not need to be smaller than the interval $|x - x_0| \leqq a$ over which conditions were imposed upon $f(x, y)$.

In rough language, the theorem states that if $f(x, y)$ is sufficiently well behaved near the point (x_0, y_0) then the differential equation

$$\frac{dy}{dx} = f(x, y) \tag{1}$$

has a solution that passes through the point (x_0, y_0) and that solution is unique near (x_0, y_0).

In Example (a) of Section 5 we can consider (x_0, y_0) to be any point in the plane, since $f(x, y) = y$ and its partial derivative $\partial f / \partial y = 1$ are continuous in any rectangle. Therefore our existence theorem assures us that through any point (x_0, y_0) there is exactly one solution, a situation that we assumed when we sketched the solution curves in Figure 5.

Again in Example (b), Section 5, the function $f(x, y) = x^2 + y^2$ and its partial derivative $\partial f / \partial y = 2y$ are continuous in any rectangle. It follows that through any point (x_0, y_0) in the plane there is exactly one solution curve, a fact that is suggested by the solution curves in Figure 6.

7. Separation of variables

We begin our study of the methods for solving first-order equations by studying an equation of the form

$$M \, dx + N \, dy = 0,$$

where M and N may be functions of both x and y. Some equations of this type are so simple that they can be put in the form

$$A(x) \, dx + B(y) \, dy = 0; \tag{1}$$

that is, the variables can be separated. Then a solution can be written at once. For it is only a matter of finding a function F whose total differential is the left member of (1). Then $F = c$, where c is an arbitrary constant, is the desired result.

EXAMPLE (a):   Solve the equation

$$\frac{dy}{dx} = \frac{2y}{x}, \qquad \text{for } x > 0 \text{ and } y > 0. \tag{2}$$

We note that for the function in equation (2), the theorem of the previous section applies and assures the existence of a unique continuous solution through any point in the first quadrant. By separating the variables we can write

$$\frac{dy}{y} = \frac{2 \, dx}{x}.$$

Hence we obtain a family of solutions

$$\ln |y| = 2 \ln |x| + c \tag{3}$$

or, because we are in the first quadrant,

$$y = e^c x^2. \tag{4}$$

If we now put $c_1 = e^c$, we can write

$$y = c_1 x^2, \qquad c_1 > 0. \tag{5}$$

EXAMPLE (b):   Solve the equation of the previous example for $x \neq 0$.

The argument now must be taken in two parts. First, if $y \neq 0$, we can proceed as before to equation (3). However, equation (5) must be written

$$|y| = c_1 x^2, \qquad c_1 > 0. \tag{6}$$

Second, if $y = 0$, we see immediately that since $x \neq 0$, $y = 0$ is a solution of the differential equation (2).

As a matter of convenience the solutions given by equation (6) are usually written

$$y = c_2 x^2, \tag{7}$$

where c_2 is taken to be an arbitrary real number. Indeed this form for the solutions incorporates the special case $y = 0$. Thus we say that the family of curves given in Figure 3, page 12, represents a family of solution curves for the differential equation (2).

We must be cautious, however. The function defined by

$$g(x) = x^2, \qquad x > 0$$
$$= -4x^2, \qquad x \leq 0,$$

obtained by piecing together two different parabolic arcs could also be considered a solution of the differential equation, even though this function is not included in the family of equation (7). The uniqueness statement in the theorem of Section 6 indicates that, as long as we restrict our attention to a point (x_0, y_0) with $x_0 \neq 0$ and consider a rectangle with center at (x_0, y_0) containing no points at which $x = 0$, then in that rectangle there is a unique solution that passes through (x_0, y_0) and is continuous in the rectangle.

EXAMPLE (c):   Solve the equation

$$(1 + y^2)\, dx + (1 + x^2)\, dy = 0, \tag{8}$$

with the "initial condition" that when $x = 0$, $y = -1$.

If we write this equation in the form

$$\frac{dy}{dx} = \frac{-(1 + y^2)}{1 + x^2}$$

we observe that the right member and its partial derivative with respect to y are continuous near $(0, -1)$. It follows that a unique solution exists for equation (8) that passes through the point $(0, -1)$.

From the differential equation we get

$$\frac{dx}{1 + x^2} + \frac{dy}{1 + y^2} = 0$$

from which it follows at once that

$$\text{arc tan } x + \text{arc tan } y = c. \tag{9}$$

In the set of solutions (9), each "arc tan" stands for the principal value of the inverse tangent and is subject to the restriction

$$-\tfrac{1}{2}\pi < \text{arc tan } x < \tfrac{1}{2}\pi.$$

The initial condition that $y = -1$ when $x = 0$ permits us to determine the value of c that must be used to obtain the particular solution desired here. Since arc tan $0 = 0$ and arc tan $(-1) = -\tfrac{1}{4}\pi$, the solution of the initial value problem is

$$\text{arc tan } x + \text{arc tan } y = -\tfrac{1}{4}\pi. \tag{10}$$

Suppose next that we wish to sketch the graph of (10). Resorting to a device of trigonometry, we take the tangent of each side of (10). Because

$$\tan (\text{arc tan } x) = x$$

and

$$\tan (A + B) = \frac{\tan A + \tan B}{1 - \tan A \tan B},$$

we are led to the equation

$$\frac{x + y}{1 - xy} = -1,$$

or

$$xy - x - y - 1 = 0. \tag{11}$$

Now (11) is the equation of an equilateral hyperbola with asymptotes $x = 1$ and $y = 1$. But if we turn to (10), we see from

$$\text{arc tan } x = -\tfrac{1}{4}\pi - \text{arc tan } y$$

that, since $(-\text{arc tan } y) < \tfrac{1}{2}\pi$,

$$\text{arc tan } x < \tfrac{1}{4}\pi.$$

Hence $x < 1$, and equation (10) represents only one branch of the hyperbola (11). In Figure 7, the solid curve is the graph of equation (10); the solid curve and the dotted curve together are the graph of equation (11).

Each branch of the hyperbola (11) represents a solution of the differential equation, one branch for $x < 1$, the other for $x > 1$. In this example we were forced onto the left branch, equation (10), by the initial condition that $y = -1$ when $x = 0$.

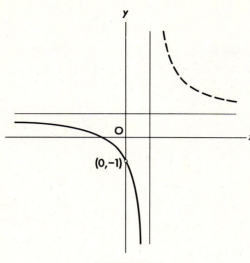

FIGURE 7

One distinction between equations (10) and (11) can be seen by noting that a computing machine, given the differential equation (8) and seeking a solution that passes through the point $(0, -1)$, would draw only the left branch of the curve in Figure 7. The barrier (asymptote) at $x = 1$ would prevent the machine from learning of the existence of the other branch of the hyperbola (11).

EXAMPLE (d): Solve the initial value problem

$$2x(y + 1)\, dx - y\, dy = 0, \tag{12}$$

where $x = 0$ and $y = -2$.

Separating the variables in equation (12), we obtain

$$2x\, dx = \left(1 - \frac{1}{y + 1}\right) dy, \qquad y \neq -1.$$

Integrating, we get a family of solutions given implicitly by

$$x^2 = y - \ln|y + 1| + c. \tag{13}$$

Since we seek a member of this family that passes through the point $(0, -2)$, we must have

$$0 = -2 - \ln|-1| + c,$$

or

$$c = 2.$$

Thus the solution to the problem is given implicitly by

$$x^2 = y - \ln|y + 1| + 2.$$

The reader should note how the theorem of Section 6 applies to this problem to indicate that we have found implicitly the unique solution to the initial value problem which is continuous for $y < -1$.

Exercises

In exercises 1 through 24 obtain the general solution.

1. $(1 - x)y' = y^2$.   ANS. $y \ln|c(1 - x)| = 1$.
2. $\sin x \sin y \, dx + \cos x \cos y \, dy = 0$.   ANS. $\sin y = c \cos x$.
3. $xy^3 \, dx + e^{x^2} \, dy = 0$.   ANS. $e^{-x^2} + y^{-2} = c$.
4. $2y \, dx = 3x \, dy$.   ANS. $x^2 = cy^3$.
5. $my \, dx = nx \, dy$.   ANS. $x^m = cy^n$.
6. $y' = xy^2$.   ANS. $y(x^2 + c) + 2 = 0$.
7. $dV/dP = -V/P$.   ANS. $PV = C$.
8. $y \, e^{2x} \, dx = (4 + e^{2x}) \, dy$.   ANS. $c^2 y^2 = 4 + e^{2x}$.
9. $dr = b(\cos \theta \, dr + r \sin \theta \, d\theta)$.   ANS. $r = c(1 - b \cos \theta)$.
10. $xy \, dx - (x + 2) \, dy = 0$.   ANS. $e^x = cy(x + 2)^2$.
11. $x^2 \, dx + y(x - 1) \, dy = 0$.   ANS. $(x + 1)^2 + y^2 + 2 \ln|c(x - 1)| = 0$.
12. $(xy + x) \, dx = (x^2 y^2 + x^2 + y^2 + 1) \, dy$.

     ANS. $\ln(x^2 + 1) = y^2 - 2y + 4 \ln|c(y + 1)|$.

13. $x \cos^2 y \, dx + \tan y \, dy = 0$.   ANS. $x^2 + \tan^2 y = c^2$.

14. $xy^3 \, dx + (y + 1)e^{-x} \, dy = 0$.   ANS. $e^x(x - 1) = \dfrac{1}{y} + \dfrac{1}{2y^2} + c$.

15. $x^2 yy' = e^y$.   ANS. $x(y + 1) = (1 + cx)e^y$.
16. $\tan^2 y \, dy = \sin^3 x \, dx$.   ANS. $\cos^3 x - 3 \cos x = 3(\tan y - y + c)$.
17. $y' = \cos^2 x \cos y$.   ANS. $4 \ln|\sec y + \tan y| = 2x + \sin 2x + c$.
18. $y' = y \sec x$.   ANS. $y = c(\sec x + \tan x)$.
19. $dx = t(1 + t^2) \sec^2 x \, dt$.   ANS. $2x + \sin 2x = c + (1 + t^2)^2$.
20. $(e^{2x} + 4)y' = y$.   ANS. $y^8(1 + 4e^{-2x}) = c^2$.
21. $\alpha \, d\beta + \beta \, d\alpha + \alpha\beta(3 \, d\alpha + d\beta) = 0$.   ANS. $c\alpha\beta = \exp(-3\alpha - \beta)$.
22. $(1 + \ln x) \, dx + (1 + \ln y) \, dy = 0$.   ANS. $x \ln x + y \ln y = c$.
23. $x \, dx - \sqrt{a^2 - x^2} \, dy = 0$.

     ANS. $y - c = -\sqrt{a^2 - x^2}$, the lower half of the circle $x^2 + (y - c)^2 = a^2$.

24. $x \, dx + \sqrt{a^2 - x^2} \, dy = 0$.

     ANS. $y - c = \sqrt{a^2 - x^2}$, the upper half of the circle $x^2 + (y - c)^2 = a^2$.

25. $a^2 \, dx = x\sqrt{x^2 - a^2} \, dy$.   ANS. $x = a \sec \dfrac{y + c}{a}$.

26. $y \ln x \ln y \, dx + dy = 0$.   ANS. $x \ln x + \ln|\ln y| = x + c$.

In exercises 27 through 32, obtain the particular solution satisfying the initial condition indicated. In each exercise interpret your answer in the light of the existence theorem of Section 6 and draw a graph of the solution.

27. $dr/dt = -4rt$; when $t = 0, r = r_0$.   ANS. $r = r_0 \exp(-2t^2)$.

28. $2xyy' = 1 + y^2$; when $x = 2, y = 3$.   ANS. $y = \sqrt{5x - 1}$.

29. $xyy' = 1 + y^2$; when $x = 2, y = 3$.   ANS. $y = \frac{1}{2}\sqrt{10x^2 - 4}$.

30. $2y\,dx = 3x\,dy$; when $x = 2, y = 1$.   ANS. $y = (x/2)^{2/3}$.

31. $2y\,dx = 3x\,dy$; when $x = -2, y = 1$.   ANS. $y = (x/2)^{2/3}$.

32. $2y\,dx = 3x\,dy$; when $x = 2, y = -1$.   ANS. $y = -(x/2)^{2/3}$.

In exercises 33 through 36, obtain the particular solution satisfying the initial condition indicated.

33. $y' = x \exp(y - x^2)$; when $x = 0, y = 0$.   ANS. $y = \ln\left[\dfrac{2}{1 + \exp(-x^2)}\right]$.

34. $xy^2\,dx + e^x\,dy = 0$; when $x \to \infty, y \to \frac{1}{2}$.   ANS. $y = e^x/(2e^x - x - 1)$.

35. $(2a^2 - r^2)\,dr = r^3 \sin\theta\,d\theta$; when $\theta = 0, r = a$.

36. $v(dv/dx) = g$; when $x = x_0, v = v_0$.   ANS. $v^2 - v_0^2 = 2g(x - x_0)$.

8. Homogeneous functions

Polynomials in which all terms are of the same degree, such as

$$x^2 - 3xy + 4y^2,$$
$$x^3 + y^3, \tag{1}$$
$$x^4y + 7y^5,$$

are called *homogeneous* polynomials. We wish now to extend the concept of homogeneity so it will apply to functions other than polynomials.

If we assign a physical dimension, say length, to each variable x and y in the polynomials in (1), then each polynomial itself also has a physical dimension, length to some power. This suggests the desired generalization. If, when certain variables are thought of as lengths, a function has the physical dimension length to the kth power, then we shall call that function homogeneous of degree k in those variables. For example, the function

$$f(x, y) = 2y^3 \exp\left(\frac{y}{x}\right) - \frac{x^4}{x + 3y} \tag{2}$$

is of dimension (length)^3 when x and y are lengths. Therefore that function is said to be homogeneous of degree 3 in x and y.

We permit the degree k to be any number. The function $\sqrt{x + 4y}$ is called homogeneous of degree $\frac{1}{2}$ in x and y. The function

$$\frac{x}{\sqrt{x^2 + y^2}}$$

is homogeneous of degree zero in x and y.

A formal definition of homogeneity is: *The function $f(x, y)$ is said to be homogeneous of degree k in x and y if, and only if,*

$$f(\lambda x, \lambda y) = \lambda^k f(x, y). \tag{3}$$

The definition is easily extended to functions of more than two variables.

For the function $f(x, y)$ of equation (2), the formal definition of homogeneity leads us to consider

$$f(\lambda x, \lambda y) = 2\lambda^3 y^3 \exp\left(\frac{\lambda y}{\lambda x}\right) - \frac{\lambda^4 x^4}{\lambda x + 3\lambda y}.$$

But we see at once that

$$f(\lambda x, \lambda y) = \lambda^3 f(x, y);$$

hence $f(x, y)$ is homogeneous of degree 3 in x and y, as stated previously.

The following theorems prove useful in the next section.

THEOREM 1: *If $M(x, y)$ and $N(x, y)$ are both homogeneous and of the same degree, the function $M(x, y)/N(x, y)$ is homogeneous of degree zero.*

Proof of Theorem 1 is left to the student.

THEOREM 2: *If $f(x, y)$ is homogeneous of degree zero in x and y, $f(x, y)$ is a function of y/x alone.*

PROOF. Let us put $y = vx$. Then Theorem 2 states that, if $f(x, y)$ is homogeneous of degree zero, $f(x, y)$ is a function of v alone. Now

$$f(x, y) = f(x, vx) = x^0 f(1, v) = f(1, v), \tag{4}$$

in which the x is now playing the role taken by λ in the definition (3) above. By (4), $f(x, y)$ depends on v alone as stated in Theorem 2.

Exercises

Determine in each exercise whether or not the function is homogeneous. If it is homogeneous, state the degree of the function.

1. $4x^2 - 3xy + y^2$.
2. $x^3 - xy + y^3$.
3. $2y + \sqrt{x^2 + y^2}$.
4. $\sqrt{x - y}$.
5. e^x.
6. $\tan x$.
7. $\exp\left(\dfrac{x}{y}\right)$.
8. $\tan\dfrac{3y}{x}$.

9. $(x^2 + y^2) \exp\left(\dfrac{2x}{y}\right) + 4xy.$ **10.** $x \sin \dfrac{y}{x} - y \sin \dfrac{x}{y}.$

11. $\dfrac{x^2 + 3xy}{x - 2y}.$ **12.** $\dfrac{x^5}{x^2 + 2y^2}.$

13. $(u^2 + v^2)^{3/2}.$ **14.** $(u^2 - 4v^2)^{-1/2}.$

15. $y^2 \tan \dfrac{x}{y}.$ **16.** $\dfrac{(x^2 + y^2)^{1/2}}{(x^2 - y^2)^{1/2}}.$

17. $\dfrac{a + 4b}{a - 4b}.$ **18.** $\ln \dfrac{x}{y}.$

19. $x \ln x - y \ln y.$ **20.** $x \ln x - x \ln y.$

ANS.   All functions are homogeneous except those of exercises 2, 5, 6, and 19.

9. Equations with homogeneous coefficients

Suppose that the coefficients M and N in an equation of order one,

$$M(x, y)\, dx + N(x, y)\, dy = 0, \tag{1}$$

are both homogeneous functions and are of the *same degree* in x and y. By Theorems 1 and 2 of Section 8, the ratio M/N is a function of y/x alone. Hence equation (1) may be put in the form

$$\frac{dy}{dx} + g\left(\frac{y}{x}\right) = 0. \tag{2}$$

This suggests the introduction of a new variable v by putting $y = vx$. Then (2) becomes

$$x\frac{dv}{dx} + v + g(v) = 0, \tag{3}$$

in which the variables are separable. We can obtain the solution of (3) by the method of Section 7, insert y/x for v, and thus arrive at the solution of (1). We have shown that the substitution $y = vx$ will transform equation (1) into an equation in v and x in which the variables are separable.

The above method would have been equally successful had we used $x = vy$ to obtain from (1) an equation in y and v. See Example (b) below.

EXAMPLE (a):   Solve the equation

$$(x^2 - xy + y^2)\, dx - xy\, dy = 0. \tag{4}$$

Since the coefficients in (4) are both homogeneous and of degree two in x and y, let us put $y = vx$. Then (4) becomes

$$(x^2 - x^2v + x^2v^2)\,dx - x^2v(v\,dx + x\,dv) = 0,$$

from which the factor x^2 should be removed at once. That done, we have to solve

$$(1 - v + v^2)\,dx - v(v\,dx + x\,dv) = 0,$$

or

$$(1 - v)\,dx - xv\,dv = 0.$$

Hence we separate variables to get

$$\frac{dx}{x} + \frac{v\,dv}{v - 1} = 0.$$

Then from

$$\frac{dx}{x} + \left[1 + \frac{1}{v - 1}\right]dv = 0$$

a family of solutions is seen to be

$$\ln|x| + v + \ln|v - 1| = \ln|c|,$$

or

$$x(v - 1)\,e^v = c.$$

In terms of the original variables, these solutions are given by

$$x\left(\frac{y}{x} - 1\right)\exp\left(\frac{y}{x}\right) = c,$$

or

$$(y - x)\exp\left(\frac{y}{x}\right) = c.$$

EXAMPLE (b): Solve the equation

$$xy\,dx + (x^2 + y^2)\,dy = 0. \tag{5}$$

Again the coefficients in the equation are homogeneous and of degree two. We could use $y = vx$, but the relative simplicity of the dx term in (5) suggests that we put $x = vy$. Then $dx = v\,dy + y\,dv$, and equation (5) is replaced by

$$vy^2(v\,dy + y\,dv) + (v^2y^2 + y^2)\,dy = 0,$$

or

$$v(v\,dy + y\,dv) + (v^2 + 1)\,dy = 0.$$

Hence we need to solve

$$vy\,dv + (2v^2 + 1)\,dy = 0, \tag{6}$$

which leads at once to

$$\ln(2v^2 + 1) + 4\ln|y| = \ln c,$$

or

$$y^4(2v^2 + 1) = c.$$

Thus the desired solutions are given by

$$y^4\left(\frac{2x^2}{y^2} + 1\right) = c;$$

that is,

$$y^2(2x^2 + y^2) = c. \tag{7}$$

Since the left member of equation (7) cannot be negative, we may, for symmetry's sake, change the arbitrary constant to c_1^4, writing

$$y^2(2x^2 + y^2) = c_1^4.$$

It is worthwhile for the student to attack equation (5) using $y = vx$. That method leads directly to the equation

$$(v^3 + 2v)\,dx + x(v^2 + 1)\,dv = 0.$$

Frequently in equations with homogeneous coefficients, it is quite immaterial whether one uses $y = vx$ or $x = vy$. However, it is sometimes easier to substitute for the variable whose differential has the simpler coefficient.

Exercises

In exercises 1 through 21, obtain a family of solutions.

1. $(x - 2y)\,dx + (2x + y)\,dy = 0.$ ANS. $\ln(x^2 + y^2) + 4\arctan(y/x) = c.$
2. $2(2x^2 + y^2)\,dx - xy\,dy = 0.$ ANS. $x^4 = c^2(4x^2 + y^2).$
3. $xy\,dx - (x^2 + 3y^2)\,dy = 0.$ ANS. $x^2 = 6y^2 \ln|y/c|.$
4. $x^2 y' = 4x^2 + 7xy + 2y^2.$ ANS. $x^2(y + 2x) = c(y + x).$
5. $3xy\,dx + (x^2 + y^2)\,dy = 0.$
6. $(x - y)(4x + y)\,dx + x(5x - y)\,dy = 0.$ ANS. $x(y + x)^2 = c(y - 2x).$
7. $(5v - u)\,du + (3v - 7u)\,dv = 0.$ ANS. $(3v + u)^2 = c(v - u).$
8. $(x^2 + 2xy - 4y^2)\,dx - (x^2 - 8xy - 4y^2)\,dy = 0.$ ANS. $x^2 + 4y^2 = c(x + y).$
9. $(x^2 + y^2)\,dx - xy\,dy = 0.$ ANS. $y^2 = 2x^2 \ln|x/c|.$

10. $x(x^2 + y^2)^2(y\,dx - x\,dy) + y^6\,dy = 0.$ ANS. $(x^2 + y^2)^3 = 6y^6 \ln |c/y|.$

11. $(x^2 + y^2)\,dx + xy\,dy = 0.$ ANS. $x^2(x^2 + 2y^2) = c^4.$

12. $xy\,dx - (x + 2y)^2\,dy = 0.$ ANS. $y^3(x + y) = ce^{x/y}.$

13. $v^2\,dx + x(x + v)\,dv = 0.$ ANS. $xv^2 = c(x + 2v).$

14. $[x \csc (y/x) - y]\,dx + x\,dy = 0.$ ANS. $\ln |x/c| = \cos (y/x).$

15. $x\,dx + \sin^2 (y/x)[y\,dx - x\,dy] = 0.$ ANS. $4x \ln |x/c| - 2y + x \sin (2y/x) = 0.$

16. $(x - y \ln y + y \ln x)\,dx + x(\ln y - \ln x)\,dy = 0.$

     ANS. $(x - y) \ln x + y \ln y = cx + y.$

17. $[x - y \arctan (y/x)]\,dx + x \arctan (y/x)\,dy = 0.$

     ANS. $2y \arctan (y/x) = x \ln [c^2(x^2 + y^2)/x^4].$

18. $y^2\,dy = x(x\,dy - y\,dx)\,e^{x/y}.$ ANS. $y \ln |y/c| = (y - x)\,e^{x/y}.$

19. $t(s^2 + t^2)\,ds - s(s^2 - t^2)\,dt = 0.$ ANS. $s^2 = -2t^2 \ln |cst|.$

20. $y\,dx = (x + \sqrt{y^2 - x^2})\,dy.$ ANS. $\arcsin (x/y) = \ln |y/c|.$

21. $(3x^2 - 2xy + 3y^2)\,dx = 4xy\,dy.$ ANS. $(y - x)(y + 3x)^3 = cx^3.$

22. Prove that, with the aid of the substitution $y = vx$, you can solve any equation of the form

$$y^n f(x)\,dx + H(x, y)(y\,dx - x\,dy) = 0,$$

where $H(x, y)$ is homogeneous in x and y.

In exercises 23 through 35, find the particular solution indicated.

23. $(x - y)\,dx + (3x + y)\,dy = 0;$ when $x = 2, y = -1.$

     ANS. $2(x + 2y) + (x + y) \ln (x + y) = 0.$

24. $(y - \sqrt{x^2 + y^2})\,dx - x\,dy = 0;$ when $x = \sqrt{3}, y = 1.$ ANS. $x^2 = 9 - 6y.$

25. $(y + \sqrt{x^2 + y^2})\,dx - x\,dy = 0;$ when $x = \sqrt{3}, y = 1.$ ANS. $x^2 = 2y + 1.$

26. $[x \cos^2 (y/x) - y]\,dx + x\,dy = 0;$ when $x = 1, y = \pi/4.$

     ANS. $\tan (y/x) = \ln (e/x).$

27. $(y^2 + 7xy + 16x^2)\,dx + x^2\,dy = 0;$ when $x = 1, y = 1.$

     ANS. $x - y = 5(y + 4x) \ln x.$

28. $y^2\,dx + (x^2 + 3xy + 4y^2)\,dy = 0;$ when $x = 2, y = 1.$

     ANS. $4(2y + x) \ln y = 2y - x.$

29. $xy\,dx + 2(x^2 + 2y^2)\,dy = 0;$ when $x = 0, y = 1.$ ANS. $y^4(3x^2 + 4y^2) = 4.$

30. $y(2x^2 - xy + y^2)\,dx - x^2(2x - y)\,dy = 0;$ when $x = 1, y = \frac{1}{2}.$

     ANS. $y^2 \ln x = 2y^2 + xy - x^2.$

31. $y(9x - 2y)\,dx - x(6x - y)\,dy = 0;$ when $x = 1, y = 1.$

     ANS. $3x^3 - x^2y - 2y^2 = 0.$

32. $y(x^2 + y^2)\,dx + x(3x^2 - 5y^2)\,dy = 0;$ when $x = 2, y = 1.$

     ANS. $2y^5 - 2x^2y^3 + 3x = 0.$

33. $(16x + 5y)\,dx + (3x + y)\,dy = 0;$ the curve to pass through the point $(1, -3).$

     ANS. $y + 3x = (y + 4x) \ln (y + 4x).$

34. $v(3x + 2v)\,dx - x^2\,dv = 0;$ when $x = 1, v = 2.$ ANS. $2x^3 + 2x^2v - 3v = 0.$

35. $(3x^2 - 2y^2)y' = 2xy;$ when $x = 0, y = -1.$ ANS. $x^2 = 2y^2(y + 1).$

36. From Theorems 1 and 2, page 26, it follows that, if F is homogeneous of degree k in x and y, F can be written in the form

$$F = x^k \varphi\left(\frac{y}{x}\right). \tag{A}$$

Use (A) to prove Euler's theorem that, if F is a homogeneous function of degree k in x and y,

$$x\frac{\partial F}{\partial x} + y\frac{\partial F}{\partial y} = kF.$$

10. Exact equations

In Section 7 it was noted that when an equation can be put in the form

$$A(x)\,dx + B(y)\,dy = 0,$$

a set of solutions can be determined by integration; that is, by finding a function whose differential is $A(x)\,dx + B(y)\,dy$.

That idea can be extended to some equations of the form

$$M(x, y)\,dx + N(x, y)\,dy = 0 \tag{1}$$

in which separation of variables may not be possible. Suppose that a function $F(x, y)$ can be found that has for its total differential the expression $M\,dx + N\,dy$; that is,

$$dF = M\,dx + N\,dy. \tag{2}$$

Then certainly

$$F(x, y) = c \tag{3}$$

defines implicitly a set of solutions of (1). For, from (3) it follows that

$$dF = 0,$$

or, in view of (2),

$$M\,dx + N\,dy = 0,$$

as desired.

Two things, then, are needed: first, to find out under what conditions on M and N a function F exists such that its total differential is exactly $M\,dx + N\,dy$; second, if those conditions are satisfied, actually to determine the function F. If a function F exists such that

$$M\,dx + N\,dy$$

is exactly the total differential of F, we call equation (1) an *exact equation*.

If the equation

$$M\,dx + N\,dy = 0 \tag{1}$$

is exact, then by definition F exists such that

$$dF = M\,dx + N\,dy.$$

But, from calculus,

$$dF = \frac{\partial F}{\partial x}\,dx + \frac{\partial F}{\partial y}\,dy,$$

so

$$M = \frac{\partial F}{\partial x}, \qquad N = \frac{\partial F}{\partial y}.$$

These two equations lead to

$$\frac{\partial M}{\partial y} = \frac{\partial^2 F}{\partial y\,\partial x} \qquad \text{and} \qquad \frac{\partial N}{\partial x} = \frac{\partial^2 F}{\partial x\,\partial y}.$$

Again from calculus

$$\frac{\partial^2 F}{\partial y\,\partial x} = \frac{\partial^2 F}{\partial x\,\partial y},$$

provided that these partial derivatives are continuous. Therefore, if (1) is an exact equation, then

$$\frac{\partial M}{\partial y} = \frac{\partial N}{\partial x}. \tag{4}$$

Thus, for (1) to be exact it is necessary that (4) be satisfied.

Let us now show that, if condition (4) is satisfied, (1) is an exact equation. Let $\phi(x, y)$ be a function for which

$$\frac{\partial \phi}{\partial x} = M.$$

The function ϕ is the result of integrating $M\,dx$ with respect to x while holding y constant. Now

$$\frac{\partial^2 \phi}{\partial y\,\partial x} = \frac{\partial M}{\partial y};$$

hence, if (4) is satisfied, then also

$$\frac{\partial^2 \phi}{\partial x\,\partial y} = \frac{\partial N}{\partial x}. \tag{5}$$

Let us integrate both sides of this last equation with respect to x, holding y fixed. In the integration with respect to x, the "arbitrary constant" may be any function of y. Let us call it $B'(y)$, for ease in indicating its integral. Then integration of (5) with respect to x yields

$$\frac{\partial \phi}{\partial y} = N + B'(y). \tag{6}$$

Now a function F can be exhibited, namely,

$$F = \phi(x, y) - B(y),$$

for which

$$dF = \frac{\partial \phi}{\partial x} dx + \frac{\partial \phi}{\partial y} dy - B'(y) dy$$

$$= M \, dx + [N + B'(y)] \, dy - B'(y) \, dy$$

$$= M \, dx + N \, dy.$$

Hence, equation (1) is exact. We have completed a proof of the theorem stated below.

THEOREM 3:   *If M, N, $\partial M / \partial y$, and $\partial N / \partial x$ are continuous functions of x and y, then a necessary and sufficient condition that*

$$M \, dx + N \, dy = 0 \tag{1}$$

be an exact equation is that

$$\frac{\partial M}{\partial y} = \frac{\partial N}{\partial x}. \tag{4}$$

Furthermore, the proof contains the germ of a method for obtaining a set of solutions, a method used in Examples (a) and (b) below.

EXAMPLE (a):   Solve the equation

$$3x(xy - 2) \, dx + (x^3 + 2y) \, dy = 0. \tag{7}$$

First, from the fact that

$$\frac{\partial M}{\partial y} = 3x^2 \quad \text{and} \quad \frac{\partial N}{\partial x} = 3x^2,$$

we conclude that equation (7) is exact. Therefore, its solution is $F = c$, where

$$\frac{\partial F}{\partial x} = M = 3x^2 y - 6x, \tag{8}$$

and

$$\frac{\partial F}{\partial y} = N = x^3 + 2y. \tag{9}$$

Let us attempt to determine F from equation (8). Integration of both sides of (8) with respect to x, holding y constant, yields

$$F = x^3 y - 3x^2 + T(y), \tag{10}$$

where the usual arbitrary constant in indefinite integration is now necessarily a function $T(y)$, as yet unknown. To determine $T(y)$, we use the fact that the function F of equation (10) must also satisfy equation (9). Hence

$$x^3 + T'(y) = x^3 + 2y,$$

$$T'(y) = 2y.$$

No arbitrary constant is needed in obtaining $T(y)$, since one is being introduced on the right in the solution $F = c$. Then

$$T(y) = y^2,$$

and from (10)

$$F = x^3y - 3x^2 + y^2.$$

Finally, a set of solutions of equation (7) is defined by

$$x^3y - 3x^2 + y^2 = c.$$

EXAMPLE (b):   Solve the equation
$$(2x^3 - xy^2 - 2y + 3)\,dx - (x^2y + 2x)\,dy = 0. \tag{11}$$

Here
$$\frac{\partial M}{\partial y} = -2xy - 2 = \frac{\partial N}{\partial x},$$

so equation (11) is exact.
   A set of solutions of (11) is $F = c$, where

$$\frac{\partial F}{\partial x} = 2x^3 - xy^2 - 2y + 3 \tag{12}$$

and

$$\frac{\partial F}{\partial y} = -x^2y - 2x. \tag{13}$$

Because (13) is simpler than (12) and, for variety's sake, let us start the determination of F from equation (13).
   At once, from (13)

$$F = -\tfrac{1}{2}x^2y^2 - 2xy + Q(x),$$

where $Q(x)$ will be determined from (12). The latter yields
$$-xy^2 - 2y + Q'(x) = 2x^3 - xy^2 - 2y + 3,$$

$$Q'(x) = 2x^3 + 3.$$

Therefore

$$Q(x) = \tfrac{1}{2}x^4 + 3x,$$

and the desired set of solutions of (11) is defined implicitly by

$$-\tfrac{1}{2}x^2 y^2 - 2xy + \tfrac{1}{2}x^4 + 3x = \tfrac{1}{2}c,$$

or

$$x^4 - x^2 y^2 - 4xy + 6x = c.$$

Exercises

Test each of the following equations for exactness and solve the equation. The equations that are not exact may, of course, be solved by methods discussed in the preceding sections.

1. $(x + 2y) dx + (2x + y) dy = 0.$ ANS. $x^2 + 4xy + y^2 = c.$
2. $(2xy - 3x^2) dx + (x^2 + 2y) dy = 0.$ ANS. $x^2 y - x^3 + y^2 = c.$
3. $(6x + y^2) dx + y(2x - 3y) dy = 0.$ ANS. $3x^2 + xy^2 - y^3 = c.$
4. $(y^2 - 2xy + 6x) dx - (x^2 - 2xy + 2) dy = 0.$
                                              ANS. $xy^2 - x^2 y + 3x^2 - 2y = c.$
5. $(2xy - y) dx + (x^2 + x) dy = 0.$ ANS. $y(x + 1)^3 = cx.$
6. $v(2uv^2 - 3) du + (3u^2 v^2 - 3u + 4v) dv = 0.$ ANS. $v(u^2 v^2 - 3u + 2v) = c.$
7. $(\cos 2y - 3x^2 y^2) dx + (\cos 2y - 2x \sin 2y - 2x^3 y) dy = 0.$
                                              ANS. $\tfrac{1}{2}\sin 2y + x \cos 2y - x^3 y^2 = c.$
8. $(1 + y^2) dx + (x^2 y + y) dy = 0.$ ANS. $2 \arctan x + \ln(1 + y^2) = c.$
9. $(1 + y^2 + xy^2) dx + (x^2 y + y + 2xy) dy = 0.$ ANS. $2x + y^2(1 + x)^2 = c.$
10. $(w^3 + wz^2 - z) dw + (z^3 + w^2 z - w) dz = 0.$ ANS. $(w^2 + z^2)^2 = 4wz + c.$
11. $(2xy - \tan y) dx + (x^2 - x \sec^2 y) dy = 0.$ ANS. $x^2 y - x \tan y = c.$
12. $(\cos x \cos y - \cot x) dx - \sin x \sin y \, dy = 0.$ ANS. $\sin x \cos y = \ln|c \sin x|.$
13. $(r + \sin \theta - \cos \theta) dr + r(\sin \theta + \cos \theta) d\theta = 0.$
                                              ANS. $r^2 + 2r(\sin \theta - \cos \theta) = c.$
14. $x(3xy - 4y^3 + 6) dx + (x^3 - 6x^2 y^2 - 1) dy = 0.$
                                              ANS. $x^3 y - 2x^2 y^3 + 3x^2 - y = c.$
15. $(\sin \theta - 2r \cos^2 \theta) dr + r \cos \theta (2r \sin \theta + 1) d\theta = 0.$
                                              ANS. $r \sin \theta - r^2 \cos^2 \theta = c.$
16. $[2x + y \cos(xy)] dx + x \cos(xy) dy = 0.$ ANS. $x^2 + \sin(xy) = c.$
17. $2xy \, dx + (y^2 + x^2) dy = 0.$ ANS. $y(3x^2 + y^2) = c.$
18. $2xy \, dx + (y^2 - x^2) dy = 0.$ ANS. $x^2 + y^2 = cy.$
19. $[2xy \cos(x^2) - 2xy + 1] dx + [\sin(x^2) - x^2] dy = 0.$
                                              ANS. $y[\sin(x^2) - x^2] = c - x.$
20. $(2x - 3y) dx + (2y - 3x) dy = 0.$ ANS. $x^2 + y^2 - 3xy = c.$
21. Do exercise 20 by a second method.
22. $(xy^2 + y - x) dx + x(xy + 1) dy = 0.$ ANS. $x^2 y^2 + 2xy - x^2 = c.$
23. $3y(x^2 - 1) dx + (x^3 + 8y - 3x) dy = 0;$ when $x = 0, y = 1.$
                                              ANS. $xy(x^2 - 3) = 4(1 - y^2).$

24. $(1 - xy)^{-2} dx + [y^2 + x^2(1 - xy)^{-2}] dy = 0$; when $x = 4$, $y = 1$.

           ANS. $xy^4 - y^3 + 3xy - 3x - 3 = 0$.

25. $(ye^{xy} - 2y^3) dx + (x e^{xy} - 6xy^2 - 2y) dy = 0$; when $x = 0$, $y = 2$.

           ANS. $e^{xy} = 2xy^3 + y^2 - 3$.

26. $(3 + y + 2y^2 \sin^2 x) dx + (x + 2xy - y \sin 2x) dy = 0$.

           ANS. $y^2 \sin 2x = c + 2x(3 + y + y^2)$.

27. $2x[3x + y - y \exp(-x^2)] dx + [x^2 + 3y^2 + \exp(-x^2)] dy = 0$.

           ANS. $x^2y + y^3 + 2x^3 + y \exp(-x^2) = c$.

28. $(xy^2 + x - 2y + 3) dx + x^2y \, dy = 2(x + y) dy$; when $x = 1$, $y = 1$.

           ANS. $(xy - 2)^2 + (x + 3)^2 = 2y^2 + 15$.

11. The linear equation of order one

In Section 10 we studied first-order differential equations that were exact. If an equation is not exact, it is natural to attempt to make it exact by the introduction of an appropriate factor, which is then called an integrating factor. Indeed, in Section 7 we multiplied by an integrating factor to separate the variables and thereby obtain an exact equation.

In general, very little can be said about the theory of integrating factors for first-order equations. In Chapter 4, we shall prove some theorems that will give some assistance in a few isolated situations. There is one important class of equations, however where the existence of an integrating factor can be demonstrated. This is the class of linear equations of order one.

An equation that is linear and of order one in the dependent variable y must by definition (Section 2) be of the form

$$A(x)\frac{dy}{dx} + B(x)y = C(x). \tag{1}$$

By dividing each member of equation (1) by $A(x)$, we obtain

$$\frac{dy}{dx} + P(x)y = Q(x), \tag{2}$$

which we choose as the standard form for the linear equation of order one.

For the moment, suppose that there exists for equation (2) a positive integrating factor $v(x) > 0$, a function of x alone. Then

$$v(x)\left[\frac{dy}{dx} + P(x)y\right] = v(x) \cdot Q(x) \tag{3}$$

must be an exact equation. But (3) is easily put into the form

$$M \, dx + N \, dy = 0$$

with

$$M = vPy - vQ,$$

and

$$N = v,$$

in which v, P, and Q are functions of x alone.

Therefore, if equation (3) is to be exact, it follows from the requirement

$$\frac{\partial M}{\partial y} = \frac{\partial N}{\partial x}$$

that v must satisfy the equation

$$vP = \frac{dv}{dx}. \tag{4}$$

From (4), v may be obtained readily, for

$$P \, dx = \frac{dv}{v},$$

so

$$\ln v = \int P \, dx,$$

or

$$v = \exp \left(\int P \, dx \right). \tag{5}$$

That is, if equation (2) has a positive integrating factor independent of y, then that factor must be as given by equation (5).

It remains to be shown that the v given by equation (5) is actually an integrating factor of

$$\frac{dy}{dx} + P(x)y = Q(x). \tag{2}$$

Let us multiply (2) by the integrating factor, obtaining

$$\exp \left(\int P \, dx \right) \frac{dy}{dx} + P \exp \left(\int P \, dx \right) y = Q \exp \left(\int P \, dx \right). \tag{6}$$

The left member of (6) is the derivative of the product

$$y \exp \left(\int P \, dx \right);$$

the right member of (6) is a function of x only. Hence equation (6) is exact, which is what we wanted to show.

Of course one integrating factor is sufficient. Hence we may use in the exponent ($\int P \, dx$) any function whose derivative is P.

Because of the great importance of the ideas just discussed and the frequent occurrence of linear equations of first order, we summarize the steps involved in solving such equations:

(a) Put the equation into standard form:

$$\frac{dy}{dx} + Py = Q.$$

(b) Obtain the integrating factor exp ($\int P \, dx$).
(c) Apply the integrating factor to the equation in its standard form.
(d) Solve the resultant exact equation.

Note in integrating the exact equation, that *the integral of the left member is always the product of the dependent variable and the integrating factor used.*

EXAMPLE (a):   Solve the equation

$$2(y - 4x^2) \, dx + x \, dy = 0.$$

The equation is linear in y. When put in standard form it becomes

$$\frac{dy}{dx} + \frac{2}{x} y = 8x, \qquad \text{when} \quad x \neq 0. \tag{7}$$

Then an integrating factor is

$$\exp\left(\int \frac{2 \, dx}{x}\right) = \exp\left(2 \ln |x|\right) = \exp\left(\ln x^2\right) = x^2.$$

Next apply the integrating factor to (7), thus obtaining the exact equation

$$x^2 \frac{dy}{dx} + 2xy = 8x^3, \tag{8}$$

which may be immediately rewritten as

$$\frac{d}{dx}(x^2 y) = 8x^3. \tag{9}$$

By integrating (9) we find that

$$x^2 y = 2x^4 + c. \tag{10}$$

This can be checked. From (10) we get (8) by differentiation. Then the original differential equation follows from (8) by a simple adjustment. Hence (10) defines a set of solutions of the original equation.

EXAMPLE (b):   Solve the equation

$$y \, dx + (3x - xy + 2) \, dy = 0.$$

Since the product $y\,dy$ occurs here, the equation is not linear in y. It is, however, linear in x. Therefore we arrange the terms as in

$$y\,dx + (3 - y)x\,dy = -2\,dy$$

and pass to the standard form,

$$\frac{dx}{dy} + \left(\frac{3}{y} - 1\right)x = \frac{-2}{y}, \qquad \text{for } y \neq 0. \tag{11}$$

Now

$$\int\left(\frac{3}{y} - 1\right)dy = 3\ln|y| - y + c_1,$$

so that an integrating factor for equation (11) is

$$\exp(3\ln|y| - y) = \exp(3\ln|y|)\cdot e^{-y}$$

$$= \exp(\ln|y|^3)\cdot e^{-y}$$

$$= |y|^3\,e^{-y}.$$

It follows that for $y > 0$, $y^3\,e^{-y}$ is an integrating factor for equation (11) and for $y < 0$, $-y^3\,e^{-y}$ serves as an integrating factor. In either case we are led to the exact equation

$$y^3\,e^{-y}\,dx + y^2(3 - y)\,e^{-y}x\,dy = -2y^2\,e^{-y}\,dy$$

from which we get

$$xy^3\,e^{-y} = -2\int y^2\,e^{-y}\,dy$$

$$= 2y^2\,e^{-y} + 4y\,e^{-y} + 4\,e^{-y} + c.$$

Thus a family of solutions is defined implicitly by

$$xy^3 = 2y^2 + 4y + 4 + c\,e^{y}.$$

12. The general solution of a linear equation

At the beginning of this chapter we stated an existence and uniqueness theorem for first-order differential equations. If the differential equation in that theorem happens to be a linear equation, we can prove a somewhat stronger statement.

Consider the linear differential equation

$$\frac{dy}{dx} + P(x)y = Q(x). \tag{1}$$

Suppose that P and Q are continuous functions on the interval $a < x < b$, and that $x = x_0$ is any number in that interval. If y_0 is an arbitrary real number, there exists a unique solution $y = y(x)$ of differential equation (1) that also satisfies the initial condition

$$y(x_0) = y_0.$$

Moreover, this solution satisfies equation (1) throughout the entire interval $a < x < b$.

The proof of this theorem has essentially been obtained in Section 11. Multiplication of equation (1) by the integrating factor $v = \exp (\int P \, dx)$ and integration gives

$$yv = \int vQ \, dx + c.$$

Since $v \neq 0$, we can write

$$y = v^{-1} \int vQ \, dx + cv^{-1}. \tag{2}$$

It is a simple matter to show that since $v \neq 0$ and v is continuous on $a < x < b$, (2) is a family of solutions of equation (1).

It is also easy to see that given any x_0 on the interval $a < x < b$ together with any number y_0, we can choose the constant c so that $y = y_0$ when $x = x_0$.

The effect of our argument is that every equation of the form of equation (1), for which P and Q have some common interval of continuity, will have a unique set of solutions containing one constant of integration that can be obtained by introducing the appropriate integrating factor. Because we are assured of the uniqueness of these solutions, we know that any other solution obtained by any other method must be one of the functions in our one-parameter family of solutions. It is for this reason that this set of solutions is called the *general solution* of equation (1). The word "general" is intended to mean that we have found all possible solutions that satisfy the differential equation on the interval $a < x < b$.

Exercises

In exercises 1 through 25, find the general solution.

1. $(x^4 + 2y) \, dx - x \, dy = 0.$ ANS. $2y = x^4 + cx^2.$
2. $(3xy + 3y - 4) \, dx + (x + 1)^2 \, dy = 0.$ ANS. $y = 2(x + 1)^{-1} + c(x + 1)^{-3}.$
3. $y' = \csc x - y \cot x.$ ANS. $y \sin x = x + c.$
4. $t(dx/dt) = 6te^{2t} + x(2t - 1).$ ANS. $xt = (3t^2 + c) e^{2t}.$
5. $dy = (x - 3y) \, dx.$ ANS. $9y = 3x - 1 + ce^{-3x}.$
6. $(3x - 1)y' = 6y - 10(3x - 1)^{1/3}.$ ANS. $y = 2(3x - 1)^{1/3} + c(3x - 1)^2.$

7. $(y - 2) dx + (3x - y) dy = 0$.   ANS. $12x = 3y + 2 + c(y - 2)^{-3}$.

8. $(2xy + x^2 + x^4) dx - (1 + x^2) dy = 0$.   ANS. $y = (1 + x^2)(c + x - \arctan x)$.

9. $y' = x - 2xy$. Solve by two methods.   ANS. $2y = 1 + ce^{-x^2}$.

10. $(y - \cos^2 x) dx + \cos x \, dy = 0$.   ANS. $y(\sec x + \tan x) = c + x - \cos x$.

11. $y' = x - 2y \cot 2x$.   ANS. $4y \sin 2x = c + \sin 2x - 2x \cos 2x$.

12. $(y - x + xy \cot x) dx + x \, dy = 0$.   ANS. $xy \sin x = c + \sin x - x \cos x$.

13. $\dfrac{dy}{dx} - my = c_1 e^{mx}$; where c_1 and m are constants.   ANS. $y = (c_1 x + c_2) e^{mx}$.

14. $\dfrac{dy}{dx} - m_2 y = c_1 e^{m_1 x}$; where c_1, m_1, m_2 are constants and $m_1 \neq m_2$.

ANS. $y = c_3 e^{m_1 x} + c_2 e^{m_2 x}$; where $c_3 = c_1/(m_1 - m_2)$.

15. $v \, dx + (2x + 1 - vx) dv = 0$.   ANS. $xv^2 = v + 1 + ce^v$.

16. $x(x^2 + 1)y' + 2y = (x^2 + 1)^3$.   ANS. $x^2 y = \frac{1}{4}(x^2 + 1)^3 + c(x^2 + 1)$.

17. $2x(y - x^2) dx + dy = 0$.   ANS. $y = x^2 - 1 + ce^{-x^2}$.

18. $(1 + xy) dx - (1 + x^2) dy = 0$.   ANS. $y = x + c(1 + x^2)^{1/2}$.

19. $2y \, dx = (x^2 - 1)(dx - dy)$.   ANS. $(x - 1)y = (x + 1)(c + x - 2 \ln |x + 1|)$.

20. $dx - (1 + 2x \tan y) dy = 0$.   ANS. $2x \cos^2 y = y + c + \sin y \cos y$.

21. $(1 + \cos x)y' = \sin x(\sin x + \sin x \cos x - y)$.

ANS. $y = (1 + \cos x)(c + x - \sin x)$.

22. $y' = 1 + 3y \tan x$.   ANS. $3y \cos^3 x = c + 3 \sin x - \sin^3 x$.

23. $(x^2 + a^2) dy = 2x[(x^2 + a^2)^2 + 3y] dx$; a is constant.

ANS. $y = (x^2 + a^2)^2[c(x^2 + a^2) - 1]$.

24. $(x + a)y' = bx - ny$; a, b, n are constants with $n \neq 0, n \neq -1$.

ANS. $n(n + 1)y = b(nx - a) + c(x + a)^{-n}$.

25. Solve the equation of exercise 24 for the exceptional cases $n = 0$ and $n = -1$.

ANS.   If $n = 0$, $y = bx + c - ab \ln |x + a|$.

If $n = -1$, $y = ab + c(x + a) + b(x + a) \ln |x + a|$.

26. In the standard form $dy + Py \, dx = Q \, dx$, put $y = vw$, thus obtaining

$$w(dv + Pv \, dx) + v \, dw = Q \, dx.$$

Then, by first choosing v so that

$$dv + Pv \, dx = 0$$

and later determining w, show how to complete the solution of

$$dy + Py \, dx = Q \, dx.$$

In exercises 27 through 33, find the particular solution indicated.

27. $(2x + 3)y' = y + (2x + 3)^{1/2}$; when $x = -1, y = 0$.

ANS. $2y = (2x + 3)^{1/2} \ln (2x + 3)$.

28. $y' = x^3 - 2xy$; when $x = 1, y = 1$.   ANS. $2y = x^2 - 1 + 2 \exp (1 - x^2)$.

29. $L\dfrac{di}{dt} + Ri = E$; where $L, R,$ and E are constants, when $t = 0, i = 0$.

ANS. $i = \dfrac{E}{R}\left[1 - \exp\left(-\dfrac{Rt}{L}\right)\right]$.

30. $L\dfrac{di}{dt} + Ri = E \sin \omega t$; when $t = 0$, $i = 0$.

ANS.   Let $Z^2 = R^2 + \omega^2 L^2$.
Then $i = EZ^{-2}[R \sin \omega t - \omega L \cos \omega t + \omega L \exp(-Rt/L)]$.

31. Find that solution of $y' = 2(2x - y)$ which passes through the point $(0, -1)$.

ANS. $y = 2x - 1$.

32. Find that solution of $y' = 2(2x - y)$ which passes through the point $(0, 1)$.

ANS. $y = 2x - 1 + 2e^{-2x}$.

33. $(1 + t^2)\,ds + 2t[st^2 - 3(1 + t^2)^2]\,dt = 0$; when $t = 0$, $s = 2$.

ANS. $s = (1 + t^2)[3 - \exp(-t^2)]$.

Miscellaneous Exercises

In each exercise, find a set of solutions, unless the statement of the exercise stipulates otherwise.

1. $y' = \exp(2x - y)$.   ANS. $2e^y = e^{2x} + c$.
2. $(x + y)\,dx + x\,dy = 0$.   ANS. $x(x + 2y) = c$.
3. $y^2\,dx - x(2x + 3y)\,dy = 0$.   ANS. $y^2(x + y) = cx$.
4. $(x^2 + 1)\,dx + x^2 y^2\,dy = 0$.   ANS. $xy^3 = 3(1 + cx - x^2)$.
5. $(x^3 + y^3)\,dx + y^2(3x + ky)\,dy = 0$; k is constant.   ANS. $ky^4 + 4xy^3 + x^4 = c$.
6. $y' = x^3 - 2xy$; when $x = 1$, $y = 2$.   ANS. $2y = x^2 - 1 + 4\exp(1 - x^2)$.
7. $dy/dx - \cos x = \cos x \tan^2 y$.   ANS. $2 \sin x = y + \sin y \cos y + c$.
8. $\cos x\,dy/dx = 1 - y - \sin x$.   ANS. $y(1 + \sin x) = (x + c)\cos x$.
9. $\sin \theta\,dr/d\theta = -1 - 2r \cos \theta$.   ANS. $r \sin^2 \theta = c + \cos \theta$.
10. $y(x + 3y)\,dx + x^2\,dy = 0$.   ANS. $x^2 y = c(2x + 3y)$.
11. $dy/dx = \sec^2 x \sec^3 y$.   ANS. $3 \tan x + c = 3 \sin y - \sin^3 y$.
12. $y(2x^3 - x^2 y + y^3)\,dx - x(2x^3 + y^3)\,dy = 0$.   ANS. $2x^2 y \ln |cx| = 4x^3 - y^3$.
13. $(1 + x^2)y' = x^4 y^4$.   ANS. $x^3 y^3 + 1 = y^3(c + 3x - 3 \arctan x)$.
14. $y(3 + 2xy^2)\,dx + 3(x^2 y^2 + x - 1)\,dy = 0$.   ANS. $x^2 y^3 = 3(c + y - xy)$.
15. $(2x^2 - 2xy - y^2)\,dx + xy\,dy = 0$.   ANS. $x^3 = c(y - x)\exp(y/x)$.
16. $y(x^2 + y^2)\,dx + x(3x^2 - 5y^2)\,dy = 0$; when $x = 2$, $y = 1$.

ANS. $2y^5 - 2x^2 y^3 + 3x = 0$.

17. $y' + ay = b$; a and b constants. Solve by two methods.   ANS. $y = b/a + ce^{-ax}$.
18. $(x - y)\,dx - (x + y)\,dy = 0$. Solve by two methods.   ANS. $x^2 - 2xy - y^2 = c$.
19. $dx/dt = \cos x \cos^2 t$.   ANS. $4 \ln |\sec x + \tan x| = 2t + \sin 2t + c$.
20. $(\sin y - y \sin x)\,dx + (\cos x + x \cos y)\,dy = 0$.   ANS. $x \sin y + y \cos x = c$.
21. $(1 + 4xy - 4x^2 y)\,dx + (x^2 - x^3)\,dy = 0$; when $x = 2$, $y = \frac{1}{4}$.

ANS. $2x^4 y = x^2 + 2x + 2 \ln(x - 1)$.

22. $3x^3 y' = 2y(y - 3)$.   ANS. $y = c(y - 3)\exp(x^{-2})$.
23. $(2y \cos x + \sin^4 x)\,dx = \sin x\,dy$; when $x = \frac{1}{2}\pi$, $y = 1$.

ANS. $y = 2 \sin^2 x \sin^2 \frac{1}{2}x$.

24. $xy(dx - dy) = x^2\,dy + y^2\,dx$.   ANS. $x = y \ln |cxy|$.
25. $a^2(dy - dx) = x^2\,dy + y^2\,dx$; a constant.

ANS. $2 \arctan(y/a) = \ln |c(x + a)/(x - a)|$.

26. $(y - \sin^2 x)\,dx + \sin x\,dy = 0$.   ANS. $y(\csc x - \cot x) = x + c - \sin x$.

27. $(x - y)\,dx + (3x + y)\,dy = 0$; when $x = 2$, $y = -1$.

ANS. $2(x + 2y) + (x + y)\ln(x + y) = 0$.

28. $y\,dx = (2x + 1)(dx - dy)$.   ANS. $3y = (2x + 1) + c(2x + 1)^{-1/2}$.

In solving exercises 29 through 33, recall that the principal value arc sin x of the inverse sine function is restricted as follows: $-\frac{1}{2}\pi \leqq \text{arc sin } x \leqq \frac{1}{2}\pi$.

29. $\sqrt{1 - y^2}\,dx + \sqrt{1 - x^2}\,dy = 0$.

ANS.   arc sin x + arc sin $y = c$, or a part of the ellipse
$x^2 + 2c_1xy + y^2 + c_1^2 - 1 = 0$; where $c_1 = \cos c$.

30. Solve the equation of exercise 29 with the added condition that, when $x = 0$, $y = \frac{1}{2}\sqrt{3}$.   ANS.   arc sin x + arc sin $y = \frac{1}{3}\pi$, or that arc of the ellipse
$x^2 + xy + y^2 = \frac{3}{4}$ that is indicated by a heavy solid line in Figure 8.

31. Solve the equation of exercise 29 with the added condition that, when $x = 0$, $y = -\frac{1}{2}\sqrt{3}$.   ANS.   arc sin x + arc sin $y = -\frac{1}{3}\pi$, or that arc of the ellipse
$x^2 + xy + y^2 = \frac{3}{4}$, that is indicated by a light solid line in Figure 8.

32. Show that after the answers to exercises 30 and 31 have been deleted, the remaining arcs of the ellipse

$$x^2 + xy + y^2 = \tfrac{3}{4}$$

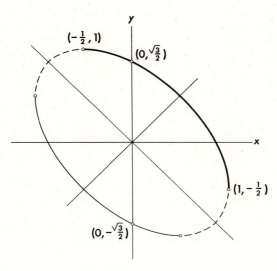

FIGURE 8

are not solutions of the differential equation

$$\sqrt{1 - y^2}\,dx + \sqrt{1 - x^2}\,dy = 0.$$

For this purpose consider the sign of the slope of the curve.

33. For the equation

$$\sqrt{1 - y^2}\, dx - \sqrt{1 - x^2}\, dy = 0$$

state and solve four problems analogous to exercises 29 through 32 above.

34. $v\, du = (e^v + 2uv - 2u)\, dv$.   ANS. $v^2 u = c\, e^{2v} - (v + 1)\, e^v$.

35. $y\, dx = (3x + y^3 - y^2)\, dy$; when $x = 1,\ y = -1$.   ANS. $x = y^2[1 + y \ln(-y)]$.

36. $y^2\, dx - (xy + 2)\, dy = 0$.   ANS. $xy = cy^2 - 1$.

37. $(x^2 - 2xy - y^2)\, dx - (x^2 + 2xy - y^2)\, dy = 0$.

   ANS. $(x + y)(x^2 - 4xy + y^2) = c^3$.

38. $y^2\, dx + (xy + y^2 - 1)\, dy = 0$; when $x = -1,\ y = 1$.

   ANS. $y^2 + 2xy + 1 = 2 \ln y$.

39. $y(y^2 - 3x^2)\, dx + x^3\, dy = 0$.   ANS. $2x^6 = y^2(x^4 + c)$.

40. $y' = \cos x - y \sec x$; when $x = 0,\ y = 1$.

   ANS. $y(1 + \sin x) = \cos x(x + 2 - \cos x)$.

41. Find that solution of $y' = 3x + y$ which passes through the point $(-1, 0)$.

   ANS. $y = -3(x + 1)$.

42. Find that solution of $y' = 3x + y$ which passes through the point $(-1, 1)$.

   ANS. $y = \exp(x + 1) - 3(x + 1)$.

43. $y' = y \tan x + \cos x$.   ANS. $2y = \sin x + (x + c) \sec x$.

44. $(x^2 - 1 + 2y)\, dx + (1 - x^2)\, dy = 0$; when $x = 2,\ y = 1$.

   ANS. $(x + 1)y = (x - 1)[x + 1 + 2 \ln(x - 1)]$.

45. $(x^3 - 3xy^2)\, dx + (y^3 - 3x^2 y)\, dy = 0$.   ANS. $x^4 - 6x^2 y^2 + y^4 = c$.

46. $(1 - x^2)y' = 1 - xy - 3x^2 + 2x^4$.   ANS. $y = x - x^3 + c(1 - x^2)^{1/2}$.

47. $(y^2 + y)\, dx - (y^2 + 2xy + x)\, dy = 0$; when $x = 3,\ y = 1$.   ANS. $2y^2 + y = x$.

48. $(y^3 - x^3)\, dx = xy(x\, dx + y\, dy)$.   ANS. $2x^2 \ln|x + y| = cx^2 + 2xy - y^2$.

49. $y' = \sec x - y \tan x$.   ANS. $y = \sin x + c \cos x$.

50. $x^2 y' = y(1 - x)$.   ANS. $x \ln|cxy| = -1$.

51. $xy' = x - y + xy \tan x$.   ANS. $xy \cos x = c + \cos x + x \sin x$.

52. $(3x^4 y - 1)\, dx + x^5\, dy = 0$; when $x = 1,\ y = 1$.   ANS. $x^4 y = 2x - 1$.

53. $y^2\, dx + x^2\, dy = 2xy\, dy$.   ANS. $y^2 = x(y + c)$.

54. $(\sin x \sin y + \tan x)\, dx - \cos x \cos y\, dy = 0$.   ANS. $\cos x \sin y = \ln|c \sec x|$.

55. $(3xy - 4y - 1)\, dx + x(x - 2)\, dy = 0$; when $x = 1,\ y = 2$.

   ANS. $2x^2(x - 2)y = x^2 - 5$.

Elementary Applications

13. Velocity of escape from the earth

Many physical problems involve differential equations of order one.

Consider the problem of determining the velocity of a particle projected in a radial direction outward from the earth and acted upon by only one force, the gravitational attraction of the earth.

We shall assume an initial velocity in a radial direction so that the motion of the particle takes place entirely on a line through the center of the earth.

According to the Newtonian law of gravitation, the acceleration of the particle will be inversely proportional to the square of the distance from the particle to the center of the earth. Let r be that variable distance, and let R be the radius of the earth. If t represents time, v the velocity of the particle, a its acceleration, and k the constant of proportionality in the Newtonian law, then

$$a = \frac{dv}{dt} = \frac{k}{r^2}.$$

The acceleration is negative because the velocity is decreasing. Hence the constant k is negative. When $r = R$, then $a = -g$, the acceleration of gravity at the surface of the earth. Thus

$$-g = \frac{k}{R^2},$$

from which

$$a = -\frac{gR^2}{r^2}.$$

We wish to express the acceleration in terms of the velocity and the distance. We have $a = dv/dt$ and $v = dr/dt$. Hence

$$a = \frac{dv}{dt} = \frac{dr}{dt}\frac{dv}{dr} = v\frac{dv}{dr},$$

so the differential equation for the velocity is now seen to be

$$v\frac{dv}{dr} = -\frac{gR^2}{r^2}. \tag{1}$$

The method of separation of variables applies to equation (1) and leads at once to the set of solutions

$$v^2 = \frac{2gR^2}{r} + C.$$

Suppose that the particle leaves the earth's surface with the velocity v_0. Then $v = v_0$ when $r = R$, from which the constant C is easily determined to be

$$C = v_0^2 - 2gR.$$

Thus, a particle projected in a radial direction outward from the earth's surface with an initial velocity v_0 will travel with a velocity v given by the equation

$$v^2 = \frac{2gR^2}{r} + v_0^2 - 2gR. \tag{2}$$

It is of considerable interest to determine whether the particle will escape from the earth. Now at the surface of the earth, at $r = R$, the velocity is positive, $v = v_0$. An examination of the right member of equation (2) shows that the velocity of the particle will remain positive if, and only if,

$$v_0^2 - 2gR \geqq 0. \tag{3}$$

If the inequality (3) is satisfied, the velocity given by equation (2) will remain positive because it cannot vanish, is continuous, and is positive at $r = R$. On the other hand, if (3) is not satisfied, then $v_0^2 - 2gR < 0$, and there will be a critical value of r for which the right member of equation (2) is zero. That is, the particle would stop, the velocity would change from positive to negative, and the particle would return to the earth.

A particle projected from the earth with a velocity v_0 such that $v_0 \geqq \sqrt{2gR}$ will escape from the earth. Hence, the minimum such velocity of projection,

$$v_e = \sqrt{2gR}, \tag{4}$$

is called the *velocity of escape.*

The radius of the earth is approximately $R = 3960$ miles. The acceleration of gravity at the surface of the earth is approximately $g = 32.16$ feet per second per second (ft/sec²), or $g = 6.09(10)^{-3}$ miles/sec². For the earth, the velocity of escape is easily found to be $v_e = 6.95$ miles/sec.

Of course, the gravitational pull of other celestial bodies, the moon, the sun, Mars, Venus, and so on has been neglected in the idealized problem treated here. It is not difficult to see that such approximations are justified, since we are interested in only the critical initial velocity v_e. Whether the particle actually recedes from the earth forever or becomes, for instance, a satellite of some heavenly body, is of no consequence in the present problem.

If in this study we happen to be thinking of the particle as an idealization of a ballistic-type rocket, then other elements must be considered. Air resistance in the first few miles may not be negligible. Methods for overcoming such difficulties are not suitable topics for discussion here.

It must be realized that the formula $v_e = \sqrt{2gR}$ applies equally well for the velocity of escape from the other members of the solar system, as long as R and g are given their appropriate values.

14. Newton's law of cooling

Experiment has shown that under certain conditions, a good approximation to the temperature of an object can be obtained by using Newton's law of cooling: The temperature of a body changes at a rate that is proportional to the difference in temperature between the outside medium and the body itself. We shall assume here that the constant of proportionality is the same whether the temperature is increasing or decreasing.

Suppose, for instance, that a thermometer, which has been at the reading 70°F inside a house, is placed outside where the air temperature is 10°F. Three minutes later it is found that the thermometer reading is 25°F. We wish to predict the thermometer reading at various later times.

Let u (°F) represent the temperature of the thermometer at time t (min), the time being measured from the instant the thermometer is placed outside. We are given that when $t = 0$, $u = 70$ and when $t = 3$, $u = 25$.

According to Newton's law, the time rate of change of temperature, du/dt, is proportional to the temperature difference $(u - 10)$. Since the thermometer temperature is decreasing, it is convenient to choose $(-k)$ as the constant of proportionality. Thus the u is to be determined from the differential equation

$$\frac{du}{dt} = -k(u - 10), \tag{1}$$

and the conditions that

$$\text{when } t = 0, u = 70 \tag{2}$$

and

$$\text{when } t = 3, u = 25. \tag{3}$$

We need to know the thermometer reading at two different times because there are two constants to be determined, k in equation (1) and the "arbitrary" constant that occurs in the solution of differential equation (1).

From equation (1) it follows at once that

$$u = 10 + C e^{-kt}.$$

Then condition (2) yields $70 = 10 + C$ from which $C = 60$, so we have

$$u = 10 + 60 e^{-kt}. \tag{4}$$

The value of k will be determined now by using condition (3). Putting $t = 3$ and $u = 25$ into equation (4) we get

$$25 = 10 + 60 e^{-3k},$$

from which $e^{-3k} = \frac{1}{4}$, so $k = \frac{1}{3} \ln 4$.

Thus, the temperature is given by the equation

$$u = 10 + 60 \exp\left(-\tfrac{1}{3}t \ln 4\right). \tag{5}$$

Since $\ln 4 = 1.39$, equation (5) may be replaced by

$$u = 10 + 60 \exp\left(-0.46t\right), \tag{6}$$

which is convenient when a table of values of e^{-x} is available.

15. Simple chemical conversion

It is known from the results of chemical experimentation that, in certain reactions in which a substance A is being converted into another substance, the time rate of change of the amount x of unconverted substance is proportional to x.

Let the amount of unconverted substance be known at some specified time; that is, let $x = x_0$ at $t = 0$. Then the amount x at any time $t > 0$ is determined by the differential equation

$$\frac{dx}{dt} = -kx \tag{1}$$

and the condition that $x = x_0$ when $t = 0$. Since the amount x is decreasing as time increases, the constant of proportionality in equation (1) is taken to be $(-k)$.

From equation (1) it follows that

$$x = C e^{-kt}.$$

But $x = x_0$ when $t = 0$. Hence $C = x_0$. Thus we have the result

$$x = x_0 e^{-kt}. \tag{2}$$

Let us now add another condition, which will enable us to determine k. Suppose it is known that at the end of half a minute, at $t = 30$ (sec), two-thirds of the original amount x_0 has already been converted. Let us determine how much unconverted substance remains at $t = 60$ (sec).

When two-thirds of the substance has been converted, one-third remains unconverted. Hence $x = \frac{1}{3}x_0$ when $t = 30$. Equation (2) now yields the relation

$$\tfrac{1}{3}x_0 = x_0 e^{-30k}$$

from which k is easily found to be $\frac{1}{30} \ln 3$. Then with t measured in seconds, the amount of unconverted substance is given by the equation

$$x = x_0 \exp\left(-\tfrac{1}{30}t \ln 3\right). \tag{3}$$

At $t = 60$,

$$x = x_0 \exp\left(-2 \ln 3\right) = x_0(3)^{-2} = \tfrac{1}{9}x_0.$$

Exercises

1. The radius of the moon is roughly 1080 miles. The acceleration of gravity at the surface of the moon is about $0.165g$, where g is the acceleration of gravity at the surface of the earth. Determine the velocity of escape for the moon.

   ANS. 1.5 miles/sec.

2. Determine, to two significant figures, the velocity of escape for each of the celestial bodies listed below. The data given are rough and g may be taken to be $6.1(10)^{-3}$ mile/sec^2.

	Acceleration of gravity at surface	Radius (miles)	Answer (miles/sec)
Venus	0.85g	3,800	6.3
Mars	0.38g	2,100	3.1
Jupiter	2.6g	43,000	37
Sun	28g	432,000	380
Ganymede	0.12g	1,780	1.6

3. A thermometer reading $18°F$ is brought into a room where the temperature is $70°F$; 1 min later the thermometer reading is $31°F$. Determine the temperature reading as a function of time and, in particular, find the temperature reading 5 min after the thermometer is first brought into the room.
   ANS. $u = 70 - 52\exp(-0.29t)$; when $t = 5, u = 58$.

4. A thermometer reading $75°F$ is taken out where the temperature is $20°F$. The reading is $30°F$ 4 min later. Find (a) the thermometer reading 7 min after the thermometer was brought outside, and (b) the time taken for the reading to drop from $75°F$ to within a half degree of the air temperature.   ANS.   (a) $23°F$; (b) 11.5 min.

5. At 1:00 P.M., a thermometer reading $70°F$ is taken outside where the air temperature is $-10°F$ (ten below zero). At 1:02 P.M., the reading is $26°F$. At 1:05 P.M., the thermometer is taken back indoors where the air is at $70°F$. What is the thermometer reading at 1:09 P.M.?   ANS. $56°F$.

6. At 9 A.M., a thermometer reading $70°F$ is taken outdoors where the temperature is $15°F$. At 9:05 A.M., the thermometer reading is $45°F$. At 9:10 A.M., the thermometer is taken back indoors where the temperature is fixed at $70°F$. Find (a) the reading at 9:20 A.M. and (b) when the reading, to the nearest degree, will show the correct ($70°F$) indoor temperature.   ANS.   (a) $58°F$; (b) 9:46 A.M.

7. At 2:00 P.M., a thermometer reading $80°F$ is taken outside where the air temperature is $20°F$. At 2:03 P.M., the temperature reading yielded by the thermometer is $42°F$. Later, the thermometer is brought inside where the air is at $80°F$. At 2:10 P.M., the reading is $71°F$. When was the thermometer brought indoors?   ANS.   At 2:05 P.M.

8. Suppose that a chemical reaction proceeds according to the law given in Section 15 above. If half the substance A has been converted at the end of 10 sec, find when nine-tenths of the substance will have been converted.   ANS.   33 sec.

9. The conversion of a substance B follows the law used in Section 15 above. If only a fourth of the substance has been converted at the end of 10 sec, find when nine-tenths of the substance will have been converted.   ANS.   80 sec.

10. For a substance C, the time rate of conversion is proportional to the square of the amount x of unconverted substance. Let k be the numerical value of the constant of proportionality and let the amount of unconverted substance be x_0 at time $t = 0$. Determine x for all $t \geq 0$.   ANS. $x = x_0/(1 + x_0kt)$.

11. Two substances, A and B, are being converted into a single compound C. In the laboratory it has been shown that, for these substances, the following law of conversion holds: the time rate of change of the amount x of compound C is proportional to the product of the amounts of unconverted substances A and B. Assume

the units of measure so chosen that one unit of compound C is formed from the combination of one unit of A with one unit of B. If at time $t = 0$ there are a units of substance A, b units of substance B, and none of compound C present, show that the law of conversion may be expressed by the equation

$$\frac{dx}{dt} = k(a - x)(b - x).$$

Solve this equation with the given initial condition.

ANS.   If $b \neq a$, $x = \dfrac{ab[\exp(b - a)kt - 1]}{b\exp(b - a)kt - a}$; if $b = a$, $x = \dfrac{a^2kt}{akt + 1}$.

12. In the solution of exercise 11 above, assume that $k > 0$ and investigate the behavior of x as $t \to \infty$.   ANS.   If $b \geq a$, $x \to a$; if $b \leq a$, $x \to b$.

13. Radium decomposes at a rate proportional to the quantity of radium present. Suppose that it is found that in 25 years approximately 1.1 % of a certain quantity of radium has decomposed. Determine approximately how long it will take for one-half the original amount of radium to decompose.   ANS.   1600 years.

14. A certain radioactive substance has a half-life of 38 hr. Find how long it takes for 90 % of the radioactivity to be dissipated.   ANS.   126 hr.

15. A bacterial population B is known to have a rate of growth proportional to B itself. If between noon and 2 P.M. the population triples, at what time, no controls being exerted, should B become 100 times what it was at noon?   ANS.   About 8:22 P.M.

16. In the motion of an object through a certain medium (air at certain pressures is an example), the medium furnishes a resisting force proportional to the square of the velocity of the moving object. Suppose a body falls, due to the action of gravity, through such a medium. Let t represent time, and v represent velocity, positive downward. Let g be the usual constant acceleration of gravity and let w be the weight of the body. Use Newton's law, force equals mass times acceleration, to conclude that the differential equation of the motion is

$$\frac{w}{g}\frac{dv}{dt} = w - kv^2,$$

where kv^2 is the magnitude of the resisting force furnished by the medium.

17. Solve the differential equation of exercise 16 with the initial condition that $v = v_0$ when $t = 0$. Introduce the constant $a^2 = w/k$ to simplify the formulas.

ANS. $\dfrac{a + v}{a - v} = \dfrac{a + v_0}{a - v_0}\exp\left(\dfrac{2gt}{a}\right)$.

18. List a consistent set of units for the dimensions of the variables and parameters of exercises 16 and 17 above.

ANS. t in sec	g in ft/sec^2
v in ft/sec	k in (lb)(sec^2)/ft^2
w in lb	a in ft/sec

19. There are mediums that resist motion through them with a force proportional to the first power of the velocity. For such a medium, state and solve problems analogous to exercises 16 through 18 above, except that for convenience a constant

$b = w/k$ may be introduced to replace the a^2 of exercise 17. Show that b has the dimensions of a velocity.

ANS. $v = b + (v_0 - b) \exp\left(-\dfrac{gt}{b}\right)$.

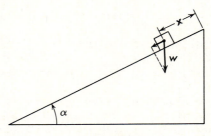

FIGURE 9

20. Figure 9 shows a weight, w pounds (lb), sliding down an inclined plane which makes an angle α with the horizontal. Assume that no force other than gravity is acting on the weight; that is, there is no friction, no air resistance, and so on. At time $t = 0$, let $x = x_0$ and let the initial velocity be v_0. Determine x for $t > 0$.

ANS. $x = \frac{1}{2}gt^2 \sin \alpha + v_0 t + x_0$.

21. A long, very smooth board is inclined at an angle of $10°$ with the horizontal. A weight starts from rest 10 ft from the bottom of the board and slides downward under the action of gravity alone. Find how long it will take the weight to reach the bottom of the board and determine the terminal speed.

ANS.   1.9 sec and 10.5 ft/sec.

22. Add to the conditions of exercise 20 above a retarding force of magnitude kv, where v is the velocity. Determine v and x under the assumption that the weight starts from rest with $x = x_0$. Use the notation $a = kg/w$.

ANS. $v = a^{-1}g \sin \alpha(1 - e^{-at}); \ x = x_0 + a^{-2}g \sin \alpha(-1 + e^{-at} + at)$.

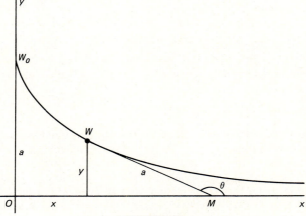

FIGURE 10

23. A man, standing at O in Figure 10, holds a rope of length a to which a weight is attached, initially at W_0. The man walks to the right dragging the weight after him. When the man is at M, the weight is at W. Find the differential equation of the path (called the tractrix) of the weight and solve the equation.

$$\text{ANS.} \quad x = a \ln \frac{a + \sqrt{a^2 - y^2}}{y} - \sqrt{a^2 - y^2}.$$

24. A tank contains 80 gallons (gal) of pure water. A brine solution with 2 lb/gal of salt enters at 2 gal/min, and the well-stirred mixture leaves at the same rate. Find (a) the amount of salt in the tank at any time, and (b) the time at which the brine leaving will contain 1 lb/gal of salt.

$$\text{ANS.} \quad \text{(a) } s = 160[1 - \exp(-t/40)]; \text{(b) } t = 40 \ln 2 \text{ min.}$$

25. For the tank in the previous exercise, determine the limiting value for the amount of salt in the tank after a long time. How much time must pass before the amount of salt in the tank reaches 80 % of this limiting value?

$$\text{ANS.} \quad \text{(a) } s = 160 \text{ lb; (b) } t = 64 \text{ min.}$$

26. A certain sum of money P draws interest compounded continuously. If at a certain time there is P_0 dollars in the account, determine the time when the principal attains the value $2P_0$ dollars, if the annual interest rate is (a) 2 %, or (b) 4 %.

$$\text{ANS.} \quad \text{(a) } 50 \ln 2 \text{ years; (b) } 25 \ln 2 \text{ years.}$$

27. A bank offers 5 % interest compounded continuously in a savings account. Determine (a) the amount of interest earned in 1 year on a deposit of $100 and (b) the equivalent rate if the compounding were done annually.

$$\text{ANS.} \quad \text{(a) } \$5.13; \text{(b) } 5.13 \%.$$

16. Logistic growth and the price of commodities

Numerous attempts have been made to develop models to study the growth of populations. One means of obtaining a simple model for that study is to assume that the average birthrate per individual is a positive constant and that the average death rate per individual is proportional to the population.

If we let $x(t)$ represent the population at time t, then the above assumptions lead to the differential equation

$$\frac{1}{x}\frac{dx}{dt} = b - ax, \tag{1}$$

where b and a are positive constants. This equation is commonly called the *logistic equation* and the growth of population determined by it is called *logistic growth*.

The variables in the logistic equation may be separated to obtain

$$\frac{dx}{x(b - ax)} = dt,$$

or

$$\left(\frac{1}{x} + \frac{a}{b - ax}\right) dx = b \, dt.$$

Integrating both sides gives us

$$\ln\left|\frac{x}{b - ax}\right| = bt + c,$$

or

$$\left|\frac{x}{b - ax}\right| = e^c e^{bt}. \tag{2}$$

To expedite the study of equation (2), let us further assume that at $t = 0$ the population is the positive number x_0. Then equation (2) may be written

$$\frac{x}{b - ax} = \frac{x_0}{b - ax_0} e^{bt},$$

and upon solving for x, we have

$$x(t) = \frac{bx_0 e^{bt}}{b - ax_0 + ax_0 e^{bt}}. \tag{3}$$

It is interesting to note that the population function obtained in equation (3) has a limiting value

$$\lim_{t \to \infty} x(t) = \lim_{t \to \infty} \frac{bx_0 e^{bt}}{b - ax_0 + ax_0 e^{bt}}$$

$$= \lim_{t \to \infty} \frac{b^2 x_0 e^{bt}}{abx_0 e^{bt}}$$

$$= \frac{b}{a},$$

where we have used l'Hospital's rule to evaluate the limit.

We should also note that the logistic equation (1) will dictate a growth or a decline in the population depending upon whether the initial population is less than or greater than b/a.

As a further example of an application in which a first-order differential equation occurs, we consider an economic model of a certain commodity market. We will assume that the price P, the supply S, and the demand D of that commodity are functions of time and that the rate of change of the price is proportional to the difference between the demand and the supply. That is,

$$\frac{dP}{dt} = k(D - S). \tag{4}$$

We further assume that the constant k is positive so that the price will increase if the demand exceeds the supply.

Different models of the commodity market will result depending upon the nature of the demand and supply functions that are indicated. If, for example, we assume that

$$D = c - dP \quad \text{and} \quad S = a + bP, \tag{5}$$

where a, b, c, and d are positive constants, we obtain a differential equation

$$\frac{dP}{dt} = k[(c - a) - (d + b)P] \tag{6}$$

that is linear in P. The assumptions (5) reflect the tendency for the demand to decrease as the price increases and the tendency for the supply to increase as the price increases, both reasonable assumptions for many commodities. We should also assume that $0 < P < c/d$, so that D is not negative.

Equation (6) may be written

$$\frac{dP}{dt} + k(d + b)P = k(c - a), \tag{7}$$

and solved by multiplying by the integrating factor $e^{k(d+b)t}$ and integrating to obtain

$$P(t) = c_1 e^{-k(d+b)t} + \frac{c - a}{d + b}.$$

If the price at $t = 0$ is $P = P_0$, we have

$$c_1 = P_0 - (c - a)/(d + b),$$

so that

$$P(t) = \left(P_0 - \frac{c - a}{d + b} \right) e^{-k(d+b)t} + \frac{c - a}{d + b}. \tag{8}$$

Equation (8) shows that under the assumptions of (4) and (5) the price will stabilize at a value $(c - a)/(d + b)$ as t becomes large.

Exercises.

1. A certain population is known to be growing at a rate given by the logistic equation $dx/dt = x(b - ax)$. Show that the maximum rate of growth will occur when the population is equal to half its equilibrium size, that is, when the population is $b/2a$.

2. A bacterial population is known to have a logistic growth pattern with initial population 1000 and an equilibrium population of 10,000. A count shows that at the end of 1 hr there are 2000 bacteria present. Determine the population as a function of time.

ANS. $x(t) = \dfrac{10{,}000\,e^{bt}}{9 + e^{bt}}$, where $b = \ln\left(\frac{9}{4}\right) \approx 0.81$.

3. For the population of Exercise 2, determine the time at which the population is increasing most rapidly and draw a sketch of the logistic curve.

ANS. $t = \dfrac{\ln 9}{\ln 9 - \ln 4} \approx 2.7\ \text{hr}$.

4. A college dormitory houses 100 students, each of whom is susceptible to a certain virus infection. A simple model of epidemics assumes that during the course of an epidemic the rate of change with respect to time of the number of infected students I is proportional to the number of infected students and also proportional to the number of uninfected students, $100 - I$. (a) If at time $t = 0$ a single student becomes infected, show that the number of infected students at time t is given by

$$I = \frac{100\,e^{100kt}}{99 + e^{100kt}}.$$

(b) If the constant of proportionality k has value 0.01 when t is measured in days, find the value of the rate of new cases $I'(t)$ at the end of each day for the first 9 days.

ANS.   3, 6, 14, 23, 24, 16, 8, 3, 1.

5. Glucose is being fed intravenously into the bloodstream of a patient at a constant rate c grams per minute. At the same time, the patient's body converts the glucose and removes it from the bloodstream at a rate proportional to the amount of glucose present. If the constant of proportionality is k, show that as time increases, the amount of glucose in the bloodstream approaches an equilibrium value of c/k.

6. The supply of food for a certain population is subject to a seasonal change that affects the growth rate of the population. The differential equation $dx/dt = cx(t)\cos t$, where c is a positive constant, provides a simple model for the seasonal growth of the population. Solve the differential equation in terms of an initial population x_0 and the constant c. Determine the maximum and the minimum populations and the time interval between maxima.   ANS.   Max $x = x_0 e^c$; Min $x = x_0 e^{-c}$.

7. Suppose that the human body dissipates a drug at a rate proportional to the amount y of drug present in the bloodstream at time t. At time $t = 0$ a first injection of y_0 grams of the drug is made into a body that was free from that drug prior to that time.

(a) Find the amount of residual drug in the bloodstream at the end of T hours.

(b) If at time T a second injection of y_0 grams is made, find the residual amount of drug at the end of $2T$ hours.

(c) If at the end of each time period of length T, an injection of y_0 grams is made, find the residual amount of drug at the end of nT hours.

(d) Find the limiting value of the answer to part (c) as n approaches infinity.

ANS.   (d) $\dfrac{y_0 e^{-kT}}{1 - e^{-kT}}$.

8. If the demand and supply functions for a commodity market are $D = c - dP$ and $S = a \sin \beta t$, determine $P(t)$ and analyze its behavior as t increases.

ANS. $P(t) = \dfrac{c}{d} + \dfrac{ka}{k^2 d^2 + \beta^2}(\beta \cos \beta t - kd \sin \beta t) + \left(P_0 - \dfrac{c}{d} - \dfrac{ka\beta}{k^2 d^2 + \beta^2} \right) e^{-kdt}$.

9. An analysis of a certain commodity market reveals that the demand and the supply functions are given by $D = c - dP$ and $S = a + bP + q \sin \beta t$, where a, b, c, d, q, and β are positive constants. Determine $P(t)$ and analyze its behavior as t increases.

ANS. $P(t) = \dfrac{c - a}{d + b} + \dfrac{qk}{Q} \cos(\beta t + \alpha) + c_1 e^{-k(d+b)t}$,

where $Q = \sqrt{k^2(d + b)^2 + \beta^2}$, $\alpha = \arccos \dfrac{\beta}{Q}$, and $c_1 = P_0 - \dfrac{c - a}{d + b} - \dfrac{\beta kq}{Q^2}$.

17. Orthogonal trajectories

Suppose that we have a family of curves given by

$$f(x, y, c) = 0, \tag{1}$$

one curve corresponding to each c in some range of values of the parameter c. In certain applications it is found desirable to know what curves have the property of intersecting a curve of the family (1) at right angles whenever they do intersect.

That is, we wish to determine a family of curves with equations

$$g(x, y, k) = 0 \tag{2}$$

such that, at any intersection of a curve of the family (2) with a curve of the family (1), the tangents to the two curves are perpendicular. The families (1) and (2) are then said to be *orthogonal trajectories** of each other.

If two curves are to be orthogonal, then at each point of intersection the slopes of the curves must be negative reciprocals of each other. That fact leads us to a method for finding orthogonal trajectories of a given family of curves. First we find the differential equation of the given family. Then, replacing dy/dx by $-dx/dy$ in that equation yields the differential equation of the orthogonal trajectories to the given curves. It remains only to solve the latter differential equation.

So far we have solved differential equations of only one form,

$$M \, dx + N \, dy = 0.$$

---

  * The word orthogonal comes from the Greek $o\rho\theta\eta$ (right) and $\gamma\omega\nu\iota\alpha$ (angle); the word trajectory comes from the Latin *trajectus* (cut across). Hence a curve that cuts across certain others at right angles is called an orthogonal trajectory of those others.

For such an equation

$$\frac{dy}{dx} = -\frac{M}{N},$$

so the differential equation of the orthogonal trajectories is

$$\frac{dy}{dx} = \frac{N}{M}$$

or

$$N\,dx - M\,dy = 0.$$

EXAMPLE: Find the orthogonal trajectories of all parabolas with vertices at the origin and foci on the x-axis.

The algebraic equation of such parabolas is

$$y^2 = 4ax. \tag{3}$$

Hence, from

$$\frac{y^2}{x} = 4a,$$

we find the differential equation of the family (3) to be

$$2x\,dy - y\,dx = 0. \tag{4}$$

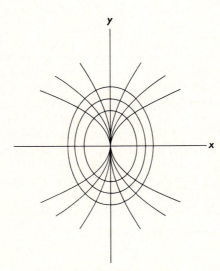

FIGURE 11

Therefore the orthogonal trajectories of the family (3) must satisfy the equation

$$2x\,dx + y\,dy = 0. \tag{5}$$

From (5) it follows that

$$2x^2 + y^2 = b^2, \tag{6}$$

where b is the arbitrary constant. Thus the orthogonal trajectories of (3) are certain ellipses (6) with centers at the origin. See Figure 11.

Exercises

In each exercise, find the orthogonal trajectories of the given family of curves. Draw a few representative curves of each family whenever a figure is requested.

1. $x - 4y = c$. Draw the figure.   ANS. $4x + y = k$.

2. $x^2 + y^2 = c^2$. Draw the figure.   ANS. $y = kx$.

3. $x^2 - y^2 = c_1$. Draw the figure.   ANS. $xy = c_2$.

4. Circles through the origin with centers on the x-axis. Draw the figure.
     ANS.   Circles through the origin with centers on the y-axis.

5. Straight lines with slope and y-intercept equal. Draw the figure.
     ANS. $(x + 1)^2 + y^2 = a^2$.

6. $y^2 = cx^3$. Draw the figure.   ANS. $2x^2 + 3y^2 = k^2$.

7. $e^x + e^{-y} = c_1$.   ANS. $e^y - e^{-x} = c_2$.

8. $y = c_1(\sec x + \tan x)$.   ANS. $y^2 = 2(c_2 - \sin x)$.

9. $x^3 = 3(y - c)$. Draw the figure.   ANS. $x(y - k) = 1$.

10. $x = c\exp(y^2)$.   ANS. $y = c_1\exp(-x^2)$.

11. $y = ce^{-mx}$; with m held fixed. Draw the figure.   ANS. $my^2 = 2(x - c_1)$.

12. Ellipses with centers at $(0,0)$ and two vertices at $(1,0)$ and $(-1,0)$.
     ANS. $x^2 + y^2 = 2\ln|cx|$.

13. $x^2 - y^2 = cx$.   ANS. $y(y^2 + 3x^2) = c_1$.

14. The cissoids, $y^2 = x^3/(a - x)$.   ANS. $(x^2 + y^2)^2 = b(2x^2 + y^2)$.

15. The trisectrices of Maclaurin, $(a + x)y^2 = x^2(3a - x)$.
     ANS. $(x^2 + y^2)^5 = cy^3(5x^2 + y^2)$.

16. $ax^2 + by^2 = c$; with a and b held fixed.   ANS. $y^a = kx^b$.

17. $ax^2 + y^2 = 2acx$; with a held fixed.   ANS.   If $a \neq 2$, $(2 - a)x^2 + y^2 = c_1y^a$.
                                                 If $a = 2$, $x^2 = -y^2\ln(c_2y)$.

18. $y(x^2 + c) + 2 = 0$.   ANS. $y^3 = -3\ln|kx|$.

19. $x^n + y^n = a^n$; with n held fixed, $n \neq 2$.   ANS. $x^{2-n} - y^{2-n} = c$.

20. $y^2 = x^2(1 - cx)$.   ANS. $x^2 + 3y^2 = c_1y$.

21. $y^2 = 4x^2(1 - cx)$.   ANS. $2x^2 = 3y^2(1 - c_1y^2)$.

22. $y^2 = ax^2(1 - cx)$; with a held fixed.
     ANS.   If $a \neq 2$, $(a - 2)x^2 = 3y^2(1 - c_1y^{a-2})$.
                                                 If $a = 2$, $x^2 = -3y^2\ln|c_2y|$.

23. $y(x^2 + 1) = cx$.   ANS. $y^2 = x^2 + 2\ln|k(x^2 - 1)|$.

24. $y = 3x - 1 + ce^{-3x}$.   ANS. $27x = 9y - 1 + ke^{-9y}$.

25. $y^2(2x^2 + y^2) = c^2$.   ANS. $y^2 = 2x^2 \ln |kx|$.

26. $y^4 = c^2(x^2 + 4y^2)$.   ANS. $x^8(2x^2 + 5y^2) = k^2$.

27. $x^4(4x^2 + 3y^2) = c^2$.   ANS. $y^8 = k^2(3x^2 + 2y^2)$.

28. For the family $x^2 + 3y^2 = cy$, find that member of the orthogonal trajectories which passes through $(1, 2)$.   ANS. $y^2 = x^2(3x + 1)$.

Additional Topics on Equations of Order One

18. Integrating factors found by inspection

In Section 11 we found that any linear equation of order one can be solved with the aid of an integrating factor. In Section 19 there is some discussion of tests for the determination of integrating factors.

At present we are concerned with equations that are simple enough to enable us to find integrating factors by inspection. The ability to do this depends largely upon recognition of certain common exact differentials and upon experience.

Below are four exact differentials that occur frequently:

$$d(xy) = x\,dy + y\,dx, \tag{1}$$

$$d\left(\frac{x}{y}\right) = \frac{y\,dx - x\,dy}{y^2}, \tag{2}$$

$$d\left(\frac{y}{x}\right) = \frac{x\,dy - y\,dx}{x^2}, \tag{3}$$

$$d\left(\arctan\frac{y}{x}\right) = \frac{x\,dy - y\,dx}{x^2 + y^2}. \tag{4}$$

Note the homogeneity of the coefficients of dx and dy in each of these differentials.

A differential involving only one variable, such as $x^{-2}\,dx$, is an exact differential.

EXAMPLE (a):   Solve the equation

$$y\,dx + (x + x^3 y^2)\,dy = 0. \tag{5}$$

Let us group the terms of like degree, writing the equation in the form

$$(y\,dx + x\,dy) + x^3 y^2\,dy = 0.$$

Now the combination $(y\,dx + x\,dy)$ attracts attention, so we rewrite the equation, obtaining

$$d(xy) + x^3 y^2\,dy = 0. \tag{6}$$

Since the differential of xy is present in equation (6), any factor that is a function of the product xy will not disturb the integrability of that term. But the other term contains the differential dy, and hence should contain a function of y alone. Therefore, let us divide by $(xy)^3$ and write

$$\frac{d(xy)}{(xy)^3} + \frac{dy}{y} = 0.$$

The equation above is integrable as it stands. A family of solutions is defined by

$$-\frac{1}{2x^2 y^2} + \ln|y| = -\ln|c|,$$

or

$$2x^2 y^2 \ln|cy| = 1.$$

EXAMPLE (b):   Solve the equation

$$y(x^3 - y)\,dx - x(x^3 + y)\,dy = 0. \tag{7}$$

Let us regroup the terms of (7) to obtain

$$x^3(y\,dx - x\,dy) - y(y\,dx + x\,dy) = 0. \tag{8}$$

Recalling that

$$d\left(\frac{x}{y}\right) = \frac{y\,dx - x\,dy}{y^2},$$

we divide the terms of equation (8) throughout by y^2 to get

$$x^3 \, d\!\left(\frac{x}{y}\right) - \frac{d(xy)}{y} = 0. \tag{9}$$

Equation (9) will be made exact by introducing a factor, if it can be found, to make the coefficient of $d(x/y)$ a function of (x/y) and the coefficient of $d(xy)$ a function of (xy). Some skill in obtaining such factors can be developed with a little practice.

There is a straightforward attack on equation (9) which has its good points. Assume that the integrating factor desired is $x^k y^n$, where k and n are to be determined. Applying that factor, we obtain

$$x^{k+3} y^n \, d\!\left(\frac{x}{y}\right) - x^k y^{n-1} \, d(xy) = 0. \tag{10}$$

Since the coefficient of $d(x/y)$ is to be a function of the ratio (x/y), the exponents of x and y in that coefficient must be numerically equal but of opposite sign. That is,

$$k + 3 = -n. \tag{11}$$

In a similar manner, from the coefficient of $d(xy)$ it follows that we must put

$$k = n - 1. \tag{12}$$

From equations (11) and (12) we conclude that $k = -2$, $n = -1$. The desired integrating factor is $x^{-2} y^{-1}$ and (10) becomes

$$\frac{x}{y} \, d\!\left(\frac{x}{y}\right) - \frac{d(xy)}{x^2 y^2} = 0$$

of which a set of solutions is given by

$$\frac{1}{2}\!\left(\frac{x}{y}\right)^2 + \frac{1}{xy} = \frac{c}{2}.$$

Finally, we may write the desired solutions of equation (7) as

$$x^3 + 2y = cxy^2.$$

EXAMPLE (c):   Solve the equation

$$3x^2 y \, dx + (y^4 - x^3) \, dy = 0.$$

Two terms in the coefficients of dx and dy are of degree three, and the other coefficient is not of degree three. Let us regroup the terms to get

$$(3x^2 y \, dx - x^3 \, dy) + y^4 \, dy = 0,$$

or

$$y \, d(x^3) - x^3 \, dy + y^4 \, dy = 0.$$

The form of the first two terms now suggests the numerator in the differential of a quotient, as in

$$d\left(\frac{u}{v}\right) = \frac{v\,du - u\,dv}{v^2}.$$

Therefore we divide each term of our equation by y^2 and obtain

$$\frac{y\,d(x^3) - x^3\,dy}{y^2} + y^2\,dy = 0,$$

or

$$d\left(\frac{x^3}{y}\right) + y^2\,dy = 0.$$

Hence a solution set of the original equation is

$$\frac{x^3}{y} + \frac{y^3}{3} = \frac{c}{3}$$

or

$$3x^3 + y^4 = cy.$$

Exercises

Except when the exercise indicates otherwise, find a set of solutions.

1. $y(2xy + 1)\,dx - x\,dy = 0.$ ANS. $x(xy + 1) = cy.$
2. $y(y^3 - x)\,dx + x(y^3 + x)\,dy = 0.$ ANS. $2xy^3 - x^2 = cy^2.$
3. $(x^3y^3 + 1)\,dx + x^4y^2\,dy = 0.$ ANS. $x^3y^3 = -3\ln|cx|.$
4. $2t\,ds + s(2 + s^2t)\,dt = 0.$ ANS. $1 + s^2t = cs^2t^2.$
5. $y(x^4 - y^2)\,dx + x(x^4 + y^2)\,dy = 0.$ ANS. $y(3x^4 + y^2) = cx^3.$
6. $y(y^2 + 1)\,dx + x(y^2 - 1)\,dy = 0.$ ANS. $x(y^2 + 1) = cy.$
7. Do exercise 6 by a second method.
8. $y(x^3 - y^5)\,dx - x(x^3 + y^5)\,dy = 0.$ ANS. $x^4 = y^4(c + 4xy).$
9. $y(x^2 - y^2 + 1)\,dx - x(x^2 - y^2 - 1)\,dy = 0.$ ANS. $x^2 + cxy + y^2 = 1.$
10. $(x^3 + xy^2 + y)\,dx + (y^3 + x^2y + x)\,dy = 0.$ ANS. $(x^2 + y^2)^2 = c - 4xy.$
11. $y(x^2 + y^2 - 1)\,dx + x(x^2 + y^2 + 1)\,dy = 0.$ ANS. $xy + \arctan(y/x) = c.$
12. $(x^3 + xy^2 - y)\,dx + (y^3 + x^2y + x)\,dy = 0.$ ANS. $2\arctan(y/x) = c - x^2 - y^2.$
13. $y(x^3e^{xy} - y)\,dx + x(y + x^3e^{xy})\,dy = 0.$ ANS. $2x^2\,e^{xy} + y^2 = cx^2.$
14. $xy(y^2 + 1)\,dx + (x^2y^2 - 2)\,dy = 0$; when $x = 1$, $y = 1.$ ANS. $x^2(y^2 + 1) = 2 + 4\ln y.$
15. $y^2(1 - x^2)\,dx + x(x^2y + 2x + y)\,dy = 0.$ ANS. $x^2y + x + y = cxy^2.$
16. $y(x^2y^2 - 1)\,dx + x(x^2y^2 + 1)\,dy = 0.$ ANS. $x^2y^2 = 2\ln|cx/y|.$

17. $x^4 y' = -x^3 y - \csc(xy)$.   ANS. $2x^2 \cos(xy) = cx^2 - 1$.

18. $[1 + y \tan(xy)]\, dx + x \tan(xy)\, dy = 0$.   ANS. $\cos(xy) = c\, e^x$.

19. $y(x^2 y^2 - m)\, dx + x(x^2 y^2 + n)\, dy = 0$.   ANS. $x^2 y^2 = 2 \ln |cx^m/y^n|$.

20. $x(x^2 - y^2 - x)\, dx - y(x^2 - y^2)\, dy = 0$; when $x = 2,\ y = 0$.
            ANS. $3(x^2 - y^2)^2 = 4(x^3 + 4)$.

21. $y(x^2 + y)\, dx + x(x^2 - 2y)\, dy = 0$; when $x = 1,\ y = 2$.
            ANS. $x^2 y - y^2 + 2x = 0$.

22. $y(x^3 y^3 + 2x^2 - y)\, dx + x^3(xy^3 - 2)\, dy = 0$; when $x = 1,\ y = 1$.
           ANS. $x^3 y^3 + 4x^2 - 7xy + 2y = 0$.

23. $y(2 - 3xy)\, dx - x\, dy = 0$.   ANS. $x^2(1 - xy) = cy$.

24. $y(2x + y^2)\, dx + x(y^2 - x)\, dy = 0$.   ANS. $x(x + y^2) = cy$.

25. $y\, dx + 2(y^4 - x)\, dy = 0$.   ANS. $y^4 + x = cy^2$.

26. $y(3x^3 - x + y)\, dx + x^2(1 - x^2)\, dy = 0$.   ANS. $y \ln |cx| = x(1 - x^2)$.

27. $2x^5 y' = y(3x^4 + y^2)$.   ANS. $x^4 = y^2(1 + cx)$.

28. $(x^n y^{n+1} + ay)\, dx + (x^{n+1} y^n + bx)\, dy = 0$.
         ANS.   If $n \neq 0,\ x^n y^n = n \ln |cx^{-a} y^{-b}|$.
                 If $n = 0,\ xy = c_1 x^{-a} y^{-b}$.

29. $(x^{n+1} y^n + ay)\, dx + (x^n y^{n+1} + ax)\, dy = 0$.
         ANS.   If $n \neq 1,\ (n - 1)(xy)^{n-1}(x^2 + y^2 - c) = 2a$.
            If $n = 1,\ x^2 + y^2 - c = -2a \ln |xy|$.

19. The determination of integrating factors

Let us see what progress can be made on the problem of the determination of an integrating factor for the equation

$$M\, dx + N\, dy = 0. \tag{1}$$

Suppose that u, possibly a function of both x and y, is to be an integrating factor of (1). Then the equation

$$uM\, dx + uN\, dy = 0 \tag{2}$$

must be exact. Therefore, by the result of Section 10,

$$\frac{\partial}{\partial y}(uM) = \frac{\partial}{\partial x}(uN).$$

Hence u must satisfy the partial differential equation

$$u \frac{\partial M}{\partial y} + M \frac{\partial u}{\partial y} = u \frac{\partial N}{\partial x} + N \frac{\partial u}{\partial x},$$

or

$$u \left(\frac{\partial M}{\partial y} - \frac{\partial N}{\partial x} \right) = N \frac{\partial u}{\partial x} - M \frac{\partial u}{\partial y}. \tag{3}$$

Furthermore, by reversing the argument above, it can be seen that, if u satisfies equation (3), u is an integrating factor for equation (1). We have "reduced" the problem of solving the ordinary differential equation (1) to the problem of obtaining a particular solution of the partial differential equation (3).

Not much has been gained because we have developed no methods for attacking an equation such as (3). Therefore, we turn the problem back into the realm of ordinary differential equations by restricting u to be a function of only one variable.

First let u be a function of x alone. Then $\partial u/\partial y = 0$ and $\partial u/\partial x$ becomes du/dx. Then (3) reduces to

$$u\left(\frac{\partial M}{\partial y} - \frac{\partial N}{\partial x}\right) = N\frac{du}{dx},$$

or

$$\frac{1}{N}\left(\frac{\partial M}{\partial y} - \frac{\partial N}{\partial x}\right)dx = \frac{du}{u}. \tag{4}$$

If the left member of the above equation is a function of x alone, then we can determine u at once. Indeed, if

$$\frac{1}{N}\left(\frac{\partial M}{\partial y} - \frac{\partial N}{\partial x}\right) = f(x), \tag{5}$$

then the desired integrating factor is $u = \exp\left(\int f(x)\,dx\right)$.

By a similar argument, assuming that u is a function of y alone, we are led to the conclusion that if

$$\frac{1}{M}\left(\frac{\partial M}{\partial y} - \frac{\partial N}{\partial x}\right) = g(y), \tag{6}$$

then an integrating factor for equation (1) is $u = \exp\left(-\int g(y)\,dy\right)$.

Our two results are expressed in the following rules:

(a) If $\dfrac{1}{N}\left(\dfrac{\partial M}{\partial y} - \dfrac{\partial N}{\partial x}\right) = f(x)$, a function of x alone, then $\exp\left(\int f(x)\,dx\right)$ is an integrating factor for the equation

$$M\,dx + N\,dy = 0. \tag{1}$$

(b) If $\dfrac{1}{M}\left(\dfrac{\partial M}{\partial y} - \dfrac{\partial N}{\partial x}\right) = g(y)$, a function of y alone, then $\exp\left(-\int g(y)\,dy\right)$

is an integrating factor for equation (1).

It should be emphasized that if neither of the preceding criteria is satisfied, we can say only that the equation does not have an integrating factor that is a function of x or y alone. For example, the student should show that the above criteria fail in the case of Example (a) of Section 18, even though $(xy)^{-3}$ is an integrating factor for the differential equation.

EXAMPLE (a): Solve the equation

$$(4xy + 3y^2 - x)\,dx + x(x + 2y)\,dy = 0. \tag{7}$$

Here $M = 4xy + 3y^2 - x$, $N = x^2 + 2xy$, so

$$\frac{\partial M}{\partial y} - \frac{\partial N}{\partial x} = 4x + 6y - (2x + 2y) = 2x + 4y.$$

Hence

$$\frac{1}{N}\left(\frac{\partial M}{\partial y} - \frac{\partial N}{\partial x}\right) = \frac{2x + 4y}{x(x + 2y)} = \frac{2}{x}.$$

Therefore an integrating factor for equation (7) is

$$\exp\left(2\int \frac{dx}{x}\right) = \exp\left(2\ln|x|\right) = x^2.$$

Returning to the original equation (7), we insert the integrating factor and obtain

$$(4x^3y + 3x^2y^2 - x^3)\,dx + (x^4 + 2x^3y)\,dy = 0, \tag{8}$$

which we know must be an exact equation. The methods of Section 10 apply. We are then led to put equation (8) in the form

$$(4x^3y\,dx + x^4\,dy) + (3x^2y^2\,dx + 2x^3y\,dy) - x^3\,dx = 0,$$

from which the solution set

$$x^4y + x^3y^2 - \tfrac{1}{4}x^4 = \tfrac{1}{4}c,$$

or

$$x^3(4xy + 4y^2 - x) = c$$

follows at once.

EXAMPLE (b): Solve the equation

$$y(x + y + 1)\,dx + x(x + 3y + 2)\,dy = 0. \tag{9}$$

First we form

$$\frac{\partial M}{\partial y} = x + 2y + 1, \qquad \frac{\partial N}{\partial x} = 2x + 3y + 2.$$

Then we see that

$$\frac{\partial M}{\partial y} - \frac{\partial N}{\partial x} = -x - y - 1,$$

so

$$\frac{1}{N}\left(\frac{\partial M}{\partial y} - \frac{\partial N}{\partial x}\right) = -\frac{x + y + 1}{x(x + 3y + 2)}$$

is not a function of x alone. But

$$\frac{1}{M}\left(\frac{\partial M}{\partial y} - \frac{\partial N}{\partial x}\right) = -\frac{x + y + 1}{y(x + y + 1)} = -\frac{1}{y}.$$

Therefore $\exp(\ln|y|) = |y|$ is the desired integrating factor for (9).

It follows that for $y > 0$, y itself is an integrating factor of equation (9) and for $y < 0$, $-y$ is an integrating factor. In either case (9) becomes

$$(xy^2 + y^3 + y^2)\,dx + (x^2y + 3xy^2 + 2xy)\,dy = 0,$$

or

$$(xy^2\,dx + x^2y\,dy) + (y^3\,dx + 3xy^2\,dy) + (y^2\,dx + 2xy\,dy) = 0.$$

Then a set of solutions of (9) is defined implicitly by

$$\tfrac{1}{2}x^2y^2 + xy^3 + xy^2 = \tfrac{1}{2}c,$$

or

$$xy^2(x + 2y + 2) = c.$$

EXAMPLE (c):   Solve the equation

$$y(x + y)\,dx + (x + 2y - 1)\,dy = 0. \tag{10}$$

From $\partial M/\partial y = x + 2y$, $\partial N/\partial x = 1$, we conclude at once that

$$\frac{1}{N}\left(\frac{\partial M}{\partial y} - \frac{\partial N}{\partial x}\right) = \frac{x + 2y - 1}{x + 2y - 1} = 1.$$

Hence e^x is an integrating factor for (10). Then

$$(xy\,e^x + y^2\,e^x)\,dx + (x\,e^x + 2y\,e^x - e^x)\,dy = 0$$

is an exact equation. Grouping the terms in the following manner,

$$[xy\,e^x\,dx + (x\,e^x - e^x)\,dy] + (y^2\,e^x\,dx + 2y\,e^x\,dy) = 0,$$

leads us at once to the family of solutions defined by

$$e^x(x - 1)y + y^2\,e^x = c,$$

or

$$y(x + y - 1) = c\,e^{-x}.$$

Exercises

Solve each of the following equations.

1. $(x^2 + y^2 + 1) dx + x(x - 2y) dy = 0.$ ANS. $x^2 - y^2 + xy - 1 = cx.$
2. $2y(x^2 - y + x) dx + (x^2 - 2y) dy = 0.$ ANS. $y(x^2 - y) = c e^{-2x}.$
3. $y(2x - y + 1) dx + x(3x - 4y + 3) dy = 0.$ ANS. $xy^3(x - y + 1) = c.$
4. $y(4x + y) dx - 2(x^2 - y) dy = 0.$ ANS. $2x^2 + xy + 2y \ln|y| = cy.$
5. $(xy + 1) dx + x(x + 4y - 2) dy = 0.$ ANS. $xy + \ln|x| + 2y^2 - 2y = c.$
6. $(2y^2 + 3xy - 2y + 6x) dx + x(x + 2y - 1) dy = 0.$
    ANS. $x^2(y^2 + xy - y + 2x) = c.$

7. $y(y + 2x - 2) dx - 2(x + y) dy = 0.$ ANS. $y(2x + y) = c e^x.$
8. $y^2 dx + (3xy + y^2 - 1) dy = 0.$ ANS. $y^2(y^2 + 4xy - 2) = c.$
9. $2y(x + y + 2) dx + (y^2 - x^2 - 4x - 1) dy = 0.$
    ANS. $x^2 + 2xy + y^2 + 4x + 1 = cy.$

10. $2(2y^2 + 5xy - 2y + 4) dx + x(2x + 2y - 1) dy = 0.$
    ANS. $x^4(y^2 + 2xy - y + 2) = c.$

11. $3(x^2 + y^2) dx + x(x^2 + 3y^2 + 6y) dy = 0.$ ANS. $x(x^2 + 3y^2) = c e^{-y}.$
12. $y(8x - 9y) dx + 2x(x - 3y) dy = 0.$ ANS. $x^3 y(2x - 3y) = c.$
13. Do exercise 12 by another method.
14. $y(2x^2 - xy + 1) dx + (x - y) dy = 0.$ ANS. $y(2x - y) = c \exp(-x^2).$
15. Euler's theorem (exercise 36, page 30) on homogeneous functions states that, if F is a homogeneous function of degree k in x and y, then

$$x\frac{\partial F}{\partial x} + y\frac{\partial F}{\partial y} = kF.$$

Use Euler's theorem to prove the result that, if M and N are homogeneous functions of the same degree, and if $Mx + Ny \neq 0$, then

$$\frac{1}{Mx + Ny}$$

is an integrating factor for the equation

$$M dx + N dy = 0. \tag{A}$$

16. In the result to be proved in exercise 15 above there is an exceptional case, namely, when $Mx + Ny = 0$. Solve equation (A) when $Mx + Ny = 0.$ ANS. $y = cx.$

Use the integrating factor in the result of exercise 15 above to solve each of the equations in exercises 17 through 20.

17. $xy dx - (x^2 + 2y^2) dy = 0.$ ANS. $x^2 = 4y^2 \ln|y/c|.$
18. $v^2 dx + x(x + v) dv = 0.$ ANS. $xv^2 = c(x + 2v).$
19. $v(u^2 + v^2) du - u(u^2 + 2v^2) dv = 0.$ ANS. $u^2 = 2v^2 \ln|cv^2/u|.$
20. $(x^2 + y^2) dx - xy dy = 0.$ (Exercise 9, page 29.)
21. Apply the method of this section to the general linear equation of order one.

20. Substitution suggested by the equation

An equation of the form

$$M\,dx + N\,dy = 0$$

may not yield at once (or at all) to the methods of Chapter 2. Even then the usefulness of those methods is not exhausted. It may be possible by some change of variables to transform the equation into a type that we know how to solve.

A natural source of suggestions for useful transformations is the differential equation itself. If a particular function of one or both variables stands out in the equation, then it is worthwhile to examine the equation after that function has been introduced as a new variable. For instance, in the equation

$$(x + 2y - 1)\,dx + 3(x + 2y)\,dy = 0 \tag{1}$$

the combination $(x + 2y)$ occurs twice and thus attracts attention. Hence we put

$$x + 2y = v,$$

and, because no other function of x and y stands out, we retain either x or y for the other variable. The solution is completed in Example (a) below.

In the equation

$$(1 + 3x \sin y)\,dx - x^2 \cos y\,dy = 0, \tag{2}$$

the presence of both $\sin y$ and its differential $\cos y\,dy$, and the fact that y appears in the equation in no other manner, leads us to put $\sin y = w$ and to obtain the differential equation in w and x. See Example (b) below.

EXAMPLE (a):   Solve the equation

$$(x + 2y - 1)\,dx + 3(x + 2y)\,dy = 0. \tag{1}$$

As suggested above, put

$$x + 2y = v.$$

Then

$$dx = dv - 2\,dy$$

and equation (1) becomes

$$(v - 1)(dv - 2\,dy) + 3v\,dy = 0,$$

or

$$(v - 1)\,dv + (v + 2)\,dy = 0.$$

Now the variables can be separated. From the equation in the form

$$\frac{v-1}{v+2}\,dv + dy = 0,$$

we get

$$\left(1 - \frac{3}{v+2}\right)dv + dy = 0$$

and then

$$v - 3\ln|v+2| + y + c = 0.$$

But $v = x + 2y$, so our final result is

$$x + 3y + c = 3\ln|x+2y+2|.$$

EXAMPLE (b):   Solve the equation

$$(1 + 3x\sin y)\,dx - x^2\cos y\,dy = 0. \tag{2}$$

Put $\sin y = w$. Then $\cos y\,dy = dw$ and (2) becomes

$$(1 + 3xw)\,dx - x^2\,dw = 0,$$

an equation linear in w. From the standard form

$$dw - \frac{3}{x}w\,dx = \frac{dx}{x^2}$$

an integrating factor is seen to be

$$\exp(-3\ln|x|) = |x|^{-3}.$$

Application of the integrating factor yields the exact equation

$$x^{-3}\,dw - 3x^{-4}w\,dx = x^{-5}\,dx,$$

for either $x > 0$ or $x < 0$, from which we get

$$x^{-3}w = -\tfrac{1}{4}x^{-4} + \tfrac{1}{4}c,$$

or

$$4xw = cx^4 - 1.$$

Hence (2) has the solution set

$$4x\sin y = cx^4 - 1.$$

21. Bernoulli's equation

A well-known equation that fits into the category of Section 20 is Bernoulli's equation,

$$y' + P(x)y = Q(x)y^n. \tag{1}$$

If $n = 1$ in (1), the variables are separable, so we concentrate on the case $n \neq 1$. Equation (1) may be put in the form

$$y^{-n} \, dy + Py^{-n+1} \, dx = Q \, dx. \tag{2}$$

But the differential of y^{-n+1} is $(1 - n)y^{-n} \, dy$, so equation (2) may be simplified by putting

$$y^{-n+1} = z,$$

from which

$$(1 - n)y^{-n} \, dy = dz.$$

Thus the equation in z and x is

$$dz + (1 - n)Pz \, dx = (1 - n)Q \, dx,$$

a linear equation in standard form. Hence any Bernoulli equation can be solved with the aid of the above change of dependent variable (unless $n = 1$, when no substitution is needed).

EXAMPLE (a):   Solve the equation

$$y(6y^2 - x - 1) \, dx + 2x \, dy = 0. \tag{3}$$

First let us group the terms according to powers of y, writing

$$2x \, dy - y(x + 1) \, dx + 6y^3 \, dx = 0.$$

Now it can be seen that the equation is a Bernoulli equation, since it involves only terms containing respectively dy, y, and y^n ($n = 3$ here). Therefore, we divide throughout by y^3, obtaining

$$2xy^{-3} \, dy - y^{-2}(x + 1) \, dx = -6 \, dx.$$

This equation is linear in y^{-2}, so we put $y^{-2} = v$, obtain $dv = -2y^{-3} \, dy$, and need to solve the equation

$$x \, dv + v(x + 1) \, dx = 6 \, dx,$$

or

$$dv + v(1 + x^{-1}) \, dx = 6x^{-1} \, dx. \tag{4}$$

Since

$$\exp{(x + \ln{|x|})} = |x|\, e^x$$

is an integrating factor for (4), the equation

$$x\, e^x\, dv + v\, e^x(x + 1)\, dx = 6\, e^x\, dx$$

is exact. Its solution set

$$xv\, e^x = 6\, e^x + c,$$

together with $v = y^{-2}$, leads us to the final result

$$y^2(6 + c\, e^{-x}) = x.$$

EXAMPLE (b): Solve the equation

$$6y^2\, dx - x(2x^3 + y)\, dy = 0. \tag{5}$$

This is a Bernoulli equation with x as the dependent variable, so it can be treated in the manner used in Example (a). That method of attack is left for the exercises.

Equation (5) can equally well be treated as follows. Note that if each member of (5) is multiplied by x^2, the equation becomes

$$6y^2x^2\, dx - x^3(2x^3 + y)\, dy = 0. \tag{6}$$

In (6), the variable x appears only in the combinations x^3 and its differential $3x^2\, dx$. Hence a reasonable choice for a new variable is $w = x^3$. The equation in w and y is

$$2y^2\, dw - w(2w + y)\, dy = 0,$$

an equation with coefficients homogeneous of degree two in y and w. The further change of variable $w = zy$ leads to the equation

$$2y\, dz - z(2z - 1)\, dy = 0,$$

$$\frac{4\, dz}{2z - 1} - \frac{2\, dz}{z} - \frac{dy}{y} = 0.$$

Therefore, we have

$$2 \ln{|2z - 1|} - 2 \ln{|z|} - \ln{|y|} = \ln{|c|},$$

or

$$(2z - 1)^2 = cyz^2.$$

But $z = w/y = x^3/y$, so the solutions we seek are determined by

$$(2x^3 - y)^2 = cyx^6.$$

Exercises

In exercises 1 through 21, solve the equation.

1. $(3x - 2y + 1) dx + (3x - 2y + 3) dy = 0.$
ANS. $5(x + y + c) = 2 \ln |15x - 10y + 11|.$

2. $\sin y(x + \sin y) dx + 2x^2 \cos y dy = 0.$
ANS. $x^3 \sin^2 y = c(3x + \sin y)^2.$

3. $dy/dx = (9x + 4y + 1)^2.$
ANS. $3 \tan (6x + c) = 2(9x + 4y + 1).$

4. $y' = y - xy^3 e^{-2x}.$
ANS. $e^{2x} = y^2(x^2 + c).$

5. $dy/dx = \sin (x + y).$
ANS. $x + c = \tan (x + y) - \sec (x + y).$

6. $xy dx + (x^2 - 3y) dy = 0.$
ANS. $x^2 y^2 = 2y^3 + c.$

7. $(3 \tan x - 2 \cos y) \sec^2 x dx + \tan x \sin y dy = 0.$
ANS. $\cos y \tan^2 x = \tan^3 x + c.$

8. $(x + 2y - 1) dx + (2x + 4y - 3) dy = 0.$ Solve by two methods.
ANS. $(x + 2y - 1)^2 = 2y + c.$

9. Solve the equation $6y^2 dx - x(2x^3 + y) dy = 0$ of Example (b) above by treating it as a Bernoulli equation in the dependent variable x.

10. $2x^3 y' = y(y^2 + 3x^2).$ Solve by two methods.
ANS. $y^2(c - x) = x^3.$

11. $(3 \sin y - 5x) dx + 2x^2 \cot y dy = 0.$
ANS. $x^3(\sin y - x)^2 = c \sin^2 y.$

12. $y' = 1 + 6x \exp (x - y).$
ANS. $\exp (y - x) = 3x^2 + c.$

13. $dv/du = (u - v)^2 - 2(u - v) - 2.$
ANS. $(u - v - 3) \exp (4u) = c(u - v + 1).$

14. $2y dx + x(x^2 \ln y - 1) dy = 0.$
ANS. $y(1 + x^2 - x^2 \ln y) = cx^2.$

15. $\cos y \sin 2x dx + (\cos^2 y - \cos^2 x) dy = 0.$
ANS. $\cos^2 x(1 + \sin y) = \cos y(y + c - \cos y).$

16. $(ke^{2v} - u) du = 2 e^{2v}(e^{2v} + ku) dv.$
ANS. $2k \arctan (u e^{-2v}) = \ln |c(u^2 + e^{4v})|.$

17. $y' \tan x \sin 2y = \sin^2 x + \cos^2 y.$
ANS. $(\sin^2 x + 3 \cos^2 y) \sin x = c.$

18. $(x + 2y - 1) dx - (x + 2y - 5) dy = 0.$

19. $y(x \tan x + \ln y) dx + \tan x dy = 0.$
ANS. $\sin x \ln y = x \cos x - \sin x + c.$

20. $xy' - y = x^k y^n$, where $n \neq 1$ and $k + n \neq 1.$
ANS. $(k + n - 1)y^{1-n} = (1 - n)x^k + cx^{1-n}.$

21. Solve the equation of exercise 20 for the values of k and n not included there.
ANS. If $n = 1$ and $k \neq 0$, $x^k = k \ln |cy/x|.$
If $n = 1$ and $k = 0$, $y = cx^2.$
If $n \neq 1$ but $k + n = 1$, $y^{1-n} = (1 - n)x^{1-n} \ln |cx|.$

In exercises 22 through 27, find the particular solution required.

22. $4(3x + y - 2) dx - (3x + y) dy = 0$; when $x = 1, y = 0.$
ANS. $7(4x - y - 4) = 8 \ln \dfrac{21x + 7y - 8}{13}.$

23. $y' = 2(3x + y)^2 - 1$; when $x = 0, y = 1.$
ANS. $4 \arctan (3x + y) = 8x + \pi.$

24. $2xyy' = y^2 - 2x^3.$ Find the solution that passes through the point $(1, 2).$
ANS. $y^2 = x(5 - x^2).$

25. $(y^4 - 2xy) dx + 3x^2 dy = 0$; when $x = 2, y = 1.$
ANS. $x^2 = y^3(x + 2).$

26. $(2y^3 - x^3)\,dx + 3xy^2\,dy = 0$; when $x = 1$, $y = 1$. Solve by two methods.

ANS. $5x^2y^3 = x^5 + 4$.

27. $(x^2 + 6y^2)\,dx - 4xy\,dy = 0$; when $x = 1$, $y = 1$. Solve by three methods.

ANS. $2y^2 = x^2(3x - 1)$.

22. Coefficients linear in the two variables

Consider the equation

$$(a_1x + b_1y + c_1)\,dx + (a_2x + b_2y + c_2)\,dy = 0, \tag{1}$$

in which the a's, b's, and c's are constants. We know already how to solve the special case in which $c_1 = 0$ and $c_2 = 0$ because then the coefficients in (1) are each homogeneous and of degree one in x and y. It is reasonable, therefore, to attempt to reduce equation (1) to that situation.

In connection with (1) consider the lines

$$a_1x + b_1y + c_1 = 0, \tag{2}$$

$$a_2x + b_2y + c_2 = 0.$$

They may be parallel or they may intersect. There will not be two lines if a_1 and b_1 are zero or if a_2 and b_2 are zero, but equation (1) will then be linear in one of its variables.

If the lines (2) intersect, let the point of intersection be (h, k). Then the translation

$$x = u + h, \tag{3}$$

$$y = v + k$$

will change the equations (2) into equations of lines through the origin of the uv coordinate system, namely,

$$a_1u + b_1v = 0, \tag{4}$$

$$a_2u + b_2v = 0.$$

Therefore, since $dx = du$ and $dy = dv$, the change of variables

$$x = u + h,$$

$$y = v + k,$$

where (h, k) is the point of intersection of the lines (2), will transform the differential equation (1) into

$$(a_1u + b_1v)\,du + (a_2u + b_2v)\,dv = 0, \tag{5}$$

an equation that we know how to solve.

If the lines (2) do not intersect, a constant k exists such that

$$a_2 x + b_2 y = k(a_1 x + b_1 y),$$

so that equation (1) appears in the form

$$(a_1 x + b_1 y + c_1)\, dx + [k(a_1 x + b_1 y) + c_2]\, dy = 0. \tag{6}$$

The recurrence of the expression $(a_1 x + b_1 y)$ in (6) suggests the introduction of a new variable $w = a_1 x + b_1 y$. Then the new equation, in w and x or in w and y, is one with variables separable, since its coefficients contain only w and constants.

EXAMPLE (a):   Solve the equation

$$(x + 2y - 4)\, dx - (2x + y - 5)\, dy = 0. \tag{7}$$

The lines

$$x + 2y - 4 = 0,$$

$$2x + y - 5 = 0$$

intersect at the point (2, 1). Hence put

$$x = u + 2,$$

$$y = v + 1.$$

Then equation (7) becomes

$$(u + 2v)\, du - (2u + v)\, dv = 0, \tag{8}$$

which has coefficients homogeneous and of degree one in u and v. Therefore let $u = vz$, which transforms (8) into

$$(z + 2)(z\, dv + v\, dz) - (2z + 1)\, dv = 0,$$

or

$$(z^2 - 1)\, dv + v(z + 2)\, dz = 0.$$

Separation of the variables v and z leads us to the equation

$$\frac{dv}{v} + \frac{(z + 2)\, dz}{z^2 - 1} = 0.$$

With the aid of partial fractions, we can write the above equation in the form

$$\frac{2\, dv}{v} + \frac{3\, dz}{z - 1} - \frac{dz}{z + 1} = 0.$$

Hence we get

$$2 \ln |v| + 3 \ln |z - 1| - \ln |z + 1| = \ln |c|$$

from which it follows that

$$v^2(z - 1)^3 = c(z + 1),$$

or

$$(vz - v)^3 = c(vz + v).$$

Now $vz = u$, so a set of solutions appears as

$$(u - v)^3 = c(u + v).$$

But $u = x - 2$ and $v = y - 1$. Therefore the desired result in terms of x and y is

$$(x - y - 1)^3 = c(x + y - 3).$$

For other methods of solution of equation (7), see exercises 23 and 30 below.

EXAMPLE (b):   Solve the equation

$$(2x + 3y - 1)\, dx + (2x + 3y + 2)\, dy = 0, \tag{9}$$

with the condition that $y = 3$ when $x = 1$.

The lines

$$2x + 3y - 1 = 0$$

and

$$2x + 3y + 2 = 0$$

are parallel. Therefore we proceed, as we should have upon first glancing at the equation, to put

$$2x + 3y = v.$$

Then $2\, dx = dv - 3\, dy$, and equation (9) is transformed into

$$(v - 1)(dv - 3\, dy) + 2(v + 2)\, dy = 0,$$

or

$$(v - 1)\, dv - (v - 7)\, dy = 0. \tag{10}$$

Equation (10) is easily solved, leading us to the relation

$$v - y + c + 6 \ln |v - 7| = 0.$$

Therefore a solution set of (9) is

$$2x + 2y + c = -6 \ln |2x + 3y - 7|.$$

But $y = 3$ when $x = 1$, so $c = -8 - 6 \ln 4$. Hence the particular solution required is given by

$$x + y - 4 = -3 \ln [\tfrac{1}{4}(2x + 3y - 7)].$$

Exercises

In exercises 1 through 18, solve the equations.

1. $(y - 2) dx - (x - y - 1) dy = 0.$ ANS. $x - 3 = (2 - y) \ln |c(y - 2)|.$
2. $(x - 4y - 9) dx + (4x + y - 2) dy = 0.$

ANS. $\ln [(x - 1)^2 + (y + 2)^2] - 8 \arc\tan \dfrac{x - 1}{y + 2} = c.$

3. $(2x - y) dx + (4x + y - 6) dy = 0.$ ANS. $(x + y - 3)^3 = c(2x + y - 4)^2.$
4. $(x - 4y - 3) dx - (x - 6y - 5) dy = 0.$ ANS. $(x - 2y - 1)^2 = c(x - 3y - 2).$
5. $(2x + 3y - 5) dx + (3x - y - 2) dy = 0.$ Solve by two methods.
6. $2 dx + (2x - y + 3) dy = 0.$ Use a change of variable.

ANS. $y + c = -\ln |2x - y + 4|.$

7. Solve the equation of exercise 6 by using the fact that the equation is linear in x.
8. $(x - y + 2) dx + 3 dy = 0.$ ANS. $x + c = 3 \ln |x - y + 5|.$
9. Solve exercise 8 by another method.
10. $(x + y - 1) dx + (2x + 2y + 1) dy = 0.$ ANS. $x + 2y + c = 3 \ln |x + y + 2|.$
11. $(3x + 2y + 7) dx + (2x - y) dy = 0.$ Solve by two methods.
12. $(x - 2) dx + 4(x + y - 1) dy = 0.$ ANS. $2(y + 1) = -(x + 2y) \ln |c(x + 2y)|.$
13. $(x - 3y + 2) dx + 3(x + 3y - 4) dy = 0.$

ANS. $\ln [(x - 1)^2 + 9(y - 1)^2] - 2 \arc\tan \dfrac{x - 1}{3(y - 1)} = c.$

14. $(6x - 3y + 2) dx - (2x - y - 1) dy = 0.$ ANS. $3x - y + c = 5 \ln |2x - y + 4|.$
15. $(9x - 4y + 4) dx - (2x - y + 1) dy = 0.$

ANS. $y - 1 = 3(y - 3x - 1) \ln |c(3x - y + 1)|.$

16. $(x + 3y - 4) dx + (x + 4y - 5) dy = 0.$

ANS. $y - 1 = (x + 2y - 3) \ln |c(x + 2y - 3)|.$

17. $(x + 2y - 1) dx - (2x + y - 5) dy = 0.$ ANS. $(x - y - 4)^3 = c(x + y - 2).$
18. $(x - 1) dx - (3x - 2y - 5) dy = 0.$ ANS. $(2y - x + 3)^2 = c(y - x + 2).$

In exercises 19 through 22 obtain the particular solution indicated.

19. $(2x - 3y + 4) dx + 3(x - 1) dy = 0$; when $x = 3$, $y = 2$.

ANS. $3(y - 2) = -2(x - 1) \ln \dfrac{x - 1}{2}.$

20. Solve the equation of exercise 19 but with the condition that when $x = -1$, $y = 2$.

ANS. $3(y - 2) = -2(x - 1) \ln \dfrac{1 - x}{2}.$

21. $(x + y - 4) dx - (3x - y - 4) dy = 0$; when $x = 4$, $y = 1$.

ANS. $2(x + 2y - 6) = 3(x - y) \ln \dfrac{x - y}{3}.$

22. Solve the equation of exercise 21 but with the condition that when $x = 3$, $y = 7$.

ANS. $y - 5x + 8 = 2(y - x) \ln \dfrac{y - x}{4}.$

23. Prove that the change of variables

$$x = \alpha_1 u + \alpha_2 v, \qquad y = u + v$$

will transform the equation

$$(a_1 x + b_1 y + c_1)\, dx + (a_2 x + b_2 y + c_2)\, dy = 0 \tag{A}$$

into an equation in which the variables u and v are separable, if α_1 and α_2 are roots of the equation

$$a_1 \alpha^2 + (a_2 + b_1)\alpha + b_2 = 0, \tag{B}$$

and if $\alpha_2 \neq \alpha_1$.

Note that this method of solution of (A) is not practical for us unless the roots of equation (B) are real and distinct.

Solve exercises 24 through 29 by the method indicated in exercise 23.

24. Do exercise 4 above. As a check, the equation (B) for this case is

$$\alpha^2 - 5\alpha + 6 = 0,$$

so we may choose $\alpha_1 = 2$ and $\alpha_2 = 3$.

The equation in u and v turns out to be

$$(v - 1)\, du - 2(u + 2)\, dv = 0.$$

25. Do exercise 3 above.   **26.** Do exercise 10 above.

27. Do exercise 17 above.   **28.** Do exercise 18 above.

29. Do Example (a) in the text of this section.

30. Prove that the change of variables

$$x = \alpha_1 u + \beta v, \qquad y = u + v$$

will transform the equation

$$(a_1 x + b_1 y + c_1)\, dx + (a_2 x + b_2 y + c_2)\, dy = 0 \tag{A}$$

into an equation that is linear in the variable u, if α_1 is a root of the equation

$$a_1 \alpha^2 + (a_2 + b_1)\alpha + b_2 = 0, \tag{B}$$

and if β is any number such that $\beta \neq \alpha_1$.

Note that this method is not practical for us unless the roots of equation (B) are real; however, they need not be distinct as they had to be in the theorem of exercise 23. The method of this exercise is particularly useful when the roots of (B) are equal.

Solve exercises 31 through 35 by the method indicated in exercise 30.

31. Do exercise 16 above. The only possible α_1 is (-2). Then β may be chosen to be anything else.

32. Do exercise 12 above.   **33.** Do exercise 15 above.   **34.** Do exercise 18 above.

35. Do exercise 4 above. As seen in exercise 24, the roots of the "α-equation" are 2 and 3. If you choose $\alpha_1 = 2$, for example, then you make β anything except 2. Of course, if you choose $\alpha_1 = 2$ and $\beta = 3$, then you are reverting to the method of exercise 23.

23. Solutions involving nonelementary integrals

In solving differential equations we frequently are confronted with the need for integrating an expression that is not the differential of any elementary* function. Following is a short list of nonelementary integrals:

$$\int \exp(-x^2)\,dx \qquad \int \frac{e^{-x}}{x}\,dx \qquad \int x \tan x\,dx$$

$$\int \sin x^2\,dx \qquad \int \frac{\sin x}{x}\,dx \qquad \int \frac{dx}{\ln x}$$

$$\int \cos x^2\,dx \qquad \int \frac{\cos x}{x}\,dx \qquad \int \frac{dx}{\sqrt{1-x^3}}.$$

Integrals involving the square root of a polynomial of degree greater than two are, in general, nonelementary. In special instances they may degenerate into elementary integrals.

The following example presents two ways of dealing with problems in which nonelementary integrals arise.

EXAMPLE: Solve the equation

$$y' - 2xy = 1$$

with the initial condition that when $x = 0$, $y = 1$.

The equation being linear in y, we write

$$dy - 2xy\,dx = dx,$$

obtain the integrating factor $\exp(-x^2)$, and prepare to solve

$$\exp(-x^2)\,dy - 2xy\exp(-x^2)\,dx = \exp(-x^2)\,dx. \tag{1}$$

The left member is, of course, the differential of $y\exp(-x^2)$. But the right member is not the differential of any elementary function; that is, $\int \exp(-x^2)\,dx$ is a nonelementary integral.

Let us turn to power series for help. From the series

$$\exp(-x^2) = \sum_{n=0}^{\infty} \frac{(-1)^n x^{2n}}{n!},$$

---

* By an elementary function we mean a function studied in the ordinary beginning calculus course. For example, polynomials, exponentials, logarithms, trigonometric, and inverse trigonometric functions are elementary. All functions obtained from them by a finite number of applications of the elementary operations of addition, subtraction, multiplication, division, extraction of roots, and raising to powers are elementary. Finally, we include such functions as $\sin(\sin x)$, in which the argument in a function previously classed as elementary is replaced by an elementary function.

obtained in calculus, it follows that

$$\int \exp(-x^2)\,dx = c + \sum_{n=0}^{\infty} \frac{(-1)^n x^{2n+1}}{n!(2n+1)}.$$

Thus the differential equation (1) has the general solution

$$y \exp(-x^2) = c + \sum_{n=0}^{\infty} \frac{(-1)^n x^{2n+1}}{n!(2n+1)}.$$

Since $y = 1$ when $x = 0$, c may be found from

$$1 = c + 0.$$

Therefore, the particular solution desired is

$$y \exp(-x^2) = 1 + \sum_{n=0}^{\infty} \frac{(-1)^n x^{2n+1}}{n!(2n+1)}. \tag{2}$$

An alternative procedure is the introduction of a definite integral. In calculus, the error function defined by

$$\operatorname{erf} x = \frac{2}{\sqrt{\pi}} \int_0^x \exp(-\beta^2)\,d\beta \tag{3}$$

is sometimes studied. Since, from (3),

$$\frac{d}{dx} \operatorname{erf} x = \frac{2}{\sqrt{\pi}} \exp(-x^2),$$

we may integrate the exact equation (1) as follows:

$$y \exp(-x^2) = \tfrac{1}{2}\sqrt{\pi}\,\operatorname{erf} x + c. \tag{4}$$

Since $\operatorname{erf} 0 = 0$, the condition that $y = 1$ when $x = 0$ yields $c = 1$. Hence, as an alternative to (2) we obtain

$$y \exp(-x^2) = 1 + \tfrac{1}{2}\sqrt{\pi}\,\operatorname{erf} x. \tag{5}$$

Equation (5) means the same as

$$y \exp(-x^2) = 1 + \int_0^x \exp(-\beta^2)\,d\beta. \tag{6}$$

Writing a solution in the form of (6) implies that the definite integral is to be evaluated by power series, approximate integration such as Simpson's rule, mechanical quadrature, or any other available tool. If it happens, as in this case, that the definite integral is itself a tabulated function, that is a great convenience, but it is not vital. The essential thing is to reduce the solution to a computable form.

Exercises

In each exercise, express the solution with the aid of power series or definite integrals.

1. $y' = y[1 - \exp(-x^2)]$.   ANS. $\ln |cy| = x - \frac{1}{2}\sqrt{\pi}\,\text{erf}\,x$.

2. $(xy - \sin x)\,dx + x^2\,dy = 0$.   ANS. $xy = c + \displaystyle\int_0^x \frac{\sin w}{w}\,dw$,

$$\text{or } y = cx^{-1} + \sum_{n=0}^{\infty}(-1)^n \frac{x^{2n}}{(2n+1)(2n+1)!}.$$

3. $y' = 1 - 4x^3 y$.   ANS. $y = \exp(-x^4)\left[c + \displaystyle\int_0^x \exp(\beta^4)\,d\beta\right]$.

4. $(y\cos^2 x - x\sin x)\,dx + \sin x\cos x\,dy = 0$.

ANS. $y\sin x = c + \displaystyle\int_0^x \beta\tan\beta\,d\beta$.

5. $(1 + xy)\,dx - x\,dy = 0$; when $x = 1$, $y = 0$.   ANS. $y = e^x \displaystyle\int_1^x \frac{e^{-\beta}}{\beta}\,d\beta$.

6. $\left[x\exp\left(\dfrac{y^2}{x^2}\right) - y\right]dx + x\,dy = 0$; when $x = 1$, $y = 2$.

ANS. $\ln x = \displaystyle\int_{y/x}^2 \exp(-\beta^2)\,d\beta$.

7. $x(2y + x)\,dx - dy = 0$; when $x = 0$, $y = 1$.

ANS. $2y = 2\exp(x^2) - x + \frac{1}{2}\sqrt{\pi}\exp(x^2)\,\text{erf}\,x$.

Miscellaneous Exercises

In each exercise, find a set of solutions unless the statement of the exercise stipulates otherwise.

1. $(y^2 - 3y - x)\,dx + (2y - 3)\,dy = 0$.   ANS. $y^2 - 3y - x + 1 = ce^{-x}$.
2. $(y^3 + y + 1)\,dx + x(x - 3y^2 - 1)\,dy = 0$.   ANS. $y^3 - xy + y + 1 = cx$.
3. $(x + 3y - 5)\,dx - (x - y - 1)\,dy = 0$.

ANS. $2(y - 1) = (x + y - 3)\ln |c(x + y - 3)|$.

4. $(x^5 - y^2)\,dx + 2xy\,dy = 0$.   ANS. $x^5 + 4y^2 = cx$.
5. $(2x + y - 4)\,dx + (x - 3y + 12)\,dy = 0$.

ANS. $2x^2 + 2xy - 3y^2 - 8x + 24y = c$.

6. Solve in two ways the equation $y' = ax + by + c$; with $b \neq 0$.

ANS. $b^2 y = c_1 e^{bx} - abx - a - cb$.

7. $y^3 \sec^2 x\,dx - (1 - 2y^2 \tan x)\,dy = 0$.   ANS. $y^2 \tan x = \ln |cy|$.
8. $x^3 y\,dx + (3x^4 - y^3)\,dy = 0$.   ANS. $15x^4 y^{12} = 4y^{15} + c$.
9. $(a_1 x + ky + c_1)\,dx + (kx + b_2 y + c_2)\,dy = 0$.

ANS. $a_1 x^2 + 2kxy + b_2 y^2 + 2c_1 x + 2c_2 y = c$.

10. $(x - 4y + 7)\,dx + (x + 2y + 1)\,dy = 0$.

ANS. $(x - y + 4)^3 = c(x - 2y + 5)^2$.

11. $xy\,dx + (y^4 - 3x^2)\,dy = 0$.   ANS. $x^2 = y^4(1 + cy^2)$.

12. $(x + 2y - 1)\,dx - (2x + y - 5)\,dy = 0$.

13. $(5x + 3\,e^y)\,dx + 2x\,e^y\,dy = 0$.   ANS. $x^3(x + e^y)^2 = c$.

14. $(3x + y - 2)\,dx + (3x + y + 4)\,dy = 0$.   ANS. $x + y + c = 3\ln|3x + y + 7|$.

15. $(x - 3y + 4)\,dx + 2(x - y - 2)\,dy = 0$.   ANS. $(x + y - 8)^4 = c(x - 2y + 1)$.

16. $(x - 2)\,dx + 4(x + y - 1)\,dy = 0$.

ANS. $2(y + 1) = -(x + 2y)\ln|c(x + 2y)|$.

17. $y\,dx = x(1 + xy^4)\,dy$.   ANS. $y(5 + xy^4) = cx$.

18. $2x\,dv + v(2 + v^2x)\,dx = 0$; when $x = 1, v = \tfrac{1}{2}$.   ANS. $xv^2(5x - 1) = 1$.

19. $2(x - y)\,dx + (3x - y - 1)\,dy = 0$.   ANS. $(x + y - 1)^4 = c(4x - 2y - 1)$.

20. $(2x - 5y + 12)\,dx + (7x - 4y + 15)\,dy = 0$.

ANS. $(x + 2y - 3)^3 = c(x - y + 3)$.

21. $y\,dx + x(x^2y - 1)\,dy = 0$.   ANS. $y^2(2x^2y - 3) = cx^2$.

22. $dy/dx = \tan y \cot x - \sec y \cos x$.   ANS. $\sin y + \sin x \ln|c \sin x| = 0$.

23. $[1 + (x + y)^2]\,dx + [1 + x(x + y)]\,dy = 0$.

ANS. $y^2 = (x + y)^2 + 2\ln|x + y| + c$.

24. $(x - 2y - 1)\,dx - (x - 3)\,dy = 0$. Solve by two methods.

ANS. $(x - 3)^2(x - 3y) = c$.

25. $(2x - 3y + 1)\,dx - (3x + 2y - 4)\,dy = 0$. Solve by two methods.

ANS. $x^2 + x - 3xy - y^2 + 4y = c$.

26. $(4x + 3y - 7)\,dx + (4x + 3y + 1)\,dy = 0$.

ANS. $x + y + c = 8\ln|4x + 3y + 25|$.

27. Find a change of variables that will reduce any equation of the form

$$xy' = yf(xy)$$

to an equation in which the variables are separable.

28. $(x + 4y + 3)\,dx - (2x - y - 3)\,dy = 0$.

ANS. $3(y + 1) = (x + y)\ln|c(x + y)|$.

29. $(3x - 3y - 2)\,dx - (x - y + 1)\,dy = 0$.

ANS. $2(y - 3x + c) = 5\ln|2x - 2y - 3|$.

30. $(x - 6y + 2)\,dx + 2(x + 2y + 2)\,dy = 0$.

ANS. $4y = -(x - 2y + 2)\ln|c(x - 2y + 2)|$.

31. $(x^4 - 4x^2y^2 - y^4)\,dx + 4x^3y\,dy = 0$; when $x = 1, y = 2$.

ANS. $y^2(5 - 3x) = x^2(5 + 3x)$.

32. $(x - y - 1)\,dx - 2(y - 2)\,dy = 0$.   ANS. $(x + y - 5)^2(x - 2y + 1) = c$.

33. $(x - 3y + 3)\,dx + (3x + y + 9)\,dy = 0$.

ANS. $\ln[(x + 3)^2 + y^2] = c + 6\arctan[(x + 3)/y]$.

34. $(2x + 4y - 1)\,dx - (x + 2y - 3)\,dy = 0$.   ANS. $\ln|x + 2y - 1| = y - 2x + c$.

35. $4y\,dx + 3(2x - 1)(dy + y^4\,dx) = 0$; when $x = 1, y = 1$.

ANS. $y^3(2x - 1)(5x - 4) = 1$.

36. $y(x - 1)\,dx - (x^2 - 2x - 2y)\,dy = 0$.   ANS. $x^2 - 2x - 4y = cy^2$.

37. $(6xy - 3y^2 + 2y)\,dx + 2(x - y)\,dy = 0$.   ANS. $y(2x - y) = c\,e^{-3x}$.

38. $y' = x - y + 2$. Solve by two methods.   ANS. $\ln|x - y + 1| = c - x$.

39. $(x + y - 2)\,dx - (x - 4y - 2)\,dy = 0$.

ANS. $\ln[(x - 2)^2 + 4y^2] + \arctan[(x - 2)/2y] = c$.

40. $4\,dx + (x - y + 2)^2\,dy = 0$.   ANS. $y + 2\arctan[(x - y + 2)/2] = c$.

Linear Differential Equations

24. The general linear equation

The general linear differential equation of order n is an equation that can be written

$$b_0(x)\frac{d^n y}{dx^n} + b_1(x)\frac{d^{n-1} y}{dx^{n-1}} + \cdots + b_{n-1}(x)\frac{dy}{dx} + b_n(x)y = R(x). \qquad (1)$$

The functions $R(x)$ and $b_i(x)$; $i = 0, 1, \ldots, n$, are to be independent of the variable y. If $R(x)$ is identically zero, equation (1) is said to be linear and homogeneous*; if $R(x)$ is not identically zero, equation (1) is called linear and nonhomogeneous. In this chapter we shall obtain some fundamental and important properties of linear equations.

First we prove that if y_1 and y_2 are solutions of the homogeneous equation

$$b_0(x)y^{(n)} + b_1(x)y^{(n-1)} + \cdots + b_{n-1}(x)y' + b_n(x)y = 0, \qquad (2)$$

---

* It is perhaps unfortunate that the word homogeneous as it is used here has a very different meaning from that in Sections 8 and 9.

and if c_1 and c_2 are constants, then

$$y = c_1 y_1 + c_2 y_2$$

is a solution of equation (2).

The statement that y_1 and y_2 are solutions of (2) means that

$$b_0 y_1^{(n)} + b_1 y_1^{(n-1)} + \cdots + b_{n-1} y_1' + b_n y_1 = 0 \tag{3}$$

and

$$b_0 y_2^{(n)} + b_1 y_2^{(n-1)} + \cdots + b_{n-1} y_2' + b_n y_2 = 0. \tag{4}$$

Now let us multiply each member of (3) by c_1, each member of (4) by c_2, and add the results. We get

$$b_0(c_1 y_1^{(n)} + c_2 y_2^{(n)}) + b_1(c_1 y_1^{(n-1)} + c_2 y_2^{(n-1)}) + \cdots$$
$$+ b_{n-1}(c_1 y_1' + c_2 y_2') + b_n(c_1 y_1 + c_2 y_2) = 0. \tag{5}$$

Since $c_1 y_1' + c_2 y_2' = (c_1 y_1 + c_2 y_2)'$, and so on, equation (5) is neither more nor less than the statement that $c_1 y_1 + c_2 y_2$ is a solution of equation (2). The proof is completed. The special case $c_2 = 0$ is worth noting; that is, for a homogeneous linear equation any constant times a solution is also a solution.

In a similar manner, or by iteration of the above result, it can be seen that if y_i, with $i = 1, 2, \ldots, k$, are solutions of equation (2), and if c_i, with $i = 1, 2, \ldots, k$, are constants, then

$$y = c_1 y_1 + c_2 y_2 + \cdots + c_k y_k \tag{6}$$

is a solution of equation (2).

The expression in equation (6) is called a linear combination of the functions $y_1, y_2, \ldots, y_k$. The theorem just proved can thus be stated as follows:

THEOREM 4: *Any linear combination of solutions of a linear homogeneous differential equation is also a solution.*

25. Linear independence

Given the functions $f_1, \ldots, f_n$, if constants $c_1, c_2, \ldots, c_n$, not all zero, exist such that

$$c_1 f_1(x) + c_2 f_2(x) + \cdots + c_n f_n(x) = 0 \tag{1}$$

identically in some interval $a \leq x \leq b$, then the functions $f_1, f_2, \ldots, f_n$ are said to be *linearly dependent*. If no such relation exists, the functions are said to be *linearly independent*. That is, the functions $f_1, f_2, \ldots, f_n$ are linearly independent when equation (1) implies that $c_1 = c_2 = \cdots = c_n = 0$.

It should be clear that, if the functions of a set are linearly dependent, at least one of them is a linear combination of the others; if they are linearly independent, then none of them is a linear combination of the others.

26. An existence and uniqueness theorem

In Section 12 we stated an existence theorem for an initial value problem involving a first-order linear differential equation. The generalization of this theorem to nth order linear equations can be stated as follows:

THEOREM 5: *Let $P_1, P_2, \ldots, P_n$, and R be functions that are continuous on an interval $a < x < b$. Suppose that x_0 is a real number in the interval and that $y_0, y_1, \ldots, y_{n-1}$ are n arbitrary real numbers. Then a unique function $y = y(x)$ exists, defined on the interval $a < x < b$, which is a solution of the equation*

$$y^{(n)} + P_1 y^{(n-1)} + \cdots + P_n y = R \tag{1}$$

on the interval and which satisfies the initial conditions

$$y(x_0) = y_0, \qquad y'(x_0) = y_1, \ldots, y^{(n-1)}(x_0) = y_{n-1}.$$

EXAMPLE: Find the unique solution of the initial value problem

$$y'' + y = 0, y(0) = 0, y'(0) = 1. \tag{2}$$

We observe that $\sin x$ and $\cos x$ are solutions of (2), so that for arbitrary c_1 and c_2

$$y = c_1 \sin x + c_2 \cos x$$

is also a solution by the theorem of Section 24.

Because of the initial conditions in (2), we are led to choose c_1 and c_2 so that $c_1 \sin 0 + c_2 \cos 0 = 0$ and $c_1 \cos 0 - c_2 \sin 0 = 1$. This can be done in only one way, namely by choosing $c_1 = 1$ and $c_2 = 0$. We find that the function $\sin x$ is a solution of the initial value problem (2). Moreover, since the problem satisfies the conditions required in Theorem 5, no matter what interval we choose, $\sin x$ is the only solution to the problem given in (2).

27. The Wronskian

With the definitions of Section 25 in mind, we shall now obtain a sufficient condition that n functions be linearly independent over an interval $a \leqq x \leqq b$.

Let us assume that each of the functions $f_1, f_2, \ldots, f_n$ is differentiable at least $(n-1)$ times in the interval $a \leq x \leq b$. Then from the equation

$$c_1 f_1 + c_2 f_2 + \cdots + c_n f_n = 0, \tag{1}$$

it follows by successive differentiation that

$$c_1 f'_1 + c_2 f'_2 + \cdots + c_n f'_n = 0,$$

$$c_1 f''_1 + c_2 f''_2 + \cdots + c_n f''_n = 0,$$

$$\vdots$$

$$c_1 f_1^{(n-1)} + c_2 f_2^{(n-1)} + \cdots + c_n f_n^{(n-1)} = 0.$$

Considered as a system of equations in $c_1, c_2, \ldots, c_n$, the n linear equations directly above will have no solution except the one with each of the c's equal to zero, if the determinant of the system does not vanish. That is, if

$$\begin{vmatrix} f_1 & f_2 & \cdots & f_n \\ f'_1 & f'_2 & \cdots & f'_n \\ f''_1 & f''_2 & \cdots & f''_n \\ & & \vdots & \\ f_1^{(n-1)} & f_2^{(n-1)} & \cdots & f_n^{(n-1)} \end{vmatrix} \neq 0, \tag{2}$$

then the functions $f_1, f_2, \ldots, f_n$ are linearly independent. The determinant in (2) is called the *Wronskian** of the n functions involved. We have shown that the nonvanishing of the Wronskian is a sufficient condition that the functions be linearly independent.

The nonvanishing of the Wronskian on an interval is not a necessary condition for linear independence. The Wronskian may vanish even when the functions are linearly independent, as exhibited in exercise 9 below.

If the n functions involved are solutions of a homogeneous linear differential equation, the situation is simplified as is shown by Theorem 6. A proof of this theorem in the case $n = 2$, is suggested in exercises 10 through 13 below.

THEOREM 6: *If, on the interval $a < x < b$, $b_0(x) \neq 0$, $b_0, b_1, \ldots, b_n$ are continuous, and $y_1, y_2, \ldots, y_n$ are solutions of the equation*

$$b_0 y^{(n)} + b_1 y^{(n-1)} + \cdots + b_{n-1} y' + b_n y = 0, \tag{3}$$

then a necessary and sufficient condition that $y_1, \ldots, y_n$ be linearly independent is the nonvanishing of the Wronskian of $y_1, \ldots, y_n$ on the interval $a < x < b$.

---

* The Wronskian determinant is named after the Polish mathematician Hoëné Wronski (1778–1853).

The functions $\cos \omega t$, $\sin \omega t$, $\sin(\omega t + \alpha)$, in which t is the variable and ω and α are constants, are linearly dependent because constants c_1, c_2, c_3 exist such that

$$c_1 \cos \omega t + c_2 \sin \omega t + c_3 \sin(\omega t + \alpha) = 0$$

for all t. Indeed, one set of such constants is $c_1 = \sin \alpha$, $c_2 = \cos \alpha$, $c_3 = -1$.

One of the best-known sets of n linearly independent functions of x is the set $1, x, x^2, \ldots, x^{n-1}$. The linear independence of the powers of x follows at once from the fact that, if $c_1, c_2, \ldots, c_n$ are not all zero, the equation

$$c_1 + c_2 x + \cdots + c_n x^{n-1} = 0$$

can have, at most, $(n - 1)$ distinct roots and so cannot vanish identically in any interval. See also exercise 1 below.

Exercises

1. Obtain the Wronskian of the functions

$$1, x, x^2, \ldots, x^{n-1} \text{ for } n > 1.$$

ANS. $W = 0! 1! 2! \cdots (n - 1)!$.

2. Show that the functions e^x, e^{2x}, e^{3x} are linearly independent.

ANS. $W = 2 e^{6x} \neq 0$.

3. Show that the functions e^x, $\cos x$, $\sin x$ are linearly independent.

ANS. $W = 2 e^x \neq 0$.

4. By determining constants c_1, c_2, c_3, c_4, which are not all zero and are such that $c_1 f_1 + c_2 f_2 + c_3 f_3 + c_4 f_4 = 0$ identically, show that the functions

$$f_1 = x, \quad f_2 = e^x, \quad f_3 = x e^x, \quad f_4 = (2 - 3x) e^x$$

are linearly dependent.

ANS.   One such set of c's is: $c_1 = 0$, $c_2 = -2$, $c_3 = 3$, $c_4 = 1$.

5. Show that $\cos(\omega t - \beta)$, $\cos \omega t$, $\sin \omega t$ are linearly dependent functions of t.

6. Show that $1, \sin x, \cos x$ are linearly independent.

7. Show that $1, \sin^2 x, \cos^2 x$ are linearly dependent.

8. Show that two nonvanishing differentiable functions of x are linearly dependent if, and only if, their Wronskian vanishes identically. This statement is not true for more than two functions.

9. Let $f_1(x) = 1 + x^3$ for $x \leq 0$, $f_1(x) = 1$ for $x \geq 0$;
    $f_2(x) = 1$ for $x \leq 0$, $f_2(x) = 1 + x^3$ for $x \geq 0$;
    $f_3(x) = 3 + x^3$ for all x.
Show that (a) f, f', f'' are continuous for all x for each of f_1, f_2, f_3;
         (b) the Wronskian of f_1, f_2, f_3 is zero for all x;
         (c) f_1, f_2, f_3 are linearly independent over the interval $-1 \leq x \leq 1$.
In part (c) you must show that if $c_1 f_1(x) + c_2 f_2(x) + c_3 f_3(x) = 0$ for all x in $-1 \leq x \leq 1$ then $c_1 = c_2 = c_3 = 0$. Use $x = -1, 0, 1$ successively to obtain three equations to solve for c_1, c_2, and c_3.

10. Given any interval $a < x < b$ with x_0 a fixed number in the interval and suppose y is a solution of the homogeneous equation

$$y'' + Py' + Qy = 0. \tag{A}$$

Further, suppose that $y(x_0) = y'(x_0) = 0$. Use the existence and uniqueness theorem of Section 26 to prove that $y(x) = 0$ for every x in the interval $a < x < b$.

11. Suppose that y_1 and y_2 are solutions of equation (A) of exercise 10 and suppose the Wronskian of y_1 and y_2 is identically zero on $a < x < b$. Show that for x_0 in the interval $a < x < b$, there must exist constants $\bar{c}_1$ and $\bar{c}_2$ not both zero such that

$$\bar{c}_1 y_1(x_0) + \bar{c}_2 y_2(x_0) = 0,$$

and

$$\bar{c}_1 y_1'(x_0) + \bar{c}_2 y_2'(x_0) = 0.$$

12. Consider the function defined by

$$y(x) = \bar{c}_1 y_1(x) + \bar{c}_2 y_2(x),$$

where $\bar{c}_1$ and $\bar{c}_2$ are the constants determined in exercise 11. Show that this function is a solution of equation (A) above and that it follows that $y(x) \equiv 0$ on $a < x < b$. (Use the results of exercise 10.)

13. Combine the results of exercises 10 to 12 to obtain a proof of the necessity condition of Theorem 6 in the case when $n = 2$. Also note that the sufficiency condition was established in the text of this section.

28. General solution of a homogeneous equation

One of the basic results of the subject of linear differential equations is contained in Theorem 7.

THEOREM 7:   *Let $\{y_1, y_2, \ldots, y_n\}$ be a linearly independent set of solutions of the homogeneous linear equation*

$$b_0(x)y^{(n)} + b_1(x)y^{(n-1)} + \cdots + b_{n-1}(x)y' + b_n(x)y = 0, \tag{1}$$

for x on the interval $a < x < b$. Suppose further that $b_0(x), \ldots, b_n(x)$ are continuous functions and that $b_0(x) \neq 0$ on $a < x < b$.

If ϕ is any solution of equation (1), valid on $a < x < b$, there exist constants $\bar{c}_1, \bar{c}_2, \ldots, \bar{c}_n$ such that

$$\phi = \bar{c}_1 y_1 + \bar{c}_2 y_2 + \cdots + \bar{c}_n y_n. \tag{2}$$

It is because of this theorem that we define the *general solution* of equation (1) to be

$$y = c_1 y_1 + c_2 y_2 + \cdots + c_n y_n, \tag{3}$$

where $c_1, c_2, \ldots, c_n$ are arbitrary constants.

In a sense each particular solution of the linear equation (1) is a special case (some choice of the c's) of the general solution (3). The basic ideas needed for a proof of this important theorem are exhibited here for an equation of order two. No additional complications occur for equations of higher order.

PROOF.   Consider the equation

$$b_0(x)y'' + b_1(x)y' + b_2(x)y = 0. \tag{4}$$

Let y_1 and y_2 be linearly independent solutions of equation (4) on the interval $a < x < b$ and choose x_0 between a and b. By Theorem 6, page 87, the Wronskian of y_1 and y_2 is not zero at x_0. That is,

$$W = \begin{vmatrix} y_1(x_0) & y_2(x_0) \\ y_1'(x_0) & y_2'(x_0) \end{vmatrix} \neq 0. \tag{5}$$

It follows that the system of equations

$$c_1 y_1(x_0) + c_2 y_2(x_0) = \phi(x_0),$$
$$c_1 y_1'(x_0) + c_2 y_2'(x_0) = \phi'(x_0),$$

has a unique solution $c_1 = \bar{c}_1, c_2 = \bar{c}_2$. That is,

$$\bar{c}_1 y_1(x_0) + \bar{c}_2 y_2(x_0) = \phi(x_0),$$
$$\bar{c}_1 y_1'(x_0) + \bar{c}_2 y_2'(x_0) = \phi'(x_0).$$

Now consider the function

$$f = \bar{c}_1 y_1 + \bar{c}_2 y_2. \tag{6}$$

Because f is a linear combination of two solutions of equation (4) on the interval $a < x < b$, it is also a solution on that interval. Moreover,

$$f(x_0) = \bar{c}_1 y_1(x_0) + \bar{c}_2 y_2(x_0),$$
$$f'(x_0) = \bar{c}_1 y_1'(x_0) + \bar{c}_2 y_2'(x_0),$$

so that $f(x_0) = \phi(x_0)$ and $f'(x_0) = \phi'(x_0)$. It follows from the uniqueness theorem of Section 26 that f and ϕ are the same solution. That is,

$$\phi = \bar{c}_1 y_1 + \bar{c}_2 y_2,$$

which completes the proof of the theorem.

It is necessary to keep in mind that the above discussion used the fact that $b_0(x) \neq 0$ in the interval $a < x < b$. It is easy to see that the linear equation

$$xy' - 2y = 0$$

has the general solution $y = cx^2$ and also such particular solutions as

$$y_1 = x^2, \qquad 0 \leqq x,$$
$$= -4x^2, \qquad x < 0.$$

The solution y_1 is not a special case of the general solution. But in any interval throughout which $b_0(x) = x \neq 0$, this particular solution is a special case of the general solution. It was, of course, made up by piecing together at $x = 0$ two parts, each drawn from the general solution.

29. General solution of a nonhomogeneous equation

Let y_p be any particular solution (not necessarily involving any arbitrary constants) of the equation

$$b_0 y^{(n)} + b_1 y^{(n-1)} + \cdots + b_{n-1} y' + b_n y = R(x) \tag{1}$$

and let y_c be a solution of the corresponding homogeneous equation

$$b_0 y^{(n)} + b_1 y^{(n-1)} + \cdots + b_{n-1} y' + b_n y = 0. \tag{2}$$

Then

$$y = y_c + y_p \tag{3}$$

is a solution of equation (1). For, using the y of equation (3) we see that

$$b_0 y^{(n)} + \cdots + b_n y = (b_0 y_c^{(n)} + \cdots + b_n y_c)$$
$$+ (b_0 y_p^{(n)} + \cdots + b_n y_p) = 0 + R(x) = R(x).$$

If $y_1, y_2, \ldots, y_n$ are linearly independent solutions of equation (2), then

$$y_c = c_1 y_1 + c_2 y_2 + \cdots + c_n y_n, \tag{4}$$

in which the c's are arbitrary constants, is the general solution of equation (2). The right member of equation (4) is called the *complementary function* for equation (1).

The general solution of the nonhomogeneous equation (1) is the sum of the complementary function and any particular solution. To justify this usage of the term "general solution" we must show that if f is any solution of equation (1) then $f \equiv y_c + y_p$ for some particular choice of the $c_1, \ldots, c_n$. We note that since f and y_p are both solutions of the nonhomogeneous equation (1), $f - y_p$ is a solution of the homogeneous equation (2). Hence by Theorem 7 of Section 28

$$f - y_p \equiv c_1 y_1 + c_2 y_2 + \cdots + c_n y_n$$

for some particular choice of the $c_1, \ldots, c_n$. This establishes what we wished to show.

EXAMPLE (a): Find the general solution of

$$y'' = 4. \tag{5}$$

We first observe that the functions 1 and x are linearly independent on any interval and are solutions of the homogeneous equation $y'' = 0$. Hence, the complementary function for equation (5) is

$$y_c = c_1 + c_2 x.$$

On the other hand, the function $2x^2$ is a particular solution of equation (5). Hence, the general solution of equation (5) is

$$y = c_1 + c_2 x + 2x^2.$$

EXAMPLE (b): Find the general solution of the equation

$$y'' - y = 4. \tag{6}$$

It is easily seen that $y = -4$ is a solution of equation (6). Therefore the y_p in equation (3) may be taken to be (-4). As we shall see later, the homogeneous equation

$$y'' - y = 0$$

has as its general solution

$$y_c = c_1 e^x + c_2 e^{-x}.$$

Thus the complementary function for equation (6) is $c_1 e^x + c_2 e^{-x}$ and a particular solution of (6) is $y_p = -4$. Hence the general solution of equation (6) is

$$y = c_1 e^x + c_2 e^{-x} - 4,$$

in which c_1 and c_2 are arbitrary constants.

30. Differential operators

Let D denote differentiation with respect to x, D^2 differentiation twice with respect to x, and so on; that is, for positive integral k,

$$D^k y = \frac{d^k y}{dx^k}.$$

The expression

$$A = a_0 D^n + a_1 D^{n-1} + \cdots + a_{n-1} D + a_n \tag{1}$$

is called a differential operator of order n. It may be defined as that operator which, when applied to any function* y, yields the result

---

* The function y is assumed to possess as many derivatives as may be encountered in whatever operations take place.

$$Ay = a_0\frac{d^n y}{dx^n} + a_1\frac{d^{n-1} y}{dx^{n-1}} + \cdots + a_{n-1}\frac{dy}{dx} + a_n y. \tag{2}$$

The coefficients $a_0, a_1, \ldots, a_n$ in the operator A may be functions of x, but in this book the only operators used will be those with constant coefficients.

Two operators A and B are said to be equal if, and only if, the same result is produced when each acts upon the function y. That is, $A = B$ if, and only if, $Ay = By$ for all functions y possessing the derivatives necessary for the operations involved.

The product AB of two operators A and B is defined as that operator which produces the same result as is obtained by using the operator B followed by the operator A. Thus $ABy = A(By)$. The product of two differential operators always exists and is a differential operator. For operators with *constant coefficients*, but not usually for those with variable coefficients, it is true that $AB = BA$.

EXAMPLE (a):   Let $A = D + 2$ and $B = 3D - 1$.
Then

$$By = (3D - 1)y = 3\frac{dy}{dx} - y$$

and

$$A(By) = (D + 2)\left(3\frac{dy}{dx} - y\right)$$

$$= 3\frac{d^2 y}{dx^2} - \frac{dy}{dx} + 6\frac{dy}{dx} - 2y$$

$$= 3\frac{d^2 y}{dx^2} + 5\frac{dy}{dx} - 2y$$

$$= (3D^2 + 5D - 2)y.$$

Hence $AB = (D + 2)(3D - 1) = 3D^2 + 5D - 2$.
Now consider BA. Acting upon y, the operator BA yields

$$B(Ay) = (3D - 1)\left(\frac{dy}{dx} + 2y\right)$$

$$= 3\frac{d^2 y}{dx^2} + 6\frac{dy}{dx} - \frac{dy}{dx} - 2y$$

$$= 3\frac{d^2 y}{dx^2} + 5\frac{dy}{dx} - 2y.$$

Hence

$$BA = 3D^2 + 5D - 2 = AB.$$

EXAMPLE (b):   Let $G = xD + 2$, and $H = D - 1$. Then

$$G(Hy) = (xD + 2)\left(\frac{dy}{dx} - y\right)$$

$$= x\frac{d^2y}{dx^2} - x\frac{dy}{dx} + 2\frac{dy}{dx} - 2y$$

$$= x\frac{d^2y}{dx^2} + (2 - x)\frac{dy}{dx} - 2y,$$

so

$$GH = xD^2 + (2 - x)D - 2.$$

On the other hand

$$H(Gy) = (D - 1)\left(x\frac{dy}{dx} + 2y\right)$$

$$= \frac{d}{dx}\left(x\frac{dy}{dx} + 2y\right) - \left(x\frac{dy}{dx} + 2y\right)$$

$$= x\frac{d^2y}{dx^2} + \frac{dy}{dx} + 2\frac{dy}{dx} - x\frac{dy}{dx} - 2y$$

$$= x\frac{d^2y}{dx^2} + (3 - x)\frac{dy}{dx} - 2y;$$

that is,

$$HG = xD^2 + (3 - x)D - 2.$$

It is worthy of notice that here we have two operators G and H (one of them with variable coefficients), whose product is dependent on the order of the factors. On this topic see also exercises 17 through 22 in the next section.

    The sum of two differential operators is obtained by expressing each in the form

$$a_0 D^n + a_1 D^{n-1} + \cdots + a_{n-1}D + a_n$$

and adding corresponding coefficients. For instance, if

$$A = 3D^2 - D + x - 2$$

and

$$B = x^2 D^2 + 4D + 7,$$

then

$$A + B = (3 + x^2)D^2 + 3D + x + 5.$$

Differential operators are linear operators; that is, if A is any differential operator, c_1 and c_2 are constants, and f_1 and f_2 are any functions of x each possessing the required number of derivatives, then

$$A(c_1 f_1 + c_2 f_2) = c_1 A f_1 + c_2 A f_2.$$

31. The fundamental laws of operation

Let A, B, and C be any differential operators as defined in Section 30. With the above definitions of addition and multiplication, it follows that differential operators satisfy the following:

(a) The commutative law of addition:

$$A + B = B + A.$$

(b) The associative law of addition:

$$(A + B) + C = A + (B + C).$$

(c) The associative law of multiplication:

$$(AB)C = A(BC).$$

(d) The distributive law of multiplication with respect to addition:

$$A(B + C) = AB + AC.$$

(e) And if A and B are operators with *constant coefficients*, then they also satisfy the commutative law of multiplication:

$$AB = BA.$$

Therefore, differential operators with constant coefficients satisfy all the laws of the algebra of polynomials with respect to the operations of addition and multiplication.

If m and n are any two positive integers, then

$$D^m D^n = D^{m+n},$$

a useful result which follows immediately from the definitions.

Since for purposes of addition and multiplication the operators with constant coefficients behave just as algebraic polynomials behave, it is legitimate to use the tools of elementary algebra. In particular, synthetic division may be used to factor operators with constant coefficients.

Exercises

Perform the indicated multiplications in exercises 1 through 4.

1. $(4D + 1)(D - 2)$.   ANS. $4D^2 - 7D - 2$.
2. $(2D - 3)(2D + 3)$.   ANS. $4D^2 - 9$.
3. $(D + 2)(D^2 - 2D + 5)$.   ANS. $D^3 + D + 10$.
4. $(D - 2)(D + 1)^2$.   ANS. $D^3 - 3D - 2$.

In exercises 5 through 16, factor each of the operators.

5. $2D^2 + 3D - 2$.   ANS. $(D + 2)(2D - 1)$.
6. $2D^2 - 5D - 12$.
7. $D^3 - 2D^2 - 5D + 6$.   ANS. $(D - 1)(D + 2)(D - 3)$.
8. $4D^3 - 4D^2 - 11D + 6$.
9. $D^4 - 4D^2$.   ANS. $D^2(D - 2)(D + 2)$.
10. $D^3 - 3D^2 + 4$.
11. $D^3 - 21D + 20$.   ANS. $(D - 1)(D - 4)(D + 5)$.
12. $2D^3 - D^2 - 13D - 6$.
13. $2D^4 + 11D^3 + 18D^2 + 4D - 8$.   ANS. $(D + 2)^3(2D - 1)$.
14. $8D^4 + 36D^3 - 66D^2 + 35D - 6$.
15. $D^4 + D^3 - 2D^2 + 4D - 24$.   ANS. $(D - 2)(D + 3)(D^2 + 4)$.
16. $D^3 - 11D - 20$.   ANS. $(D - 4)(D^2 + 4D + 5)$.

Perform the indicated multiplications in exercises 17 through 22.

17. $(D - x)(D + x)$.   ANS. $D^2 + 1 - x^2$.
18. $(D + x)(D - x)$.   ANS. $D^2 - 1 - x^2$.
19. $D(xD - 1)$.   ANS. xD^2.
20. $(xD - 1)D$.   ANS. $xD^2 - D$.
21. $(xD + 2)(xD - 1)$.   ANS. $x^2D^2 + 2xD - 2$.
22. $(xD - 1)(xD + 2)$.   ANS. $x^2D^2 + 2xD - 2$.

32. Some properties of differential operators

Since for constant m and positive integral k,

$$D^k e^{mx} = m^k e^{mx}, \tag{1}$$

it is easy to find the effect an operator has upon e^{mx}. Let $f(D)$ be a polynomial in D,

$$f(D) = a_0 D^n + a_1 D^{n-1} + \cdots + a_{n-1}D + a_n. \tag{2}$$

Then

$$f(D) e^{mx} = a_0 m^n e^{mx} + a_1 m^{n-1} e^{mx} + \cdots + a_{n-1} m e^{mx} + a_n e^{mx},$$

so

$$f(D) e^{mx} = e^{mx} f(m). \tag{3}$$

If m is a root of the equation $f(m) = 0$, then in view of equation (3),

$$f(D) e^{mx} = 0.$$

Next consider the effect of the operator $D - a$ on the product of e^{ax} and a function y. We have

$$(D - a)(e^{ax} y) = D(e^{ax} y) - a e^{ax} y$$

$$= e^{ax} Dy,$$

and

$$(D - a)^2 (e^{ax} y) = (D - a)(e^{ax} Dy)$$

$$= e^{ax} D^2 y.$$

Repeating the operation, we are led to

$$(D - a)^n (e^{ax} y) = e^{ax} D^n y. \tag{4}$$

Using the linearity of differential operators, we conclude that when $f(D)$ is a polynomial in D with constant coefficients, then

$$e^{ax} f(D) y = f(D - a)[e^{ax} y]. \tag{5}$$

The relation (5) shows us how to shift an exponential factor from the left of a differential operator to the right of the operator. This relation has many uses, some of which we will examine in Chapter 6.

EXAMPLE (a):   Let $f(D) = 2D^2 + 5D - 12$. Then the equation $f(m) = 0$ is

$$2m^2 + 5m - 12 = 0,$$

or

$$(m + 4)(2m - 3) = 0,$$

of which the roots are $m_1 = -4$ and $m_2 = \frac{3}{2}$.

With the aid of equation (3) above it can be seen that

$$(2D^2 + 5D - 12) e^{-4x} = 0$$

and that

$$(2D^2 + 5D - 12) \exp\left(\tfrac{3}{2} x\right) = 0.$$

In other words, $y_1 = e^{-4x}$ and $y_2 = \exp(\frac{3}{2}x)$ are solutions of

$$(2D^2 + 5D - 12)y = 0.$$

EXAMPLE (b): Show that

$$(D - m)^n(x^k e^{mx}) = 0 \qquad \text{for } k = 0, 1, \ldots, (n - 1). \tag{6}$$

In equation (5) we let $f(D) = (D - m)^n$ and $y = x^k$. Then using the exponential shift we obtain

$$(D - m)^n(x^k e^{mx}) = e^{mx}D^n x^k.$$

But $D^n x^k = 0$ for $k = 0, 1, 2, \ldots, n - 1$, which gives us equation (6) directly.

The results obtained in equations (3), (5), and (6) are of fundamental importance to the solving of linear differential equations with constant coefficients which we consider in Chapter 6.

EXAMPLE (c): As an illustration of the use of the exponential shift we solve the differential equation

$$(D + 3)^4 y = 0. \tag{7}$$

First we multiply equation (7) by e^{3x} to obtain

$$e^{3x}(D + 3)^4 y = 0.$$

Applying the exponential shift as in equation (5) leads to

$$D^4(e^{3x}y) = 0.$$

Integrating four times gives us

$$e^{3x}y = c_1 + c_2 x + c_3 x^2 + c_4 x^3,$$

and finally,

$$y = (c_1 + c_2 x + c_3 x^2 + c_4 x^3) e^{-3x}. \tag{8}$$

Note that each of the four functions e^{-3x}, $x e^{-3x}$, $x^2 e^{-3x}$, and $x^3 e^{-3x}$ is a solution of equation (7). This of course is assured by the theorem of equation (6) of Example (b).

If we now show that the four functions are linearly independent, equation (8) gives the general solution of equation (7). See exercise 5 below.

Exercises

In exercises 1 through 4, use the exponential shift as in Example (c) above to find the general solution.

1. $(D - 2)^3 y = 0.$ ANS. $y = (c_1 + c_2 x + c_3 x^2) e^{2x}.$

2. $(D + 1)^2 y = 0.$

3. $(2D - 1)^2 y = 0.$ ANS. $y = (c_1 + c_2 x) \exp(\tfrac{1}{2}x).$

4. $(D + 7)^6 y = 0.$

5. To show that the four functions in Example (c) above are linearly independent on any interval, assume that they are linearly dependent and show that this leads to a contradiction of the results obtained in exercise 1 of Section 27.

6. Prove that the set of functions

$$e^{ax}, x\, e^{ax}, x^2\, e^{ax}, \ldots, x^{n-1}\, e^{ax}$$

is a linearly independent set on any interval. See exercise 5.

Linear Equations with Constant Coefficients

33. Introduction

Several methods for solving differential equations with constant coefficients are presented in this book. A classical technique is treated in this and the next chapter. Chapters 11 and 12 contain a development of the Laplace transform and its use in solving linear differential equations. Chapter 13 studies matrix techniques for solving linear equations with constant coefficients. Each method has its advantages and its disadvantages. Each is theoretically sufficient: all are necessary for maximum efficiency.

34. The auxiliary equation; distinct roots

Any linear homogeneous differential equation with constant coefficients,

$$a_0 \frac{d^n y}{dx^n} + a_1 \frac{d^{n-1} y}{dx^{n-1}} + \cdots + a_{n-1} \frac{dy}{dx} + a_n y = 0, \tag{1}$$

may be written in the form

$$f(D)y = 0, \tag{2}$$

where $f(D)$ is a linear differential operator. As we saw in the preceding chapter, if m is any root of the algebraic equation $f(m) = 0$, then

$$f(D)\, e^{mx} = 0,$$

which means simply that $y = e^{mx}$ is a solution of equation (2). The equation

$$f(m) = 0 \tag{3}$$

is called the *auxiliary equation* associated with (1) or (2).

The auxiliary equation for (1) is of degree n. Let its roots be $m_1, m_2, \ldots, m_n$. If these roots are all real and distinct, then the n solutions

$$y_1 = \exp{(m_1 x)}, y_2 = \exp{(m_2 x)}, \ldots, y_n = \exp{(m_n x)}$$

are linearly independent and the general solution of (1) can be written at once. It is

$$y = c_1 \exp{(m_1 x)} + c_2 \exp{(m_2 x)} + \cdots + c_n \exp{(m_n x)},$$

in which $c_1, c_2, \ldots, c_n$ are arbitrary constants.

Repeated roots of the auxiliary equation will be treated in the next section. Imaginary roots will be avoided until Section 37, where the corresponding solutions will be put into a desirable form.

EXAMPLE (a):   Solve the equation

$$\frac{d^3 y}{dx^3} - 4\frac{d^2 y}{dx^2} + \frac{dy}{dx} + 6y = 0.$$

First write the auxiliary equation

$$m^3 - 4m^2 + m + 6 = 0,$$

whose roots $m = -1, 2, 3$ may be obtained by synthetic division. Then the general solution is seen to be

$$y = c_1 e^{-x} + c_2 e^{2x} + c_3 e^{3x}.$$

EXAMPLE (b):   Solve the equation

$$(3D^3 + 5D^2 - 2D)y = 0.$$

The auxiliary equation is

$$3m^3 + 5m^2 - 2m = 0$$

and its roots are $m = 0, -2, \frac{1}{3}$. Using the fact that $e^{0x} = 1$, the desired solution may be written

$$y = c_1 + c_2 e^{-2x} + c_3 \exp\left(\tfrac{1}{3}x\right).$$

EXAMPLE (c): Solve the equation

$$\frac{d^2x}{dt^2} - 4x = 0$$

with the conditions that when $t = 0$, $x = 0$ and $dx/dt = 3$.
    The auxiliary equation is

$$m^2 - 4 = 0,$$

with roots $m = 2, -2$. Hence the general solution of the differential equation
is

$$x = c_1 e^{2t} + c_2 e^{-2t}.$$

It remains to enforce the conditions at $t = 0$. Now

$$\frac{dx}{dt} = 2c_1 e^{2t} - 2c_2 e^{-2t}.$$

Thus the condition that $x = 0$ when $t = 0$ requires that

$$0 = c_1 + c_2,$$

and the condition that $dx/dt = 3$ when $t = 0$ requires that

$$3 = 2c_1 - 2c_2.$$

From the simultaneous equations for c_1 and c_2 we conclude that $c_1 = \tfrac{3}{4}$ and
$c_2 = -\tfrac{3}{4}$. Therefore

$$x = \tfrac{3}{4}(e^{2t} - e^{-2t}),$$

which can also be put in the form

$$x = \tfrac{3}{2}\sinh(2t).$$

Exercises

In exercises 1 through 22, find the general solution. When the operator D is used, it
is implied that the independent variable is x.

1. $(D^2 - D - 2)y = 0.$ ANS. $y = c_1 e^{-x} + c_2 e^{2x}.$
2. $(D^2 + 3D)y = 0.$ ANS. $y = c_1 + c_2 e^{-3x}.$
3. $(D^2 - D - 6)y = 0.$ ANS. $y = c_1 e^{-2x} + c_2 e^{3x}.$
4. $(D^2 + 5D + 6)y = 0.$ ANS. $y = c_1 e^{-2x} + c_2 e^{-3x}.$
5. $(D^3 + 2D^2 - 15D)y = 0.$ ANS. $y = c_1 + c_2 e^{3x} + c_3 e^{-5x}.$
6. $(D^3 + 2D^2 - 8D)y = 0.$ ANS. $y = c_1 + c_2 e^{2x} + c_3 e^{-4x}.$

7. $(D^3 - D^2 - 4D + 4)y = 0.$ ANS. $y = c_1 e^{-2x} + c_2 e^x + c_3 e^{2x}.$

8. $(D^3 - 3D^2 - D + 3)y = 0.$ ANS. $y = c_1 e^{3x} + c_2 e^x + c_3 e^{-x}.$

9. $(4D^3 - 13D + 6)y = 0.$ ANS. $y = c_1 e^{x/2} + c_2 e^{3x/2} + c_3 e^{-2x}.$

10. $(4D^3 - 49D - 60)y = 0.$ ANS. $y = c_1 e^{4x} + c_2 e^{-5x/2} + c_3 e^{-3x/2}.$

11. $\dfrac{d^3 x}{dt^3} - 2\dfrac{d^2 x}{dt^2} - 3\dfrac{dx}{dt} = 0.$ ANS. $x = c_1 + c_2 e^{-t} + c_3 e^{3t}.$

12. $\dfrac{d^3 x}{dt^3} - 7\dfrac{dx}{dt} + 6x = 0.$ ANS. $x = c_1 e^t + c_2 e^{2t} + c_3 e^{-3t}.$

13. $(10D^3 + D^2 - 7D + 2)y = 0.$ ANS. $y = c_1 e^{-x} + c_2 e^{x/2} + c_3 e^{2x/5}.$

14. $(4D^3 - 13D - 6)y = 0.$ ANS. $y = c_1 e^{2x} + c_2 e^{-3x/2} + c_3 e^{-x/2}.$

15. $(D^3 - 5D - 2)y = 0.$ ANS. $y = c_1 e^{-2x} + c_2 e^{(1+\sqrt{2})x} + c_3 e^{(1-\sqrt{2})x}.$

16. $(D^3 - 3D^2 - 3D + 1)y = 0.$ ANS. $y = c_1 e^{-x} + c_2 e^{(2+\sqrt{3})x} + c_3 e^{(2-\sqrt{3})x}.$

17. $(4D^4 - 15D^2 + 5D + 6)y = 0.$ ANS. $y = c_1 e^{-2x} + c_2 e^{-x/2} + c_3 e^{3x/2} + c_4 e^x.$

18. $(D^4 - 2D^3 - 13D^2 + 38D - 24)y = 0.$

         ANS. $y = c_1 e^x + c_2 e^{2x} + c_3 e^{3x} + c_4 e^{-4x}.$

19. $(6D^4 + 23D^3 + 28D^2 + 13D + 2)y = 0.$

20. $(4D^4 - 45D^2 - 70D - 24)y = 0.$

         ANS. $y = c_1 e^{4x} + c_2 e^{-2x} + c_3 e^{-x/2} + c_4 e^{-3x/2}.$

21. $(D^2 - 4aD + 3a^2)y = 0;\ a$ real $\neq 0.$

22. $[D^2 - (a + b)D + ab]y = 0;\ a$ and b real and unequal. ANS. $y = c_1 e^{ax} + c_2 e^{bx}.$

In exercises 23 and 24, find the particular solution indicated.

23. $(D^2 - 2D - 3)y = 0;$ when $x = 0,\ y = 0,\ y' = -4.$ ANS. $y = e^{-x} - e^{3x}.$

24. $(D^2 - D - 6)y = 0;$ when $x = 0,\ y = 0,$ and when $x = 1,\ y = e^3.$

         ANS. $y = (e^{3x} - e^{-2x})/(1 - e^{-5}).$

In exercises 25 through 29, find for $x = 1$ the y value for the particular solution required.

25. $(D^2 - 2D - 3)y = 0;$ when $x = 0,\ y = 4,\ y' = 0.$

         ANS. When $x = 1,\ y = e^3 + 3e^{-1} = 21.2.$

26. $(D^3 - 4D)y = 0;$ when $x = 0,\ y = 0,\ y' = 0,\ y'' = 2.$

         ANS. When $x = 1,\ y = \sinh^2 1.$

27. $(D^2 - D - 6)y = 0;$ when $x = 0,\ y = 3,\ y' = -1.$ ANS. When $x = 1,\ y = 20.4.$

28. $(D^2 + 3D - 10)y = 0;$ when $x = 0,\ y = 0,$ and when $x = 2,\ y = 1.$

         ANS. When $x = 1,\ y = 0.135.$

29. $(D^3 - 2D^2 - 5D + 6)y = 0;$ when $x = 0,\ y = 1,\ y' = -7,\ y'' = -1.$

         ANS. When $x = 1,\ y = -19.8.$

35. The auxiliary equation; repeated roots

Suppose that in the equation

$$f(D)y = 0 \qquad\qquad (1)$$

the operator $f(D)$ has repeated factors; that is, the auxiliary equation $f(m) = 0$ has repeated roots. Then the method of the previous section does not yield the general solution. Let the auxiliary equation have three equal roots $m_1 = b$, $m_2 = b$, $m_3 = b$. The corresponding part of the solution yielded by the method of Section 34 is

$$y = c_1 e^{bx} + c_2 e^{bx} + c_3 e^{bx},$$

$$y = (c_1 + c_2 + c_3) e^{bx}. \tag{2}$$

Now (2) can be replaced by

$$y = c_4 e^{bx} \tag{3}$$

with $c_4 = c_1 + c_2 + c_3$. Thus, corresponding to the three roots under consideration, this method has yielded only the solution (3). The difficulty is present, of course, because the three solutions corresponding to the roots $m_1 = m_2 = m_3 = b$ are not linearly independent.

What is needed is a method for obtaining n linearly independent solutions corresponding to n equal roots of the auxiliary equation. Suppose that the auxiliary equation $f(m) = 0$ has the n equal roots

$$m_1 = m_2 = \cdots = m_n = b.$$

Then the operator $f(D)$ must have a factor $(D - b)^n$. We wish to find n linearly independent y's for which

$$(D - b)^n y = 0. \tag{4}$$

Turning to the result (6) near the end of Section 32 and writing $m = b$, we find that

$$(D - b)^n (x^k e^{bx}) = 0 \qquad \text{for} \quad k = 0, 1, 2, \dots, (n - 1). \tag{5}$$

The functions $y_k = x^k e^{bx}$ where $k = 0, 1, 2, \dots, (n - 1)$ are linearly independent because, aside from the common factor e^{bx}, they contain only the respective powers $x^0, x^1, x^2, \dots, x^{n-1}$. (See exercise 5, Section 32.)

The general solution of equation (4) is

$$y = c_1 e^{bx} + c_2 x e^{bx} + \cdots + c_n x^{n-1} e^{bx}. \tag{6}$$

Furthermore, if $f(D)$ contains the factor $(D - b)^n$, then the equation

$$f(D)y = 0 \tag{1}$$

can be written

$$g(D)(D - b)^n y = 0 \tag{7}$$

where $g(D)$ contains all the factors of $f(D)$ except $(D - b)^n$. Then any solution of

$$(D - b)^n y = 0 \tag{4}$$

is also a solution of (7) and therefore of (1).

Now we are in a position to write the solution of equation (1) whenever the auxiliary equation has only real roots. Each root of the auxiliary equation is either distinct from all the other roots or it is one of a set of equal roots. Corresponding to a root m_i distinct from all others, there is the solution

$$y_i = c_i \exp(m_i x), \tag{8}$$

and corresponding to n equal roots $m_1, m_2, \ldots, m_n$, each equal to b, there are the solutions

$$c_1 e^{bx}, c_2 x e^{bx}, \ldots, c_n x^{n-1} e^{bx}. \tag{9}$$

The collection of solutions (9) has the proper number of elements, a number equal to the order of the differential equation, because there is one solution corresponding to each root of the auxiliary equation. The solutions thus obtained can be proved to be linearly independent.

EXAMPLE (a): Solve the equation

$$(D^4 - 7D^3 + 18D^2 - 20D + 8)y = 0. \tag{10}$$

With the aid of synthetic division, it is easily seen that the auxiliary equation

$$m^4 - 7m^3 + 18m^2 - 20m + 8 = 0$$

has the roots $m = 1, 2, 2, 2$. Then the general solution of equation (10) is

$$y = c_1 e^x + c_2 e^{2x} + c_3 x e^{2x} + c_4 x^2 e^{2x},$$

or

$$y = c_1 e^x + (c_2 + c_3 x + c_4 x^2) e^{2x}.$$

EXAMPLE (b): Solve the equation

$$\frac{d^4 y}{dx^4} + 2\frac{d^3 y}{dx^3} + \frac{d^2 y}{dx^2} = 0.$$

The auxiliary equation is

$$m^4 + 2m^3 + m^2 = 0,$$

with roots $m = 0, 0, -1, -1$. Hence the desired solution is

$$y = c_1 + c_2 x + c_3 e^{-x} + c_4 x e^{-x}.$$

Exercises

In exercises 1 through 20 find the general solution.

1. $(4D^2 - 4D + 1)y = 0.$ ANS. $y = (c_1 + c_2 x) e^{x/2}.$

2. $(D^2 + 6D + 9)y = 0.$

3. $(D^3 - 4D^2 + 4D)y = 0$.   ANS. $y = c_1 + (c_2 + c_3 x) e^{2x}$.

4. $(9D^3 + 6D^2 + D)y = 0$.

5. $(2D^4 - 3D^3 - 2D^2)y = 0$.   ANS. $y = c_1 + c_2 x + c_3 e^{2x} + c_4 e^{-x/2}$.

6. $(2D^4 - 5D^3 - 3D^2)y = 0$.

7. $(D^3 + 3D^2 - 4)y = 0$.   ANS. $y = c_1 e^x + (c_2 + c_3 x) e^{-2x}$.

8. $(4D^3 - 27D + 27)y = 0$.

9. $(D^3 + 3D^2 + 3D + 1)y = 0$.   ANS. $y = (c_1 + c_2 x + c_3 x^2) e^{-x}$.

10. $(D^3 + 6D^2 + 12D + 8)y = 0$.

11. $(D^5 - D^3)y = 0$.   ANS. $y = c_1 + c_2 x + c_3 x^2 + c_4 e^x + c_5 e^{-x}$;

or $y = c_1 + c_2 x + c_3 x^2 + c_6 \cosh x + c_7 \sinh x$.

12. $(D^5 - 16D^3)y = 0$.

13. $(4D^4 + 4D^3 - 3D^2 - 2D + 1)y = 0$.

ANS. $y = (c_1 + c_2 x) e^{x/2} + (c_3 + c_4 x) e^{-x}$.

14. $(4D^4 - 4D^3 - 23D^2 + 12D + 36)y = 0$.

ANS. $y = (c_1 + c_2 x) e^{-3x/2} + (c_3 + c_4 x) e^{2x}$.

15. $(D^4 + 3D^3 - 6D^2 - 28D - 24)y = 0$.

ANS. $y = c_1 e^{3x} + (c_2 + c_3 x + c_4 x^2) e^{-2x}$.

16. $(27D^4 - 18D^2 + 8D - 1)y = 0$.

17. $(4D^5 - 23D^3 - 33D^2 - 17D - 3)y = 0$.

ANS. $y = c_1 e^{3x} + (c_2 + c_3 x) e^{-x} + (c_4 + c_5 x) e^{-x/2}$.

18. $(4D^5 - 15D^3 - 5D^2 + 15D + 9)y = 0$.

19. $(D^4 - 5D^2 - 6D - 2)y = 0$.

ANS. $y = (c_1 + c_2 x) e^{-x} + c_3 e^{(1+\sqrt{3})x} + c_4 e^{(1-\sqrt{3})x}$.

20. $(D^5 - 5D^4 + 7D^3 + D^2 - 8D + 4)y = 0$.

In exercises 21 through 26, find the particular solution indicated.

21. $(D^2 + 4D + 4)y = 0$; when $x = 0$, $y = 1$, $y' = -1$.   ANS. $y = (1 + x) e^{-2x}$.

22. The equation of exercise 21 with the condition that the graph of the solution pass through the points $(0, 2)$ and $(2, 0)$.   ANS. $y = (2 - x) e^{-2x}$.

23. $(D^3 - 3D - 2)y = 0$; when $x = 0$, $y = 0$, $y' = 9$, $y'' = 0$.

ANS. $y = 2 e^{2x} + (3x - 2) e^{-x}$.

24. $(D^4 + 3D^3 + 2D^2)y = 0$; when $x = 0$, $y = 0$, $y' = 4$, $y'' = -6$, $y''' = 14$.

ANS. $y = 2(x + e^{-x} - e^{-2x})$.

25. The equation of exercise 24 with the conditions that when $x = 0$, $y = 0$, $y' = 3$, $y'' = -5$, $y''' = 9$.   ANS. $y = 2 - e^{-x} - e^{-2x}$.

26. $(D^3 + D^2 - D - 1)y = 0$; when $x = 0$, $y = 1$, when $x = 2$, $y = 0$, and also as $x \to \infty$, $y \to 0$.   ANS. $y = \frac{1}{2}(2 - x) e^{-x}$.

In exercises 27 through 29, find for $x = 2$ the y value for the particular solution required.

27. $(4D^2 - 4D + 1)y = 0$; when $x = 0$, $y = -2$, $y' = 2$.

ANS.   When $x = 2$, $y = 4 e$.

28. $(D^3 + 2D^2)y = 0$; when $x = 0$, $y = -3$, $y' = 0$, $y'' = 12$.

ANS.   When $x = 2$, $y = 3 e^{-4} + 6$.

29. $(D^3 + 5D^2 + 3D - 9)y = 0$; when $x = 0$, $y = -1$, when $x = 1$, $y = 0$, and also as $x \to \infty$, $y \to 0$.   ANS.   When $x = 2$, $y = e^{-6}$.

36. A definition of exp z for imaginary z

Since the auxiliary equation may have imaginary roots, we need now to lay down a definition of exp z for imaginary z.

Let $z = \alpha + i\beta$ with α and β real. Since it is desirable to have the ordinary laws of exponents remain valid, it is wise to require that

$$\exp(\alpha + i\beta) = e^\alpha \cdot e^{i\beta}. \tag{1}$$

To e^α with α real, we attach the usual meaning.

Now consider $e^{i\beta}$, β real. In calculus it is shown that for all real x

$$e^x = 1 + \frac{x}{1!} + \frac{x^2}{2!} + \frac{x^3}{3!} + \cdots + \frac{x^n}{n!} + \cdots, \tag{2}$$

or

$$e^x = \sum_{n=0}^{\infty} \frac{x^n}{n!}. \tag{2}$$

If we now tentatively put $x = i\beta$ in (2) as a definition of $e^{i\beta}$, we get

$$e^{i\beta} = 1 + \frac{i\beta}{1!} + \frac{i^2\beta^2}{2!} + \frac{i^3\beta^3}{3!} + \frac{i^4\beta^4}{4!} + \cdots + \frac{i^n\beta^n}{n!} + \cdots. \tag{3}$$

Separating the even powers of β from the odd powers of β in (3) yields

$$e^{i\beta} = 1 + \frac{i^2\beta^2}{2!} + \frac{i^4\beta^4}{4!} + \cdots + \frac{i^{2k}\beta^{2k}}{(2k)!} + \cdots$$
$$+ \frac{i\beta}{1!} + \frac{i^3\beta^3}{3!} + \cdots + \frac{i^{2k+1}\beta^{2k+1}}{(2k+1)!} + \cdots, \tag{4}$$

or

$$e^{i\beta} = \sum_{k=0}^{\infty} \frac{i^{2k}\beta^{2k}}{(2k)!} + \sum_{k=0}^{\infty} \frac{i^{2k+1}\beta^{2k+1}}{(2k+1)!}. \tag{4}$$

Now $i^{2k} = (-1)^k$, so we may write

$$e^{i\beta} = 1 - \frac{\beta^2}{2!} + \frac{\beta^4}{4!} + \cdots + \frac{(-1)^k\beta^{2k}}{(2k)!} + \cdots$$
$$+ i\left[\frac{\beta}{1!} - \frac{\beta^3}{3!} + \cdots + \frac{(-1)^k\beta^{2k+1}}{(2k+1)!} + \cdots\right], \tag{5}$$

or

$$e^{i\beta} = \sum_{k=0}^{\infty} \frac{(-1)^k\beta^{2k}}{(2k)!} + i\sum_{k=0}^{\infty} \frac{(-1)^k\beta^{2k+1}}{(2k+1)!}. \tag{5}$$

But the series on the right in (5) are precisely those for $\cos \beta$ and $\sin \beta$ as developed in calculus. Hence we are led to the tentative result

$$e^{i\beta} = \cos \beta + i \sin \beta. \tag{6}$$

The student should realize that the manipulations above have no meaning in themselves at this stage (assuming that infinite series with complex terms are not a part of the content of elementary mathematics). What has been accomplished is this: the formal manipulations above have suggested the meaningful definition (6). Combining (6) with (1) we now put forward a reasonable *definition* of $\exp(\alpha + i\beta)$, namely,

$$\exp(\alpha + i\beta) = e^{\alpha}(\cos \beta + i \sin \beta), \qquad \text{when } \alpha \text{ and } \beta \text{ are real.} \tag{7}$$

Replacing β by $(-\beta)$ in (7) yields a result that is of value to us in the next section,

$$\exp(\alpha - i\beta) = e^{\alpha}(\cos \beta - i \sin \beta).$$

It is interesting and important that, with the definition (7), the function e^z for complex z retains many of the properties possessed by the function e^x for real x. Such matters are often studied in detail in books on complex variables.* Here we need in particular to know that if

$$y = \exp(a + ib)x,$$

with a, b, and x real, then

$$(D - a - ib)y = 0.$$

The result desired follows at once by differentiation, with respect to x, of the function

$$y = e^{ax}(\cos bx + i \sin bx).$$

37. The auxiliary equation; imaginary roots

Consider a differential equation $f(D)y = 0$ for which the auxiliary equation $f(m) = 0$ has real coefficients. From elementary algebra we know that if the auxiliary equation has any imaginary roots those roots must occur in conjugate pairs. Thus if

$$m_1 = a + ib$$

is a root of the equation $f(m) = 0$, with a and b real and $b \neq 0$, then

$$m_2 = a - ib$$

*For example, R. V. Churchill, J. W. Brown, and R. F. Verhey, *Complex Variables and Applications*, 3rd ed. (New York: McGraw-Hill Book Company, 1976), pp. 52–56.

is also a root of $f(m) = 0$. It must be kept in mind that this result is a consequence of the reality of the coefficients in the equation $f(m) = 0$. Imaginary roots do not necessarily appear in pairs in an algebraic equation whose coefficients involve imaginaries.

We can now construct in usable form solutions of

$$f(D)y = 0 \tag{1}$$

corresponding to imaginary roots of $f(m) = 0$. For, since $f(m)$ is assumed to have real coefficients, any imaginary roots appear in conjugate pairs

$$m_1 = a + ib \qquad \text{and} \qquad m_2 = a - ib.$$

Then, according to the preceding section, equation (1) is satisfied by

$$y = c_1 \exp\left[(a + ib)x\right] + c_2 \exp\left[(a - ib)x\right]. \tag{2}$$

Taking x to be real along with a and b, we get from (2) the result

$$y = c_1 e^{ax}(\cos bx + i \sin bx) + c_2 e^{ax}(\cos bx - i \sin bx). \tag{3}$$

Now (3) may be written

$$y = (c_1 + c_2) e^{ax} \cos bx + i(c_1 - c_2) e^{ax} \sin bx.$$

Finally, let $c_1 + c_2 = c_3$, and $i(c_1 - c_2) = c_4$, where c_3 and c_4 are new arbitrary constants. Then equation (1) is seen to have the solutions

$$y = c_3 e^{ax} \cos bx + c_4 e^{ax} \sin bx, \tag{4}$$

corresponding to the two roots $m_1 = a + ib$ and $m_2 = a - ib$ ($b \neq 0$) of the auxiliary equation.

The reduction of the solution (2) above to the desirable form (4) has been done once and that is enough. Whenever a pair of conjugate imaginary roots of the auxiliary equation appears, we write down at once, in the form given on the right in equation (4), the particular solution corresponding to those two roots.

EXAMPLE (a):   Solve the equation

$$(D^3 - 3D^2 + 9D + 13)y = 0.$$

For the auxiliary equation

$$m^3 - 3m^2 + 9m + 13 = 0,$$

one root, $m_1 = -1$, is easily found. When the factor $(m + 1)$ is removed by synthetic division, it is seen that the other two roots are solutions of the quadratic equation

$$m^2 - 4m + 13 = 0.$$

Those roots are found to be $m_2 = 2 + 3i$ and $m_3 = 2 - 3i$. The auxiliary equation has the roots $m = -1, 2 \pm 3i$. Hence the general solution of the differential equation is

$$y = c_1 e^{-x} + c_2 e^{2x} \cos 3x + c_3 e^{2x} \sin 3x.$$

Repeated imaginary roots lead to solutions analogous to those brought in by repeated real roots. For instance, if the roots $m = a \pm ib$ occur three times, then the corresponding six linearly independent solutions of the differential equation are those appearing in the expression

$$(c_1 + c_2 x + c_3 x^2) e^{ax} \cos bx + (c_4 + c_5 x + c_6 x^2) e^{ax} \sin bx.$$

EXAMPLE (b):   Solve the equation

$$(D^4 + 8D^2 + 16)y = 0.$$

The auxiliary equation $m^4 + 8m^2 + 16 = 0$ may be written

$$(m^2 + 4)^2 = 0,$$

so its roots are seen to be $m = \pm 2i, \pm 2i$. The roots $m_1 = 2i$ and $m_2 = -2i$ occur twice each. Thinking of $2i$ as $0 + 2i$ and recalling that $e^{0x} = 1$, we write the solution of the differential equation as

$$y = (c_1 + c_2 x) \cos 2x + (c_3 + c_4 x) \sin 2x.$$

In such exercises as those below a fine check can be obtained by direct substitution of the result and its appropriate derivatives into the differential equation. The verification is particularly effective because the operations performed in the check are so different from those performed in obtaining the solution.

38. A note on hyperbolic functions

Two particular linear combinations of exponential functions appear with such frequency in both pure and applied mathematics that it has been worth-while to use special symbols for those combinations. The hyperbolic sine of x, written $\sinh x$, is defined by

$$\sinh x = \frac{e^x - e^{-x}}{2}; \tag{1}$$

the hyperbolic cosine of x, written $\cosh x$, is defined by

$$\cosh x = \frac{e^x + e^{-x}}{2}. \tag{2}$$

From the definitions of $\sinh x$ and $\cosh x$ it follows that

$$\sinh^2 x = \tfrac{1}{4}(e^{2x} - 2 + e^{-2x})$$

and

$$\cosh^2 x = \tfrac{1}{4}(e^{2x} + 2 + e^{-2x}),$$

so

$$\cosh^2 x - \sinh^2 x = 1, \tag{3}$$

an identity similar to the well-known identity $\cos^2 x + \sin^2 x = 1$ in trigonometry.

Directly from the definition we find that

$$y = \sinh u$$

is equivalent to

$$y = \tfrac{1}{2}(e^u - e^{-u}).$$

Hence, if u is a function of x, then

$$\frac{dy}{dx} = \tfrac{1}{2}(e^u + e^{-u})\frac{du}{dx},$$

that is,

$$\frac{d}{dx}\sinh u = \cosh u \frac{du}{dx}. \tag{4}$$

The same method yields the result

$$\frac{d}{dx}\cosh u = \sinh u \frac{du}{dx}. \tag{5}$$

The graphs of $y = \cosh x$ and $y = \sinh x$ are exhibited in Figure 12. Note the important properties:

(a) $\cosh x \geq 1$ for all real x;
(b) the only real value of x for which $\sinh x = 0$ is $x = 0$;
(c) $\cosh(-x) = \cosh x$; that is, $\cosh x$ is an even function of x;
(d) $\sinh(-x) = -\sinh x$; $\sinh x$ is an odd function of x.

The hyperbolic functions have no real period. Corresponding to the period 2π possessed by the circular functions, there is a period $2\pi i$ for the hyperbolic functions.

The hyperbolic cosine curve is that in which a transmission line, cable, piece of string, watch chain, etc., hangs between two points at which it is suspended. This result is obtained in Chapter 16.

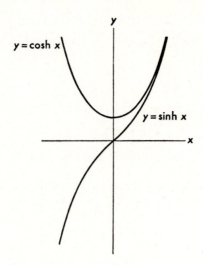

FIGURE 12

Since $D^2 \cosh ax = a^2 \cosh ax$ and $D^2 \sinh ax = a^2 \sinh ax$, it follows that both $\cosh ax$ and $\sinh ax$ are solutions of

$$(D^2 - a^2)y = 0, \qquad a \neq 0. \tag{6}$$

Furthermore the Wronskian of these two functions,

$$W(x) = \begin{vmatrix} \cosh ax & \sinh ax \\ a \sinh ax & a \cosh ax \end{vmatrix} = a,$$

is not zero, so that $\cosh ax$ and $\sinh ax$ are linearly independent solutions of equation (6). Hence the general solution of (6) may be written

$$y = c_1 \cosh ax + c_2 \sinh ax$$

instead of using the form

$$y = c_3 e^{ax} + c_4 e^{-ax}.$$

It is often very convenient to use this alternative form for representing the general solution of (6).

EXAMPLE: Find the solution of the problem

$$(D^2 - 4)y = 0; \qquad \text{when } x = 0, \ y = 0, \ y' = 2. \tag{7}$$

The general solution of the differential equation (7) may be written

$$y = c_1 \cosh 2x + c_2 \sinh 2x,$$

from which

$$y' = 2c_1 \sinh 2x + 2c_2 \cosh 2x.$$

The initial conditions now require that $0 = c_1$ and $2 = 2c_2$, so that finally

$$y = \sinh 2x.$$

Note that if we were to choose the alternative form

$$y = c_3 e^{2x} + c_4 e^{-2x}$$

for the general solution of (7), we would obtain the same result with a little more fuss in determining c_3 and c_4. Indeed, one major reason for using the hyperbolic functions is that $\cosh ax$ and $\sinh ax$ have values 1 and 0 when $x = 0$, a fact that is particularly useful in solving initial value problems.

Exercises

Find the general solution except when the exercise stipulates otherwise.

1. Verify directly that the relation

$$y = c_3 e^{ax} \cos bx + c_4 e^{ax} \sin bx \tag{A}$$

satisfies the equation

$$[(D - a)^2 + b^2]y = 0.$$

2. $(D^2 - 2D + 5)y = 0$. Verify your answer. ANS. $y = c_1 e^x \cos 2x + c_2 e^x \sin 2x$.
3. $(D^2 - 2D + 2)y = 0$.
4. $(D^2 + 9)y = 0$. Verify your answer. ANS. $y = c_1 \cos 3x + c_2 \sin 3x$.
5. $(D^2 - 9)y = 0$. ANS. $y = c_1 \cosh 3x + c_2 \sinh 3x$.
6. $(D^2 + 6D + 13)y = 0$. Verify your answer.
7. $(D^2 - 4D + 7)y = 0$. ANS. $y = c_1 e^{2x} \cos \sqrt{3}x + c_2 e^{2x} \sin \sqrt{3}x$.
8. $(D^3 + 2D^2 + D + 2)y = 0$. Verify your answer.
9. $(D^2 - 1)y = 0$; when $x = 0$, $y = y_0$, $y' = 0$. ANS. $y = y_0 \cosh x$.
10. $(D^2 + 1)y = 0$; when $x = 0$, $y = y_0$, $y' = 0$. ANS. $y = y_0 \cos x$.
11. $(D^4 + 2D^3 + 10D^2)y = 0$. ANS. $y = c_1 + c_2 x + c_3 e^{-x} \cos 3x + c_4 e^{-x} \sin 3x$.
12. $(D^3 + 7D^2 + 19D + 13)y = 0$; when $x = 0$, $y = 0$, $y' = 2$, $y'' = -12$.
    ANS. $y = e^{-3x} \sin 2x$.
13. $(D^5 + D^4 - 7D^3 - 11D^2 - 8D - 12)y = 0$.
    ANS. $y = c_1 \cos x + c_2 \sin x + c_3 e^{-2x} + c_4 x e^{-2x} + c_5 e^{3x}$.
14. $(D^4 - 2D^3 + 2D^2 - 2D + 1)y = 0$. Verify your answer.
15. $(D^4 + 18D^2 + 81)y = 0$. ANS. $y = (c_1 + c_2 x) \cos 3x + (c_3 + c_4 x) \sin 3x$.
16. $(2D^4 + 11D^3 - 4D^2 - 69D + 34)y = 0$. Verify your answer.
17. $(D^6 + 9D^4 + 24D^2 + 16)y = 0$.
    ANS. $y = c_1 \cos x + c_2 \sin x + (c_3 + c_4 x) \cos 2x + (c_5 + c_6 x) \sin 2x$.
18. $(2D^3 - D^2 + 36D - 18)y = 0$.
    ANS. $y = c_1 e^{x/2} + c_2 \cos (3\sqrt{2}x) + c_3 \sin (3\sqrt{2}x)$.

19. $\dfrac{d^2x}{dt^2} + k^2x = 0$, k real; when $t = 0$, $x = 0$, $\dfrac{dx}{dt} = v_0$. Verify your result completely.

ANS. $x = (v_0/k) \sin kt$.

20. $(D^3 + D^2 + 4D + 4)y = 0$; when $x = 0$, $y = 0$, $y' = -1$, $y'' = 5$.

ANS. $y = e^{-x} - \cos 2x$.

21. $\dfrac{d^2x}{dt^2} + 2b\dfrac{dx}{dt} + k^2x = 0$, $k > b > 0$; when $t = 0$, $x = 0$, $\dfrac{dx}{dt} = v_0$.

ANS. $x = (v_0/a)\,e^{-bt} \sin at$; where $a = \sqrt{k^2 - b^2}$.

Miscellaneous Exercises

Obtain the general solution unless otherwise instructed.

1. $(D^2 + 3D)y = 0$.   ANS. $y = c_1 + c_2 e^{-3x}$.
2. $(9D^4 + 6D^3 + D^2)y = 0$.   ANS. $y = c_1 + c_2x + (c_3 + c_4x)\exp\left(-\tfrac{1}{3}x\right)$.
3. $(D^2 + D - 6)y = 0$.   ANS. $y = c_1 e^{2x} + c_2 e^{-3x}$.
4. $(D^3 + 2D^2 + D + 2)y = 0$.   ANS. $y = c_1 e^{-2x} + c_2 \cos x + c_3 \sin x$.
5. $(D^3 - 3D^2 + 4)y = 0$.   ANS. $y = c_1 e^{-x} + e^{2x}(c_2 + c_3x)$.
6. $(D^3 - 2D^2 - 3D)y = 0$.   ANS. $y = c_1 + c_2 e^{3x} + c_3 e^{-x}$.
7. $(4D^3 - 3D + 1)y = 0$.   ANS. $y = c_1 e^{-x} + (c_2 + c_3x)\exp\left(\tfrac{1}{2}x\right)$.
8. $(D^3 + 3D^2 - 4D - 12)y = 0$.   ANS. $y = c_1 \cosh 2x + c_2 \sinh 2x + c_3 e^{-3x}$.
9. $(D^3 + 3D^2 + 3D + 1)y = 0$.   ANS. $y = e^{-x}(c_1 + c_2x + c_3x^2)$.
10. $(4D^3 - 21D - 10)y = 0$.   ANS. $y = c_1 e^{-2x} + c_2 \exp\left(\tfrac{5}{2}x\right) + c_3 \exp\left(-\tfrac{1}{2}x\right)$.
11. $(4D^3 - 7D + 3)y = 0$.   ANS. $y = c_1 e^x + c_2 \exp\left(\tfrac{1}{2}x\right) + c_3 \exp\left(-\tfrac{3}{2}x\right)$.
12. $(D^2 - D - 6)y = 0$; when $x = 0$, $y = 2$, $y' = 1$.   ANS. $y = e^{3x} + e^{-2x}$.
13. $(D^4 + 6D^3 + 9D^2)y = 0$; when $x = 0$, $y = 0$, $y' = 0$, $y'' = 6$, and as $x \to \infty$, $y' \to 1$.
    For this particular solution, find the value of y when $x = 1$.   ANS. $y = 1 - e^{-3}$.
14. $(D^3 + 6D^2 + 12D + 8)y = 0$; when $x = 0$, $y = 1$, $y' = -2$, $y'' = 2$.

ANS. $y = e^{-2x}(1 - x^2)$.

15. $(D^3 - 14D + 8)y = 0$.

ANS. $y = c_1 e^{-4x} + c_2 \exp\left[(2 + \sqrt{2})x\right] + c_3 \exp\left[(2 - \sqrt{2})x\right]$.

16. $(8D^3 - 4D^2 - 2D + 1)y = 0$.   ANS. $y = (c_1 + c_2x)\exp\left(\tfrac{1}{2}x\right) + c_3 \exp\left(-\tfrac{1}{2}x\right)$.
17. $(D^4 + D^3 - 4D^2 - 4D)y = 0$.
18. $(D^4 - 2D^3 + 5D^2 - 8D + 4)y = 0$.

ANS. $y = e^x(c_1 + c_2x) + c_3 \cos 2x + c_4 \sin 2x$.

19. $(D^4 + 2D^2 + 1)y = 0$.
20. $(D^4 + 5D^2 + 4)y = 0$.   ANS. $y = c_1 \cos x + c_2 \sin x + c_3 \cos 2x + c_4 \sin 2x$.
21. $(D^4 + 3D^3 - 4D)y = 0$.
22. $(D^5 + D^4 - 9D^3 - 13D^2 + 8D + 12)y = 0$.

ANS. $y = c_1 e^x + c_2 e^{3x} + c_3 e^{-x} + e^{-2x}(c_4 + c_5x)$.

23. $(D^4 - 11D^3 + 36D^2 - 16D - 64)y = 0$.
24. $(D^2 + 2D + 5)y = 0$.   ANS. $y = e^{-x}(c_1 \cos 2x + c_2 \sin 2x)$.
25. $(D^4 + 4D^3 + 2D^2 - 8D - 8)y = 0$.
26. $(4D^4 - 24D^3 + 35D^2 + 6D - 9)y = 0$.

ANS. $y = e^{3x}(c_1 + c_2x) + c_3 \cosh \tfrac{1}{2}x + c_4 \sinh \tfrac{1}{2}x$.

27. $(4D^4 + 20D^3 + 35D^2 + 25D + 6)y = 0$.

28. $(D^4 - 7D^3 + 11D^2 + 5D - 14)y = 0.$
29. $(D^3 + 5D^2 + 7D + 3)y = 0.$
30. $(D^3 - 2D^2 + D - 2)y = 0.$ ANS. $y = c_1 e^{2x} + c_2 \cos x + c_3 \sin x.$
31. $(D^3 - D^2 + D - 1)y = 0.$
32. $(D^3 + 4D^2 + 5D)y = 0.$
33. $(D^4 - 13D^2 + 36)y = 0.$
34. $(D^4 - 5D^3 + 5D^2 + 5D - 6)y = 0.$
                            ANS. $y = c_1 \cosh x + c_2 \sinh x + c_3 e^{2x} + c_4 e^{3x}.$
35. $(4D^3 + 8D^2 - 11D + 3)y = 0.$
36. $(D^3 + D^2 - 16D - 16)y = 0.$
37. $(D^4 - D^3 - 3D^2 + D + 2)y = 0.$ ANS. $y = c_1 e^x + c_2 e^{2x} + e^{-x}(c_3 + c_4 x).$
38. $(D^3 - 2D^2 - 3D + 10)y = 0.$
39. $(D^5 + D^4 - 6D^3)y = 0.$
40. $(4D^3 + 28D^2 + 61D + 37)y = 0.$ ANS. $y = c_1 e^{-x} + e^{-3x}(c_2 \cos \frac{1}{2}x + c_3 \sin \frac{1}{2}x).$
41. $(4D^3 + 12D^2 + 13D + 10)y = 0.$
42. $(18D^3 - 33D^2 + 20D - 4)y = 0.$
43. $(D^5 - 2D^3 - 2D^2 - 3D - 2)y = 0.$
                      ANS. $y = e^{-x}(c_1 + c_2 x) + c_3 e^{2x} + c_4 \cos x + c_5 \sin x.$
44. $(D^4 - 2D^3 + 2D^2 - 2D + 1)y = 0.$
45. $(4D^5 + 4D^4 - 9D^3 - 11D^2 + D + 3)y = 0.$
46. $(D^5 - 15D^3 + 10D^2 + 60D - 72)y = 0.$
47. $(D^4 + 2D^3 - 6D^2 - 16D - 8)y = 0.$
                  ANS. $y = e^{-2x}(c_1 + c_2 x) + e^x(c_3 \cos \sqrt{3}x + c_4 \sin \sqrt{3}x).$

Nonhomogeneous Equations: Undetermined Coefficients

39. Construction of a homogeneous equation from a specified solution

In Section 29 we saw that the general solution of the equation

$$(b_0 D^n + b_1 D^{n-1} + \cdots + b_{n-1} D + b_n)y = R(x) \tag{1}$$

is

$$y = y_c + y_p,$$

where y_c, the complementary function, is the general solution of the homogeneous equation

$$(b_0 D^n + b_1 D^{n-1} + \cdots + b_{n-1} D + b_n)y = 0 \tag{2}$$

and y_p is any particular solution of the original equation (1).

Various methods for getting a particular solution of (1) when the b_0, $b_1, \ldots, b_n$ are constants will be presented. In preparation for the method of undetermined coefficients it is wise to obtain proficiency in writing a homogeneous differential equation of which a given function of proper form is a solution.

Recall that in solving homogeneous equations with constant coefficients, a term such as $c_1 e^{ax}$ occurred only when the auxiliary equation $f(m) = 0$ had a root $m = a$, and then the operator $f(D)$ had a factor $(D - a)$. In like manner, $c_2 x e^{ax}$ appeared only when $f(D)$ contained the factor $(D - a)^2$, $c_3 x^2 e^{ax}$ only when $f(D)$ contained $(D - a)^3$, and so on.   Such terms as $c e^{ax} \cos bx$ or $c e^{ax} \sin bx$ correspond to roots $m = a \pm ib$, or to a factor $[(D - a)^2 + b^2]$.

EXAMPLE (a):   Find a homogeneous linear equation, with constant coefficients, that has as a particular solution

$$y = 7 e^{3x} + 2x.$$

First note that the coefficients (7 and 2) are quite irrelevant for the present problem, so long as they are not zero. We shall obtain an equation satisfied by $y = c_1 e^{3x} + c_2 x$, no matter what the constants c_1 and c_2 may be.

A term $c_1 e^{3x}$ occurs along with a root $m = 3$ of the auxiliary equation. The term $c_2 x$ will appear if the auxiliary equation has $m = 0, 0$; that is, a double root $m = 0$. We have recognized that the equation

$$D^2(D - 3)y = 0,$$

or

$$(D^3 - 3D^2)y = 0,$$

has $y = c_1 e^{3x} + c_2 x + c_3$ as its general solution, and therefore that it also has $y = 7 e^{3x} + 2x$ as a particular solution.

EXAMPLE (b):   Find a homogeneous linear equation with real, constant coefficients that is satisfied by

$$y = 6 + 3x e^x - \cos x. \tag{3}$$

The term 6 is associated with $m = 0$, the term $3x e^x$ with a double root $m = 1, 1$, and the term $(- \cos x)$ with the pair of imaginary roots $m = 0 \pm i$. Hence the auxiliary equation is

$$m(m - 1)^2(m^2 + 1) = 0,$$

or

$$m^5 - 2m^4 + 2m^3 - 2m^2 + m = 0.$$

Therefore the function in (3) is a solution of the differential equation

$$(D^5 - 2D^4 + 2D^3 - 2D^2 + D)y = 0. \tag{4}$$

That is, from the general solution

$$y = c_1 + (c_2 + c_3 x) e^x + c_4 \cos x + c_5 \sin x$$

of equation (4), the relation (3) follows by an appropriate choice of the constants: $c_1 = 6$, $c_2 = 0$, $c_3 = 3$, $c_4 = -1$, $c_5 = 0$.

EXAMPLE (c):   Find a homogeneous linear equation with real, constant coefficients that is satisfied by

$$y = 4x\, e^x \sin 2x.$$

The desired equation must have its auxiliary equation with roots $m = 1 \pm 2i, 1 \pm 2i$. The roots $m = 1 \pm 2i$ correspond to factors $(m - 1)^2 + 4$, so the auxiliary equation must be

$$[(m - 1)^2 + 4]^2 = 0,$$

or

$$m^4 - 4m^3 + 14m^2 - 20m + 25 = 0.$$

Hence the desired equation is

$$(D^4 - 4D^3 + 14D^2 - 20D + 25)y = 0.$$

Note that in all such problems, a correct (but undesirable) solution may be obtained by inserting additional roots of the auxiliary equation.

Exercises

In exercises 1 through 14, obtain in factored form a linear differential equation with real, constant coefficients that is satisfied by the given function.

1. $y = 4\,e^{2x} + 3\,e^{-x}$.   ANS. $(D - 2)(D + 1)y = 0$.
2. $y = 7 - 2x + \frac{1}{2}e^{4x}$.   ANS. $D^2(D - 4)y = 0$.
3. $y = -2x + \frac{1}{2}\,e^{4x}$.   ANS. $D^2(D - 4)y = 0$.
4. $y = x^2 - 5 \sin 3x$.   ANS. $D^3(D^2 + 9)y = 0$.
5. $y = 2\,e^x \cos 3x$.   ANS. $(D - 1 - 3i)(D - 1 + 3i)y = 0$; or $[(D - 1)^2 + 9]y = 0$; or $(D^2 - 2D + 10)y = 0$.
6. $y = 3\,e^{2x} \sin 3x$.   ANS. $(D^2 - 4D + 13)y = 0$.
7. $y = -2\,e^{3x} \cos x$.   ANS. $(D^2 - 6D + 10)y = 0$.
8. $y = e^{-x} \sin 2x$.   ANS. $(D^2 + 2D + 5)y = 0$.
9. $y = xe^{-x} \sin 2x + 3e^{-x} \cos 2x$.   ANS. $(D^2 + 2D + 5)^2 y = 0$.
10. $y = \sin 2x + 3 \cos 2x$.   ANS. $(D^2 + 4)y = 0$.
11. $y = \cos kx$.   ANS. $(D^2 + k^2)y = 0$.
12. $y = x \sin 2x$.   ANS. $(D^2 + 4)^2 y = 0$.
13. $y = 4 \sinh x$.   ANS. $(D^2 - 1)y = 0$.
14. $y = 2 \cosh 2x - \sinh 2x$.   ANS. $(D^2 - 4)y = 0$.

In exercises 15 through 34, list the roots of the auxiliary equation for a homogeneous linear equation with real, constant coefficients that has the given function as a particular solution.

15. $y = 3x\,e^{2x}$.   ANS. $m = 2, 2$.
16. $y = x^2\,e^{-x} + 4\,e^x$.   ANS. $m = -1, -1, -1, 1$.

17. $y = e^{-x} \cos 4x$.   ANS. $m = -1 \pm 4i$.
18. $y = 3 e^{-x} \cos 4x + 15 e^{-x} \sin 4x$.   ANS. $m = -1 \pm 4i$.
19. $y = x(e^{2x} + 4)$.   ANS. $m = 0, 0, 2, 2$.
20. $y = 4 + 2x^2 - e^{-3x}$.   ANS. $m = 0, 0, 0, -3$.
21. $y = x e^x$.   ANS. $m = 1, 1$.
22. $y = x e^x + 5 e^x$.   ANS. $m = 1, 1$.
23. $y = 4 \cos 2x$.   ANS. $m = \pm 2i$.
24. $y = 4 \cos 2x - 3 \sin 2x$.   ANS. $m = \pm 2i$.
25. $y = x \cos 2x$.   ANS. $m = 2i, 2i, -2i, -2i$.
26. $y = e^{-2x} \cos 3x$.   ANS. $m = -2 \pm 3i$.
27. $y = x \cos 2x - 3 \sin 2x$.   ANS. $m = 2i, 2i, -2i, -2i$.
28. $y = e^{-2x}(\cos 3x + \sin 3x)$.   ANS. $m = -2 \pm 3i$.
29. $y = \sin^3 x$. Use the fact that $\sin^3 x = \frac{1}{4}(3 \sin x - \sin 3x)$.   ANS. $m = \pm i, \pm 3i$.
30. $y = \cos^2 x$.   ANS. $m = 0, \pm 2i$.
31. $y = x^2 - x + e^{-x}(x + \cos x)$.
32. $y = x^2 \sin x$.
33. $y = x^2 \sin x + x \cos x$.
34. $y = 8 \cos 4x + \sin 3x$.

40. Solution of a nonhomogeneous equation

Before proceeding to the theoretical basis and the actual working technique of the useful method of undetermined coefficients, let us examine the underlying ideas as applied to a simple numerical example.

Consider the equation

$$D^2(D - 1)y = 3 e^x + \sin x. \tag{1}$$

The complementary function may be determined at once from the roots

$$m = 0, 0, 1 \tag{2}$$

of the auxiliary equation. The complementary function is

$$y_c = c_1 + c_2 x + c_3 e^x. \tag{3}$$

Since the general solution of (1) is

$$y = y_c + y_p,$$

where y_c is as given in (3) and y_p is any particular solution of (1), all that remains for us to do is to find a particular solution of (1).

The right-hand member of (1),

$$R(x) = 3 e^x + \sin x, \tag{4}$$

is a particular solution of a homogeneous linear differential equation whose auxiliary equation has the roots

$$m' = 1, \pm i. \tag{5}$$

Therefore the function R is a particular solution of the equation

$$(D - 1)(D^2 + 1)R = 0. \tag{6}$$

We wish to convert (1) into a homogeneous linear differential equation with constant coefficients, because we know how to solve any such equation. But, by (6), the operator $(D - 1)(D^2 + 1)$ will annihilate the right member of (1). Therefore, we apply that operator to both sides of equation (1) and get

$$(D - 1)(D^2 + 1)D^2(D - 1)y = 0. \tag{7}$$

Any solution of (1) must be a particular solution of (7). The general solution of (7) can be written at once from the roots of its auxiliary equation, those roots being the values $m = 0, 0, 1$ from (2) and the values $m' = 1, \pm i$ from (5). Thus the general solution of (7) is

$$y = c_1 + c_2 x + c_3 e^x + c_4 x e^x + c_5 \cos x + c_6 \sin x. \tag{8}$$

But the desired general solution of (1) is

$$y = y_c + y_p, \tag{9}$$

where

$$y_c = c_1 + c_2 x + c_3 e^x,$$

the c_1, c_2, c_3 being arbitrary constants as in (8). Thus there must exist a particular solution of (1) containing at most the remaining terms in (8). Using different letters as coefficients to emphasize that they are not arbitrary, we conclude that (1) has a particular solution

$$y_p = Ax e^x + B \cos x + C \sin x. \tag{10}$$

We now have only to determine the numerical coefficients A, B, C by direct use of the original equation

$$D^2(D - 1)y = 3 e^x + \sin x. \tag{1}$$

From (10) it follows that

$$Dy_p = A(x e^x + e^x) - B \sin x + C \cos x,$$

$$D^2 y_p = A(x e^x + 2 e^x) - B \cos x - C \sin x,$$

$$D^3 y_p = A(x e^x + 3 e^x) + B \sin x - C \cos x.$$

Substitution of y_p into (1) then yields

$$A e^x + (B + C) \sin x + (B - C) \cos x = 3 e^x + \sin x. \tag{11}$$

Because (11) is to be an identity and because e^x, $\sin x$, and $\cos x$ are linearly independent, the corresponding coefficients in the two members of (11) must be equal; that is,

$$A = 3$$

$$B + C = 1$$

$$B - C = 0.$$

Therefore $A = 3$, $B = \frac{1}{2}$, $C = \frac{1}{2}$. Returning to (10), we find that a particular solution of equation (1) is

$$y_p = 3x\, e^x + \tfrac{1}{2}\cos x + \tfrac{1}{2}\sin x.$$

The general solution of the original equation

$$D^2(D - 1)y = 3\, e^x + \sin x \tag{1}$$

is therefore obtained by adding to the complementary function the y_p found above:

$$y = c_1 + c_2 x + c_3\, e^x + 3x\, e^x + \tfrac{1}{2}\cos x + \tfrac{1}{2}\sin x. \tag{12}$$

A careful analysis of the ideas behind the process used shows that to arrive at the solution (12), we need perform only the following steps:

(a) From (1) find the values of m and m' as exhibited in (2) and (5).
(b) From the values of m and m' write y_c and y_p as in (3) and (10).
(c) Substitute y_p into (1), equate corresponding coefficients, and obtain the numerical values of the coefficients in y_p.
(d) Write the general solution of (1).

41. The method of undetermined coefficients

Let us examine the general problem of the type treated in the preceding section. Let $f(D)$ be a polynomial in the operator D. Consider the equation

$$f(D)y = R(x). \tag{1}$$

Let the roots of the auxiliary equation $f(m) = 0$ be

$$m = m_1, m_2, \ldots, m_n. \tag{2}$$

The general solution of (1) is

$$y = y_c + y_p \tag{3}$$

where y_c can be obtained at once from the values of m in (2) and where $y = y_p$ is any particular solution (yet to be obtained) of (1).

Now suppose that the right member $R(x)$ of (1) is *itself a particular solution of some homogeneous linear differential equation with constant coefficients,*

$$g(D)R = 0, \tag{4}$$

whose auxiliary equation has the roots

$$m' = m'_1, m'_2, \ldots, m'_k. \tag{5}$$

Recall that the values of m' in (5) can be obtained by inspection from $R(x)$.

The differential equation

$$g(D)f(D)y = 0 \tag{6}$$

has as the roots of its auxiliary equation the values of m from (2) and m' from (5). Hence the general solution of (6) contains the y_c of (3) and so is of the form

$$y = y_c + y_q.$$

But also any particular solution of (1) must satisfy (6). Now, if

$$f(D)(y_c + y_q) = R(x),$$

then $f(D)y_q = R(x)$ because $f(D)y_c = 0$. Then deleting the y_c from the general solution of (6) leaves a function y_q that for some numerical values of its coefficients must satisfy (1); that is, the coefficients in y_q can be determined so that $y_q = y_p$. The determination of those numerical coefficients may be accomplished as in the examples below.

It must be kept in mind that the method of this section is applicable when, and only when, the right member of the equation is itself a particular solution of some homogeneous linear differential equation with constant coefficients.

EXAMPLE (a):   Solve the equation

$$(D^2 + D - 2)y = 2x - 40\cos 2x. \tag{7}$$

Here we have

$$m = 1, -2$$

and

$$m' = 0, 0, \pm 2i.$$

Therefore we may write

$$y_c = c_1 e^x + c_2 e^{-2x},$$

$$y_p = A + Bx + C\cos 2x + E\sin 2x,$$

in which c_1 and c_2 are arbitrary constants, whereas A, B, C, and E are to be determined numerically so that y_p will satisfy the equation (7).

Since
$$Dy_p = B - 2C \sin 2x + 2E \cos 2x$$
and
$$D^2 y_p = -4C \cos 2x - 4E \sin 2x,$$

direct substitution of y_p into (7) yields

$$-4C \cos 2x - 4E \sin 2x + B - 2C \sin 2x + 2E \cos 2x - 2A$$
$$-2Bx - 2C \cos 2x - 2E \sin 2x = 2x - 40 \cos 2x. \qquad (8)$$

But (8) is to be an identity in x, so we must equate coefficients of each of the set of linearly independent functions $\cos 2x$, $\sin 2x$, x, and 1 appearing in the identity. Thus it follows that

$$-6C + 2E = -40,$$
$$-6E - 2C = 0,$$
$$-2B = 2,$$
$$B - 2A = 0.$$

The above equations determine A, B, C, and E. Indeed, they lead to

$$A = -\tfrac{1}{2}, \qquad C = 6,$$
$$B = -1, \qquad E = -2.$$

Since the general solution of (7) is $y = y_c + y_p$, we can now write the desired result,

$$y = c_1 e^x + c_2 e^{-2x} - \tfrac{1}{2} - x + 6 \cos 2x - 2 \sin 2x.$$

EXAMPLE (b):   Solve the equation

$$(D^2 + 1)y = \sin x. \qquad (9)$$

At once $m = \pm i$ and $m' = \pm i$. Therefore

$$y_c = c_1 \cos x + c_2 \sin x,$$
$$y_p = Ax \cos x + Bx \sin x.$$

Now

$$y_p'' = A(-x \cos x - 2 \sin x) + B(-x \sin x + 2 \cos x),$$

so the requirement that y_p satisfy equation (9) yields

$$-2A \sin x + 2B \cos x = \sin x,$$

from which $A = -\tfrac{1}{2}$ and $B = 0$.

The general solution of (9) is

$$y = c_1 \cos x + c_2 \sin x - \tfrac{1}{2}x \cos x.$$

EXAMPLE (c): Determine y so that it will satisfy the equation

$$y''' - y' = 4\,e^{-x} + 3\,e^{2x} \tag{10}$$

with the conditions that when $x = 0$, $y = 0$, $y' = -1$, and $y'' = 2$.
First we note that $m = 0, 1, -1$, and $m' = -1, 2$. Thus

$$y_c = c_1 + c_2\,e^x + c_3\,e^{-x},$$

$$y_p = Ax\,e^{-x} + B\,e^{2x}.$$

Now

$$y'_p = A(-x\,e^{-x} + e^{-x}) + 2B\,e^{2x},$$

$$y''_p = A(x\,e^{-x} - 2\,e^{-x}) + 4B\,e^{2x},$$

$$y'''_p = A(-x\,e^{-x} + 3\,e^{-x}) + 8B\,e^{2x}.$$

Then

$$y'''_p - y'_p = 2A\,e^{-x} + 6B\,e^{2x},$$

so that from (10) we may conclude that $A = 2$ and $B = \tfrac{1}{2}$.
The general solution of (10) is therefore

$$y = c_1 + c_2\,e^x + c_3\,e^{-x} + 2x\,e^{-x} + \tfrac{1}{2}e^{2x}. \tag{11}$$

We must determine c_1, c_2, c_3 so (11) will satisfy the conditions that when $x = 0$, $y = 0$, $y' = -1$, and $y'' = 2$.
From (11) it follows that

$$y' = c_2\,e^x - c_3\,e^{-x} - 2x\,e^{-x} + 2\,e^{-x} + e^{2x} \tag{12}$$

and

$$y'' = c_2\,e^x + c_3\,e^{-x} + 2x\,e^{-x} - 4\,e^{-x} + 2\,e^{2x}. \tag{13}$$

We put $x = 0$ in each of (11), (12), and (13) to get the equations for the determination of c_1, c_2, and c_3. These are

$$0 = c_1 + c_2 + c_3 + \tfrac{1}{2},$$

$$-1 = c_2 - c_3 + 3,$$

$$2 = c_2 + c_3 - 2,$$

from which $c_1 = -\tfrac{9}{2}$, $c_2 = 0$, $c_3 = 4$. Therefore, the final result is

$$y = -\tfrac{9}{2} + 4\,e^{-x} + 2x\,e^{-x} + \tfrac{1}{2}e^{2x}.$$

An important point, sometimes overlooked by students, is that it is the general solution, the y of (11), that must be made to satisfy the initial conditions.

Exercises

In exercises 1 through 35, obtain the general solution.

1. $(D^2 - 3D + 2)y = 2x^3 - 9x^2 + 6x.$ ANS. $y = c_1 e^x + c_2 e^{2x} + x^3.$
2. $(D^2 + 4)y = 5e^x - 4x.$ ANS. $y = c_1 \sin 2x + c_2 \cos 2x + e^x - x.$
3. $(D^2 + 4)y = 5e^x - 4x^2.$ ANS. $y = c_1 \sin 2x + c_2 \cos 2x + e^x - x^2 + \frac{1}{2}.$
4. $(D^2 + D)y = \sin x.$ ANS. $y = c_1 + c_2 e^{-x} - \frac{1}{2}\sin x - \frac{1}{2}\cos x.$
5. $(D^2 - 4D + 4)y = e^x.$ ANS. $y = (c_1 + c_2 x)e^{2x} + e^x.$
6. $(D^2 - 3D + 2)y = 2x^2 + 1.$ ANS. $y = c_1 e^x + c_2 e^{2x} + x^2 + 3x + 4.$
7. $y'' - 3y' - 4y = 6e^x.$ ANS. $y = c_1 e^{4x} + c_2 e^{-x} - e^x.$
8. $y'' - 3y' - 4y = 5e^{4x}.$ ANS. $y = (c_1 + x)e^{4x} + c_2 e^{-x}.$
9. $(D^2 - 4)y = 8e^{2x} - 12.$ ANS. $y = (c_1 + 2x)e^{2x} + c_2 e^{-2x} + 3.$
10. $(D^2 - D - 2)y = 1 - 2x - 9e^{-x}.$ ANS. $y = (c_1 + 3x)e^{-x} + c_2 e^{2x} + x - 1.$
11. $y'' - 4y' + 3y = 20\cos x.$ ANS. $y = c_1 e^x + c_2 e^{3x} + 2\cos x - 4\sin x.$
12. $y'' - 4y' + 3y = 2\cos x + 4\sin x.$ ANS. $y = c_1 e^x + c_2 e^{3x} + \cos x.$
13. $y'' + 2y' + y = 7 + 75\sin 2x.$

     ANS. $y = e^{-x}(c_1 + c_2 x) + 7 - 12\cos 2x - 9\sin 2x.$
14. $(D^2 + 4D + 5)y = 50x + 13e^{3x}.$

     ANS. $y = e^{-2x}(c_1 \cos x + c_2 \sin x) + 10x - 8 + \frac{1}{2}e^{3x}.$
15. $(D^2 + 1)y = \cos x.$ ANS. $y = c_1 \cos x + c_2 \sin x + \frac{1}{2}x \sin x.$
16. $(D^2 - 4D + 4)y = e^{2x}.$ ANS. $y = e^{2x}(c_1 + c_2 x + \frac{1}{2}x^2).$
17. $(D^2 - 1)y = e^{-x}(2\sin x + 4\cos x).$
18. $(D^2 - 1)y = 8xe^x.$ ANS. $y = c_1 e^{-x} + e^x(c_2 - 2x + 2x^2).$
19. $(D^3 - D)y = x.$ ANS. $y = c_1 + c_2 e^x + c_3 e^{-x} - \frac{1}{2}x^2.$
20. $(D^3 - D^2 + D - 1)y = 4\sin x.$

     ANS. $y = c_1 e^x + (c_2 + x)\cos x + (c_3 - x)\sin x.$
21. $(D^3 + D^2 - 4D - 4)y = 3e^{-x} - 4x - 6.$

     ANS. $y = c_1 e^{2x} + c_2 e^{-2x} + (c_3 - x)e^{-x} + x + \frac{1}{2}.$
22. $(D^4 - 1)y = 7x^2.$
23. $(D^4 - 1)y = e^{-x}.$ ANS. $y = c_1 e^x + (c_2 - \frac{1}{4}x)e^{-x} + c_3 \cos x + c_4 \sin x.$
24. $(D^2 - 1)y = 10\sin^2 x.$ Use the identity $\sin^2 x = \frac{1}{2}(1 - \cos 2x).$

     ANS. $y = c_1 e^x + c_2 e^{-x} - 5 + \cos 2x.$
25. $(D^2 + 1)y = 12\cos^2 x.$ ANS. $y = c_1 \cos x + c_2 \sin x + 6 - 2\cos 2x.$
26. $(D^2 + 4)y = 4\sin^2 x.$ ANS. $y = c_1 \cos 2x + c_2 \sin 2x + \frac{1}{2}(1 - x\sin 2x).$
27. $y'' - 3y' - 4y = 16x - 50\cos 2x.$

     ANS. $y = c_1 e^{4x} + c_2 e^{-x} + 3 - 4x + 4\cos 2x + 3\sin 2x.$
28. $(D^3 - 3D - 2)y = 100\sin 2x.$
29. $y'' + 4y' + 3y = 15e^{2x} + e^{-x}.$ ANS. $y = c_1 e^{-x} + c_2 e^{-3x} + e^{2x} + \frac{1}{2}xe^{-x}.$
30. $y'' - y = e^x - 4.$
31. $y'' - y' - 2y = 6x + 6e^{-x}.$ ANS. $y = c_1 e^{2x} + c_2 e^{-x} + \frac{3}{2} - 3x - 2xe^{-x}.$
32. $y'' + 6y' + 13y = 60\cos x + 26.$

33. $(D^3 - 3D^2 + 4)y = 6 + 80 \cos 2x$.

ANS. $y = c_1 e^{-x} + e^{2x}(c_2 + c_3 x) + \frac{3}{2} + 4 \cos 2x - 2 \sin 2x$.

34. $(D^3 + D - 10)y = 29 e^{4x}$.

35. $(D^3 + D^2 - 4D - 4)y = 8x + 8 + 6e^{-x}$.

ANS. $y = c_1 \cosh 2x + c_2 \sinh 2x + c_3 e^{-x} - 2x - 2xe^{-x}$.

In exercises 36 through 44, find the particular solution indicated.

36. $(D^2 + 1)y = 10 e^{2x}$; when $x = 0, y = 0, y' = 0$.

ANS. $y = 2(e^{2x} - \cos x - 2 \sin x)$.

37. $(D^2 - 4)y = 2 - 8x$; when $x = 0, y = 0, y' = 5$.

ANS. $y = e^{2x} - \frac{1}{2}e^{-2x} + 2x - \frac{1}{2}$.

38. $(D^2 + 3D)y = -18x$; when $x = 0, y = 0, y' = 5$.

ANS. $y = 1 + 2x - 3x^2 - e^{-3x}$.

39. $(D^2 + 4D + 5)y = 10 e^{-3x}$; when $x = 0, y = 4, y' = 0$.

40. $\dfrac{d^2x}{dt^2} + 4\dfrac{dx}{dt} + 5x = 10$; when $t = 0, x = 0, \dfrac{dx}{dt} = 0$.

ANS. $x = 2(1 - e^{-2t} \cos t - 2 e^{-2t} \sin t)$.

41. $\ddot{x} + 4\dot{x} + 5x = 8 \sin t$; when $t = 0, x = 0, \dot{x} = 0$. Note that $\dot{x} = dx/dt, \ddot{x} = d^2x/dt^2$ is a common notation when the independent variable is time.

ANS. $x = (1 + e^{-2t}) \sin t - (1 - e^{-2t}) \cos t$.

42. $y'' + 9y = 81x^2 + 14 \cos 4x$; when $x = 0, y = 0, y' = 3$.

43. $(D^3 + 4D^2 + 9D + 10)y = -24e^x$; when $x = 0, y = 0, y' = -4, y'' = 10$.

44. $y'' + 2y' + 5y = 8e^{-x}$; when $x = 0, y = 0, y' = 8$.

ANS. $y = e^{-x}(2 + 4 \sin 2x - 2 \cos 2x)$.

In exercises 45 through 48 obtain, from the particular solution indicated, the value of y and the value of y' at $x = 2$.

45. $y'' + 2y' + y = x$; at $x = 0, y = -3$, and at $x = 1, y = -1$.

ANS.   At $x = 2, y = e^{-2}, y' = 1$.

46. $y'' + 2y' + y = x$; at $x = 0, y = -2, y' = 2$.

ANS.   At $x = 2, y = 2 e^{-2}, y' = 1 - e^{-2}$.

47. $4y'' + y = 2$; at $x = \pi, y = 0, y' = 1$.

ANS.   At $x = 2, y = -0.7635, y' = +0.3012$.

48. $2y'' - 5y' - 3y = -9x^2 - 1$; at $x = 0, y = 1, y' = 0$.

ANS.   At $x = 2, y = 5.64, y' = 5.68$.

49. $(D^2 + D)y = x + 1$; when $x = 0, y = 1$, and when $x = 1, y = \frac{1}{2}$. Compute the value of y at $x = 4$.   ANS.   At $x = 4, y = 8 - e^{-1} - e^{-2} - e^{-3}$.

50. $(D^2 + 1)y = x^3$; when $x = 0, y = 0$, and when $x = \pi, y = 0$. Show that this boundary value problem has no solution.

51. $(D^2 + 1)y = 2 \cos x$; when $x = 0, y = 0$, and when $x = \pi, y = 0$. Show that this boundary value problem has an unlimited number of solutions and obtain them.

ANS. $y = (c + x) \sin x$.

52. For the equation $(D^3 + D^2)y = 4$, find the solution whose graph has at the origin a point of inflection with a horizontal tangent line.

ANS. $y = 4 - 4x + 2x^2 - 4e^{-x}$.

53. For the equation $(D^2 - D)y = 2 - 2x$, find a particular solution that has at some point (to be determined) on the x-axis an inflection point with a horizontal tangent line.   ANS.   The point is $(1, 0)$; the solution is $y = x^2 + 1 - 2\exp(x - 1)$.

42. Solution by inspection

It is frequently easy to obtain a particular solution of a nonhomogeneous equation

$$(b_0 D^n + b_1 D^{n-1} + \cdots + b_{n-1}D + b_n)y = R(x) \tag{1}$$

by inspection.

For example, if $R(x)$ is a constant R_0 and if $b_n \neq 0$,

$$y_p = \frac{R_0}{b_n} \tag{2}$$

is a solution of

$$(b_0 D^n + b_1 D^{n-1} + \cdots + b_n)y = R_0, \qquad b_n \neq 0, R_0 \text{ constant}, \tag{3}$$

because all derivatives of y_p are zero, so

$$(b_0 D^n + b_1 D^{n-1} + \cdots + b_n)y_p = b_n R_0/b_n = R_0.$$

Suppose that $b_n = 0$ in equation (3). Let $D^k y$ be the lowest-ordered derivative that actually appears in the differential equation. Then the equation may be written

$$(b_0 D^n + \cdots + b_{n-k}D^k)y = R_0, \qquad b_{n-k} \neq 0, R_0 \text{ constant}. \tag{4}$$

Now $D^k x^k = k!$, a constant, so that all higher derivatives of x^k are zero. Thus it becomes evident that (4) has a solution

$$y_p = \frac{R_0 x^k}{k! b_{n-k}}, \tag{5}$$

for then $(b_0 D^n + \cdots + b_{n-k}D^k)y_p = b_{n-k}R_0 k!/k! b_{n-k} = R_0$.

EXAMPLE (a):   Solve the equation

$$(D^2 - 3D + 2)y = 16. \tag{6}$$

By the methods of Chapter 6 we obtain the complementary function,

$$y_c = c_1 e^x + c_2 e^{2x}.$$

By inspection a particular solution of the original equation is

$$y_p = \tfrac{16}{2} = 8.$$

Hence the general solution of (6) is

$$y = c_1 e^x + c_2 e^{2x} + 8.$$

EXAMPLE (b):   Solve the equation

$$\frac{d^5 y}{dx^5} + 4\frac{d^3 y}{dx^3} = 7. \tag{7}$$

From the auxiliary equation $m^5 + 4m^3 = 0$ we get $m = 0, 0, 0, \pm 2i$. Hence

$$y_c = c_1 + c_2 x + c_3 x^2 + c_4 \cos 2x + c_5 \sin 2x.$$

A particular solution of (7) is

$$y_p = \frac{7x^3}{3! \cdot 4} = \frac{7x^3}{24}.$$

As a check, note that

$$(D^5 + 4D^3)\frac{7x^3}{24} = 0 + 4 \cdot \frac{7 \cdot 6}{24} = 7.$$

The general solution of equation (7) is

$$y = c_1 + c_2 x + c_3 x^2 + \tfrac{7}{24}x^3 + c_4 \cos 2x + c_5 \sin 2x,$$

in which the $c_1, \ldots, c_5$ are arbitrary constants.

Examination of

$$(D^2 + 4)y = \sin 3x \tag{8}$$

leads us to search for a solution proportional to $\sin 3x$ because, if y is proportional to $\sin 3x$, so is $D^2 y$. Indeed, from

$$y = A \sin 3x \tag{9}$$

we get

$$D^2 y = -9A \sin 3x,$$

so (9) is a solution of (8) if

$$(-9 + 4)A = 1$$

$$A = -\tfrac{1}{5}.$$

Thus (8) has the general solution

$$y = c_1 \cos 2x + c_2 \sin 2x - \tfrac{1}{5}\sin 3x,$$

a result easily obtained mentally.

For equation (8), the general method of undetermined coefficients leads us to write

$$m = \pm 2i, \qquad m' = \pm 3i,$$

and so to write

$$y_p = A \sin 3x + B \cos 3x. \tag{10}$$

When the y_p of (10) is substituted into (8), it is found, of course, that

$$A = -\tfrac{1}{5} \qquad B = 0.$$

In contrast, consider the equation

$$(D^2 + 4D + 4)y = \sin 3x. \tag{11}$$

Here any attempt to find a solution proportional to $\sin 3x$ is doomed to failure because, although $D^2 y$ will also be proportional to $\sin 3x$, the term Dy will involve $\cos 3x$. There is no other term on either side of (11) to compensate for this cosine term, so no solution of the form $y = A \sin 3x$ is possible. For this equation, $m = -2, -2, m' = \pm 3i$, and, in the particular solution

$$y_p = A \sin 3x + B \cos 3x,$$

it must turn out that $B \neq 0$. No labor has been saved by the inspection.
   In more complicated situations such as

$$(D^2 + 4)y = x \sin 3x - 2 \cos 3x,$$

the method of inspection will save no work.
   For the equation

$$(D^2 + 4)y = e^{5x}, \tag{12}$$

we see, since $(D^2 + 4) e^{5x} = 29 e^{5x}$, that

$$y_p = \tfrac{1}{29} e^{5x}$$

is a solution.
   Finally, note that if y_1 is a solution of

$$f(D)y = R_1(x)$$

and y_2 is a solution of

$$f(D)y = R_2(x),$$

then

$$y_p = y_1 + y_2$$

is a solution of

$$f(D)y = R_1(x) + R_2(x).$$

It follows readily that the task of obtaining a particular solution of

$$f(D)y = R(x)$$

may be split into parts by treating separate terms of $R(x)$ independently, if convenient. See the examples below. This is the basis of the "method of superposition," which plays a useful role in applied mathematics.

EXAMPLE (c): Find a particular solution of

$$(D^2 - 9)y = 3\,e^x + x - \sin 4x. \tag{13}$$

Since $(D^2 - 9)\,e^x = -8\,e^x$, we see by inspection that

$$y_1 = -\tfrac{3}{8}e^x$$

is a particular solution of

$$(D^2 - 9)y_1 = 3\,e^x.$$

In a similar manner, we see that $y_2 = -\tfrac{1}{9}x$ satisfies

$$(D^2 - 9)y_2 = x$$

and that

$$y_3 = \tfrac{1}{25}\sin 4x$$

satisfies

$$(D^2 - 9)y_3 = -\sin 4x.$$

Hence

$$y_p = -\tfrac{3}{8}e^x - \tfrac{1}{9}x + \tfrac{1}{25}\sin 4x$$

is a solution of equation (13).

EXAMPLE (d): Find a particular solution of

$$(D^2 + 4)y = \sin x + \sin 2x. \tag{14}$$

At once we see that $y_1 = \tfrac{1}{3}\sin x$ is a solution of

$$(D^2 + 4)y_1 = \sin x.$$

Then we seek a solution of

$$(D^2 + 4)y_2 = \sin 2x \tag{15}$$

by the method of undetermined coefficients. Because $m = \pm 2i$ and $m' = \pm 2i$, we put

$$y_2 = Ax \sin 2x + Bx \cos 2x$$

into (15) and easily determine that

$$4A \cos 2x - 4B \sin 2x = \sin 2x,$$

from which $A = 0, B = -\frac{1}{4}$.

Thus a particular solution of (14) is

$$y_p = \tfrac{1}{3} \sin x - \tfrac{1}{4}x \cos 2x.$$

EXAMPLE (e): Find a particular solution of

$$(D^2 + a^2)y = \cos bx. \tag{16}$$

If $b \neq a$, then a particular solution of the form $y = A \cos bx$ will exist. It follows from (16) that

$$(-b^2A + a^2A) \cos bx = \cos bx$$

and $A = (a^2 - b^2)^{-1}$. A particular solution of (16) is

$$y = (a^2 - b^2)^{-1} \cos bx.$$

If $b = a$, then equation (16) becomes

$$(D^2 + a^2)y = \cos ax, \tag{17}$$

and no function of the form $A \cos ax$ is a particular solution since the operator $D^2 + a^2$ will annihilate $A \cos ax$. However, a solution of the form $Ax \cos ax + Bx \sin ax$ exists. Upon substitution into (17) we require that

$$-2aA \sin ax + 2aB \cos ax = \cos ax,$$

an equation that is satisfied only if $A = 0$ and $B = 1/2a$. Therefore

$$y = \frac{x}{2a} \sin ax \tag{18}$$

is a particular solution of (17).

We have seen in this example an important distinction between the cases $b \neq a$ and $b = a$. In a physical application considered in Chapter 10, the presence of a solution of the form given in (18) results in a phenomenon called resonance. At this point we need only notice that the solution in (18) will be oscillatory in character, but the amplitudes of the oscillation will become increasingly large as x increases.

Exercises

1. Show that if $b \neq a$, then

$$(D^2 + a^2)y = \sin bx$$

has the particular solution $y = (a^2 - b^2)^{-1} \sin bx$.

2. Show that the equation

$$(D^2 + a^2)y = \sin ax$$

has no solution of the form $y = A \sin ax$, with A constant. Find a particular solution of the equation.

    ANS. $y = -\dfrac{x}{2a} \cos ax.$

In exercises 3 through 50 find a particular solution by inspection. Verify your solution.

3. $(D^2 + 4)y = 12.$ ANS. $y = 3.$

4. $(D^2 + 9)y = 18.$ ANS. $y = 2.$

5. $(D^2 + 4D + 4)y = 8.$ ANS. $y = 2.$

6. $(D^2 + 2D - 3)y = 6.$ ANS. $y = -2.$

7. $(D^3 - 3D + 2)y = -7.$ 8. $(D^4 + 4D^2 + 4)y = -20.$

9. $(D^2 + 4D)y = 12.$ 10. $(D^3 - 9D)y = 27.$

11. $(D^3 + 5D)y = 15.$ ANS. $y = 3x.$

12. $(D^3 + D)y = -8.$ ANS. $y = -8x.$

13. $(D^4 - 4D^2)y = 24.$ ANS. $y = -3x^2.$

14. $(D^4 + D^2)y = -12.$ ANS. $y = -6x^2.$

15. $(D^5 - D^3)y = 24.$ ANS. $y = -4x^3.$

16. $(D^5 - 9D^3)y = 27.$ ANS. $y = \frac{1}{2}x^3.$

17. $(D^2 + 4)y = 6 \sin x.$ ANS. $y = 2 \sin x.$

18. $(D^2 + 4)y = 10 \cos 3x.$ ANS. $y = -2 \cos 3x.$

19. $(D^2 + 4)y = 8x + 1 - 15 e^x.$ ANS. $y = 2x + \frac{1}{4} - 3 e^x.$

20. $(D^2 + D)y = 6 + 3 e^{2x}.$ ANS. $y = 6x + \frac{1}{2} e^{2x}.$

21. $(D^2 + 3D - 4)y = 18 e^{2x}.$ ANS. $y = 3 e^{2x}.$

22. $(D^2 + 2D + 5)y = 4 e^x - 10.$ ANS. $y = \frac{1}{2} e^x - 2.$

23. $(D^2 - 1)y = 2 e^{3x}.$ 24. $(D^2 - 1)y = 2x + 3.$

25. $(D^2 - 1)y = \cos 2x.$ 26. $(D^2 - 1)y = \sin 2x.$

27. $(D^2 + 1)y = e^x + 3x.$ 28. $(D^2 + 1)y = 5 e^{-3x}.$

29. $(D^2 + 1)y = -2x + \cos 2x.$ 30. $(D^2 + 1)y = 4 e^{-2x}.$

31. $(D^2 + 1)y = 10 \sin 4x.$ 32. $(D^2 + 1)y = -6 e^{-3x}.$

33. $(D^2 + 2D + 1)y = 12 e^x.$ ANS. $y = 3 e^x.$

34. $(D^2 + 2D + 1)y = 7 e^{-2x}.$ ANS. $y = 7 e^{-2x}.$

35. $(D^2 - 2D + 1)y = 12 e^{-x}.$ ANS. $y = 3 e^{-x}.$

36. $(D^2 - 2D + 1)y = 6 e^{-2x}.$ ANS. $y = \frac{2}{3} e^{-2x}.$

37. $(D^2 - 2D - 3)y = e^x.$ ANS. $y = -\frac{1}{4} e^x.$

38. $(D^2 - 2D - 3)y = e^{2x}.$ ANS. $y = -\frac{1}{3} e^{2x}.$

39. $(4D^2 + 1)y = 12 \sin x.$ ANS. $y = -4 \sin x.$

40. $(4D^2 + 1)y = -12 \cos x.$ ANS. $y = 4 \cos x.$

41. $(4D^2 + 4D + 1)y = 18 e^x - 5.$ 42. $(4D^2 + 4D + 1)y = 7 e^{-x} + 2.$

43. $(D^3 - 1)y = e^{-x}.$ 44. $(D^3 - 1)y = 4 - 3x^2.$

45. $(D^3 - D)y = e^{2x}.$ 46. $(D^4 + 4)y = 5 e^{2x}.$

47. $(D^4 + 4)y = 6 \sin 2x.$ 48. $(D^4 + 4)y = \cos 2x.$

49. $(D^3 - D)y = 5 \sin 2x.$ 50. $(D^3 - D)y = 5 \cos 2x.$

8

Variation of Parameters

43. Introduction

In Chapter 7 we solved the nonhomogeneous linear equation with constant coefficients

$$(b_0 D^n + b_1 D^{n-1} + \cdots + b_{n-1} D + b_n)y = R(x) \tag{1}$$

by the method of undetermined coefficients. We saw that this method would be applicable only for a certain class of differential equations: those for which $R(x)$ itself was a solution of a homogeneous linear equation with constant coefficients.

In this chapter we shall study two methods that carry no such restrictions. In fact, much of what we do will be applicable to linear equations with variable coefficients.

We begin with a procedure by D'Alembert that is often called the method of reduction of order.

133

44. Reduction of order

Consider the general second-order linear equation

$$y'' + py' + qy = R. \tag{1}$$

Suppose that we know a solution $y = y_1$ of the corresponding homogeneous equation

$$y'' + py' + qy = 0. \tag{2}$$

Then the introduction of a new dependent variable v by the substitution

$$y = y_1 v \tag{3}$$

will lead to a solution of equation (1) in the following way.

From (3) it follows that

$$y' = y_1 v' + y_1' v,$$
$$y'' = y_1 v'' + 2y_1' v' + y_1'' v,$$

so substitution of (3) into (1) yields

$$y_1 v'' + 2y_1' v' + y_1'' v + py_1 v' + py_1' v + qy_1 v = R,$$

or

$$y_1 v'' + (2y_1' + py_1)v' + (y_1'' + py_1' + qy_1)v = R. \tag{4}$$

But $y = y_1$ is a solution of (2). That is,

$$y_1'' + py_1' + qy_1 = 0$$

and equation (4) reduces to

$$y_1 v'' + (2y_1' + py_1)v' = R. \tag{5}$$

Now let $v' = w$ so equation (5) becomes

$$y_1 w' + (2y_1' + py_1)w = R, \tag{6}$$

a linear equation of first order in w.

By the usual method (integrating factor) we can find w from (6). Then we can get v from $v' = w$ by an integration. Finally $y = y_1 v$.

Note that the method is not restricted to equations with constant coefficients. It depends only upon our knowing a particular solution of equation (2); that is, upon our knowledge of the complementary function. For practical purposes, the method depends also upon our being able to effect the integrations.

EXAMPLE (a): Solve the equation

$$y'' - y = e^x. \tag{7}$$

The complementary function of (7) is

$$y_c = c_1 e^x + c_2 e^{-x}.$$

We shall take the particular solution e^x and use the method of reduction of order by setting

$$y = v e^x.$$

Then

$$y' = v e^x + v' e^x,$$

and

$$y'' = v e^x + 2v' e^x + v'' e^x.$$

Substituting into equation (7) gives

$$v'' + 2v' = 1. \tag{8}$$

Equation (8) is a first-order linear equation in the variable v'. Applying the integrating factor e^{2x} yields

$$e^{2x}(v'' + 2v') = e^{2x}.$$

Thus

$$e^{2x}v' = \tfrac{1}{2} e^{2x} + c, \tag{9}$$

where c is an arbitrary constant. Equation (9) readily gives

$$v' = \tfrac{1}{2} + c e^{-2x},$$

and hence

$$v = c_1 e^{-2x} + c_2 + \tfrac{1}{2}x,$$

where c_1 and c_2 are arbitrary constants.

Remembering that $y = v e^x$, we finally have

$$y = c_1 e^{-x} + c_2 e^x + \tfrac{1}{2}x e^x.$$

Of course, the solution to equation (7) could have been obtained by the method of undetermined coefficients. Let us now solve a problem not solvable by that method.

EXAMPLE (b): Solve the equation

$$(D^2 + 1)y = \csc x. \tag{10}$$

The complementary function is

$$y_c = c_1 \cos x + c_2 \sin x. \tag{11}$$

We may use any special case of (11) as the y_1 in the theory above. Let us then put

$$y = v \sin x.$$

We find that

$$y' = v' \sin x + v \cos x$$

and

$$y'' = v'' \sin x + 2v' \cos x - v \sin x.$$

The equation for v is

$$v'' \sin x + 2v' \cos x = \csc x,$$

or

$$v'' + 2v' \cot x = \csc^2 x. \tag{12}$$

Put $v' = w$; then equation (12) becomes

$$w' + 2w \cot x = \csc^2 x,$$

for which an integrating factor is $\sin^2 x$. Thus

$$\sin^2 x \, dw + 2w \sin x \cos x \, dx = dx \tag{13}$$

is exact. From (13) we get

$$w \sin^2 x = x,$$

and if we seek only a particular solution, we have

$$w = x \csc^2 x$$

or

$$v' = x \csc^2 x.$$

Hence

$$v = \int x \csc^2 x \, dx,$$

or

$$v = -x \cot x + \ln |\sin x|,$$

a result easily obtained using integration by parts.
   Now

$$y = v \sin x,$$

so the particular solution which we sought is

$$y_p = -x \cos x + \sin x \ln |\sin x|.$$

Finally, the complete solution of (10) is seen to be

$$y = c_1 \cos x + c_2 \sin x - x \cos x + \sin x \ln |\sin x|.$$

Exercises

Use the method of reduction of order to solve the following equations.

1. $(D^2 - 1)y = x - 1$.   ANS. $y = c_1 e^x + c_2 e^{-x} - x + 1$.
2. $(D^2 - 5D + 6)y = 2 e^x$.   ANS. $y = c_1 e^{2x} + c_2 e^{3x} + e^x$.
3. $(D^2 - 4D + 4)y = e^x$.   ANS.   See exercise 5, Section 41.
4. $(D^2 + 4)y = \sin x$.   ANS. $y = c_1 \sin 2x + c_2 \cos 2x + \frac{1}{3} \sin x$.
5. Use the substitution $y = v \cos x$ to solve the equation of Example (b) above.
6. Use $y = v e^{-x}$ to solve the equation of Example (a) above.
7. $(D^2 + 1)y = \sec x$. $c_1 \cos + c \sin \theta$
8. $(D^2 + 1)y = \sec^3 x$. Use $y = v \sin x$.
9. $(D^2 + 1)y = \csc^3 x$. Take a hint from exercise 8.
10. $(D^2 + 2D + 1)y = (e^x - 1)^{-2}$.   ANS. $y = e^{-x}(c_1 + c_2 x - \ln |1 - e^{-x}|)$.
11. $(D^2 - 3D + 2)y = (1 + e^{2x})^{-1/2}$.
12. Verify that $y = e^x$ is a solution of the equation

$$(x - 1)y'' - xy' + y = 0.$$

Use this fact to find the general solution of

$$(x - 1)y'' - xy' + y = 1.$$

ANS. $y = c_1 x + c_2 e^x + 1$.

13. Observe that $y = x$ is a particular solution of the equation

$$2x^2 y'' + xy' - y = 0$$

and find the general solution. For what values of x is the solution valid?

ANS. $y = c_1 x + c_2 x^{-1/2}$.

14. In Chapter 19 we shall study Bessel's differential equation of index zero

$$xy'' + y' + xy = 0.$$

Suppose that one solution of this equation is given the name $J_0(x)$. Show that a second solution takes the form

$$J_0(x) \int \frac{dx}{x[J_0(x)]^2}.$$

15. One solution of the Legendre differential equation

$$(1 - x^2)y'' - 2xy' + 2y = 0$$

is $y = x$. Find a second solution.   ANS. $y = -2 + x \ln \left| \dfrac{1 + x}{1 - x} \right|$.

45. Variation of parameters

In the previous section we saw that if y_1 is a solution of the homogeneous equation

$$y'' + p(x)y' + q(x)y = 0 \tag{1}$$

then we can use it to determine the general solution of the nonhomogeneous equation

$$y'' + p(x)y' + q(x)y = R(x). \tag{2}$$

In using the method of reduction of order we proceeded as follows. Because y_1 is a solution of (1), the function c_1y_1 is also a solution for an arbitrary constant c_1. We replaced the constant c_1 by a function $v(x)$ and considered the possibility of the existence of a solution of equation (2) of the form $v \cdot y_1$. This led us to a first-order linear equation in the variable v' that we were able to solve.

Suppose now that we know the general solution of the homogeneous equation (1). That is, suppose

$$y_c = c_1y_1 + c_2y_2 \tag{3}$$

is a solution of (1), where y_1 and y_2 are linearly independent on an interval $a < x < b$. Let us see what happens if we replace both of the constants in (3) with functions of x. That is, we consider

$$y = Ay_1 + By_2 \tag{4}$$

and try to determine $A(x)$ and $B(x)$ so that $Ay_1 + By_2$ is a solution of equation (2).

Note that we are involved with two unknown functions $A(x)$ and $B(x)$ and that we have only insisted that these functions satisfy one condition: the function in (4) is to be a solution of equation (2). We may therefore expect to impose a second condition on $A(x)$ and $B(x)$ in some way which would be to our advantage. Indeed, if we simply impose the condition $B(x) \equiv 0$, then we will be dealing with the method of reduction of order. Actually we impose a somewhat different condition on A and B.

From (4) it follows that

$$y' = Ay_1' + By_2' + A'y_1 + B'y_2. \tag{5}$$

Rather than become involved with derivatives of A and B of higher order than the first, we now choose some particular function for the expression

$$A'y_1 + B'y_2.$$

Technically, we could let this function be $\sin x$, e^x, or any other suitable function. For simplicity we choose

$$A'y_1 + B'y_2 = 0. \tag{6}$$

It then follows from (5) that

$$y'' = Ay_1'' + By_2'' + A'y_1' + B'y_2'. \tag{7}$$

Because y was to be a solution of (2), we substitute from (4), (5), and (7) into equation (2) to obtain

$$A(y_1'' + py_1' + qy_1) + B(y_2'' + py_2' + qy_2) + A'y_1' + B'y_2' = R(x).$$

But y_1 and y_2 are solutions of the homogeneous equation (1), so that finally

$$A'y_1' + B'y_2' = R(x). \tag{8}$$

Equations (6) and (8) now give us two equations that we wish to solve for A' and B'. This solution exists providing the determinant

$$\begin{vmatrix} y_1 & y_2 \\ y_1' & y_2' \end{vmatrix}$$

does not vanish. But this determinant is precisely the Wronskian of the functions y_1 and y_2, which were presumed to be linearly independent on the interval $a < x < b$. Therefore, the Wronskian does not vanish on that interval and we can find A' and B'. By integration we can now find A and B. Once A and B are known, equation (4) gives us the desired y.

This argument can easily be extended to equations of order higher than two, but no essentially new ideas appear. Moreover, there is nothing in the method that prohibits the linear differential equation involved from having variable coefficients.

EXAMPLE (a):   Solve the equation

$$(D^2 + 1)y = \sec x \tan x. \tag{9}$$

Of course,

$$y_c = c_1 \cos x + c_2 \sin x.$$

Let us seek a particular solution by variation of parameters. Put

$$y = A \cos x + B \sin x, \tag{10}$$

from which

$$y' = -A \sin x + B \cos x + A' \cos x + B' \sin x.$$

Next set

$$A' \cos x + B' \sin x = 0, \tag{11}$$

so that

$$y' = -A \sin x + B \cos x.$$

Then

$$y'' = -A \cos x - B \sin x - A' \sin x + B' \cos x. \tag{12}$$

Next we eliminate y by combining equations (10) and (12) with the original equation (9). Thus we get the relation

$$-A' \sin x + B' \cos x = \sec x \tan x. \tag{13}$$

From (13) and (11), A' is easily eliminated. The result is

$$B' = \tan x,$$

so that

$$B = \ln |\sec x|, \tag{14}$$

in which the arbitrary constant has been disregarded because we are seeking only a particular solution to add to our previously determined complementary function y_c.

From equations (13) and (11) it also follows easily that

$$A' = -\sin x \sec x \tan x,$$

or

$$A' = -\tan^2 x.$$

Then

$$A = -\int \tan^2 x \, dx = \int (1 - \sec^2 x) \, dx,$$

so that

$$A = x - \tan x, \tag{15}$$

again disregarding the arbitrary constant.

Returning to equation (10) with the known A from (15) and the known B from (14), we write the particular solution

$$y_p = (x - \tan x) \cos x + \sin x \ln |\sec x|,$$

or

$$y_p = x \cos x - \sin x + \sin x \ln |\sec x|.$$

Then the general solution of (9) is

$$y = c_1 \cos x + c_3 \sin x + x \cos x + \sin x \ln |\sec x|, \tag{16}$$

where the term $(-\sin x)$ in y_p has been absorbed in the complementary function term $c_3 \sin x$, since c_3 is an arbitrary constant.

The solution (16) can, as usual, be verified by direct substitution into the original differential equation.

EXAMPLE (b):   Solve the equation

$$(D^2 - 3D + 2)y = \frac{1}{1 + e^{-x}}. \tag{17}$$

Here

$$y_c = c_1 e^x + c_2 e^{2x},$$

so we put

$$y = A e^x + B e^{2x}. \tag{18}$$

Because

$$y' = A e^x + 2B e^{2x} + A' e^x + B' e^{2x},$$

we impose the condition

$$A' e^x + B' e^{2x} = 0. \tag{19}$$

Then

$$y' = A e^x + 2B e^{2x}, \tag{20}$$

from which it follows that

$$y'' = A e^x + 4B e^{2x} + A' e^x + 2B' e^{2x}. \tag{21}$$

Combining (18), (20), (21), and the original equation (17), we find that

$$A' e^x + 2B' e^{2x} = \frac{1}{1 + e^{-x}}. \tag{22}$$

Elimination of B' from equations (19) and (22) yields

$$A' e^x = -\frac{1}{1 + e^{-x}},$$

$$A' = -\frac{e^{-x}}{1 + e^{-x}}.$$

Then

$$A = \ln(1 + e^{-x}).$$

Similarly,

$$B' e^{2x} = \frac{1}{1 + e^{-x}}$$

so that

$$B = \int \frac{e^{-2x}}{1 + e^{-x}} dx = \int \left(e^{-x} - \frac{e^{-x}}{1 + e^{-x}} \right) dx,$$

or

$$B = -e^{-x} + \ln(1 + e^{-x}).$$

Then, from (18),

$$y_p = e^x \ln(1 + e^{-x}) - e^x + e^{2x} \ln(1 + e^{-x}).$$

The term $(-e^x)$ in y_p can be absorbed into the complementary function. The general solution of equation (17) is

$$y = c_3 e^x + c_2 e^{2x} + (e^x + e^{2x}) \ln(1 + e^{-x}).$$

46. Solution of $y'' + y = f(x)$

Consider next the equation

$$(D^2 + 1)y = f(x), \tag{1}$$

in which all that we require of $f(x)$ is that it be integrable in the interval on which we seek a solution. For instance, $f(x)$ may be any continuous function or any function with only a finite number of finite discontinuities on the interval $a \leq x \leq b$.

The method of variation of parameters will now be applied to the solution of (1). Put

$$y = A \cos x + B \sin x. \tag{2}$$

Then

$$y' = -A \sin x + B \cos x + A' \cos x + B' \sin x,$$

and if we choose

$$A' \cos x + B' \sin x = 0, \tag{3}$$

we obtain

$$y'' = -A \cos x - B \sin x - A' \sin x + B' \cos x. \tag{4}$$

From (1), (2), and (4) it follows that

$$-A' \sin x + B' \cos x = f(x). \tag{5}$$

Equations (3) and (5) may be solved for A' and B', yielding

$$A' = -f(x) \sin x \qquad \text{and} \qquad B' = f(x) \cos x.$$

We may now write

$$A = -\int_a^x f(\beta) \sin \beta \, d\beta, \tag{6}$$

$$B = \int_a^x f(\beta) \cos \beta \, d\beta, \tag{7}$$

for any x in $a \leq x \leq b$. It is here that we use the integrability of $f(x)$ on the interval $a \leq x \leq b$.

The A and B of (6) and (7) may be inserted in (2) to give us the particular solution

$$y_p = -\cos x \int_a^x f(\beta) \sin \beta \, d\beta + \sin x \int_a^x f(\beta) \cos \beta \, d\beta$$

$$= \int_a^x f(\beta)(\sin x \cos \beta - \cos x \sin \beta) \, d\beta. \tag{8}$$

Hence we have

$$y_p = \int_a^x f(\beta) \sin (x - \beta) \, d\beta, \tag{9}$$

and we can now write the general solution of equation (1):

$$y = c_1 \cos x + c_2 \sin x + \int_a^x f(\beta) \sin (x - \beta) \, d\beta. \tag{10}$$

Exercises

In exercises 1 through 18 use variation of parameters.

1. $(D^2 - 1)y = e^x + 1$.   ANS. $y = c_1 e^x + c_2 e^{-x} + \frac{1}{2}x e^x - \frac{1}{4}e^x - 1$.
2. $(D^2 + 1)y = \csc x \cot x$.

   ANS. $y = c_1 \cos x + c_2 \sin x - x \sin x - \cos x \ln |\sin x|$.
3. $(D^2 + 1)y = \csc x$.   ANS. $y = c_1 \sin x + c_2 \cos x - x \cos x + \sin x \ln |\sin x|$.
4. $(D^2 + 2D + 2)y = e^{-x} \csc x$.   ANS. $y_p = -xe^{-x} \cos x + e^{-x} \sin x \ln |\sin x|$.
5. $(D^2 + 1)y = \sec^3 x$.   ANS. $y = y_c + \frac{1}{2} \sec x$.
6. $(D^2 + 1)y = \sec^4 x$.   ANS. $y = y_c - \frac{1}{2} + \frac{1}{6} \sec^2 x + \frac{1}{2} \sin x \ln |\sec x + \tan x|$.
7. $(D^2 + 1)y = \tan x$.   ANS. $y = y_c - \cos x \ln |\sec x + \tan x|$.
8. $(D^2 + 1)y = \tan^2 x$.   ANS. $y = y_c - 2 + \sin x \ln |\sec x + \tan x|$.
9. $(D^2 + 1)y = \sec x \csc x$.

   ANS. $y = y_c - \cos x \ln |\sec x + \tan x| - \sin x \ln |\csc x + \cot x|$.
10. $(D^2 + 1)y = \sec^2 x \csc x$.   ANS. $y = y_c - \sin x \ln |\csc 2x + \cot 2x|$.
11. $(D^2 - 2D + 1)y = e^{2x}(e^x + 1)^{-2}$.   ANS. $y = y_c + e^x \ln (1 + e^x)$.
12. $(D^2 - 3D + 2)y = e^{2x}/(1 + e^{2x})$.

   ANS. $y = y_c + e^x \arctan (e^{-x}) - \frac{1}{2} e^{2x} \ln (1 + e^{-2x})$.
13. $(D^2 - 3D + 2)y = \cos (e^{-x})$.   ANS. $y = y_c - e^{2x} \cos (e^{-x})$.

14. $(D^2 - 1)y = 2(1 - e^{-2x})^{-1/2}$.

ANS. $y = c_1 e^x + c_2 e^{-x} - e^x \arcsin(e^{-x}) - (1 - e^{-2x})^{1/2}$.

15. $(D^2 - 1)y = e^{-2x} \sin e^{-x}$.

ANS. $y = y_c - \sin e^{-x} - e^x \cos e^{-x}$.

16. $(D - 1)(D - 2)(D - 3)y = e^x$.

ANS. $y = y_c + \frac{1}{2}x\, e^x$.

17. $y''' - y' = x$.

18. $y''' + y' = \tan x$.

19. Observe that x and e^x are solutions of the homogeneous equation associated with

$$(1 - x)y'' + xy' - y = 2(x - 1)^2 e^{-x}.$$

Use this fact to solve the nonhomogeneous equation.

20. Solve the equation

$$y'' - y = e^x$$

by the method of variation of parameters, but instead of setting $A'y_1 + B'y_2 = 0$ as in equation (6) Section 45, choose $A'y_1 + B'y_2 = k$, for constant k.

21. Apply the suggestion of exercise 20 to exercise 5 above.

22. Let y_1 and y_2 be solutions of the homogeneous equation associated with

$$y'' + p(x)y' + q(x)y = f(x). \tag{A}$$

Let $W(x)$ be the Wronskian of y_1 and y_2, and assume $W(x) \neq 0$ on the interval $a < x < b$. Show that a particular solution of equation (A) is given by

$$y_p = \int_a^x \frac{f(\beta)[y_1(\beta)y_2(x) - y_1(x)y_2(\beta)]\, d\beta}{W(\beta)}. \tag{B}$$

23. The conditions of exercise 22 imply that

$$y_1'' + py_1' + qy_1 = 0 \tag{C}$$

and

$$y_2'' + py_2' + qy_2 = 0. \tag{D}$$

If we multiply equation (C) by y_2 and equation (D) by y_1 and then subtract the two equations, we obtain

$$(y_2 y_1'' - y_1 y_2'') + p(y_2 y_1' - y_1 y_2') = 0.$$

From this equation show that the Wronskian of y_1 and y_2 can be written

$$W(x) = c \exp\left(-\int p\, dx\right), \tag{E}$$

where c is constant. Equation (E) is known as Abel's formula.

24. Conclude from exercise 23 that if $W(x_0) = 0$ for some x_0 on the interval $a < x < b$, then $W(x) \equiv 0$ for all $a < x < b$.

25. Solve the initial value problem

$$y'' + y = f(x); \qquad \text{when } x = x_0, y = y_0, y' = y_0'.$$

Hint: Show that the constant a in equations (6) and (7) of Section 46 could have been chosen to be x_0. Determine the c_1 and c_2 of equation (10) by using the form of y_p in equation (8).

ANS. $y = y_0 \cos (x - x_0) + y_0' \sin (x - x_0) + \displaystyle\int_{x_0}^{x} f(\beta) \sin (x - \beta) \, d\beta.$

Miscellaneous Exercises

1. $(D^2 - 1)y = 2 e^{-x}(1 + e^{-2x})^{-2}.$ ANS. $y = y_c - x e^{-x} - \frac{1}{2} e^{-x} \ln (1 + e^{-2x}).$

2. $(D^2 - 1)y = (1 - e^{2x})^{-3/2}.$ ANS. $y = y_c - (1 - e^{2x})^{1/2}.$

3. $(D^2 - 1)y = e^{2x}(3 \tan e^x + e^x \sec^2 e^x).$ ANS. $y = y_c + e^x \ln |\sec e^x|.$

4. $(D^2 + 1)y = \sec^2 x \tan x.$ ANS. $y = y_c + \frac{1}{2} \tan x + \frac{1}{2} \cos x \ln |\sec x + \tan x|.$

5. Do exercise 4 by another method.

6. $(D^2 + 1)y = \cot x.$ ANS. $y = y_c - \sin x \ln |\csc x + \cot x|.$

7. $(D^2 + 1)y = \sec x.$ ANS. $y = c_1 \cos x + c_2 \sin x + x \sin x + \cos x \ln |\cos x|.$

8. Do exercise 7 by another method.

9. $(D^2 - 1)y = 2/(1 + e^x).$ ANS. $y = y_c - 1 - x e^x + (e^x - e^{-x}) \ln (1 + e^x).$

10. $(D^3 + D)y = \sec^2 x.$ *Hint:* integrate once first.

ANS. $y = c_1 + c_2 \cos x + c_3 \sin x - \cos x \ln |\sec x + \tan x|.$

11. $(D^2 - 1)y = 2/(e^x - e^{-x}).$ ANS. $y = y_c - x e^{-x} + \frac{1}{2}(e^x - e^{-x}) \ln |1 - e^{-2x}|.$

12. $(D^2 - 3D + 2)y = \sin e^{-x}.$ ANS. $y = y_c - e^{2x} \sin e^{-x}.$

13. $(D^2 - 1)y = 1/(e^{2x} + 1).$ ANS. $y = y_c - \frac{1}{2} - \cosh x \arctan e^{-x}.$

14. $y'' + y = \sec^3 x \tan x.$ ANS. $y = y_c + \frac{1}{6} \sec x \tan x.$

15. $y'' + y = \sec x \tan^2 x.$ Verify your answer.

16. $y'' + 4y' + 3y = \sin e^x.$ ANS. $y = y_c - e^{-2x} \sin e^x - e^{-3x} \cos e^x.$

17. $y'' + y = \csc^3 x \cot x.$ ANS. $y = y_c + \frac{1}{6} \cot x \csc x.$

Inverse Differential Operators

47. The exponential shift

In Chapter 5 we studied some of the properties of the algebra of linear differential operators with constant coefficients. We found this algebra useful in finding solutions of homogeneous linear equations. In this chapter we show briefly how differential operators may be used to find particular solutions for nonhomogeneous linear equations.

As a first illustration we make use of the exponential shift theorem that was derived in Section 32:

$$e^{ax}f(D)y = f(D - a)[e^{ax}y], \tag{1}$$

where $f(D)$ is a linear differential operator with constant coefficients.

EXAMPLE (a):   Solve the equation

$$(D^2 - 2D + 5)y = 16x^3 e^{3x}. \tag{2}$$

Note that the complementary function is

$$y_c = c_1 e^x \cos 2x + c_2 e^x \sin 2x. \tag{3}$$

We can conclude also that there is a particular solution,

$$y_p = Ax^3 e^{3x} + Bx^2 e^{3x} + Cx e^{3x} + E e^{3x}, \tag{4}$$

which can be obtained by the method of Chapter 7. But the task of obtaining the derivatives of y_p and finding the numerical values of A, B, C, and E is a little tedious. It can be made easier by using the exponential shift (1).

Let us write (2) in the form

$$e^{-3x}(D^2 - 2D + 5)y = 16x^3,$$

and then apply the relation (1), with $a = -3$. In shifting the exponential e^{-3x} from the left to the right of the differential operator, we must replace D by $(D + 3)$ throughout, thus obtaining

$$[(D + 3)^2 - 2(D + 3) + 5](e^{-3x}y) = 16x^3,$$

or

$$(D^2 + 4D + 8)(e^{-3x}y) = 16x^3. \tag{5}$$

In equation (5), the dependent variable is $(e^{-3x}y)$. We know at once that (5) has a particular solution of the form

$$e^{-3x}y_p = Ax^3 + Bx^2 + Cx + E. \tag{6}$$

Successive differentiations of (6) are simple. Indeed,

$$D(e^{-3x}y_p) = 3Ax^2 + 2Bx + C,$$

$$D^2(e^{-3x}y_p) = 6Ax + 2B,$$

thus, from (5) we get

$$6Ax + 2B + 12Ax^2 + 8Bx + 4C + 8Ax^3 + 8Bx^2 + 8Cx + 8E = 16x^3.$$

Hence

$$8A = 16,$$

$$12A + 8B = 0,$$

$$6A + 8B + 8C = 0,$$

$$2B + 4C + 8E = 0,$$

from which $A = 2, B = -3, C = \frac{3}{2}, E = 0$.

Therefore

$$e^{-3x}y_p = 2x^3 - 3x^2 + \tfrac{3}{2}x,$$

or

$$y_p = (2x^3 - 3x^2 + \tfrac{3}{2}x)\, e^{3x},$$

and the general solution of the original equation (2) is

$$y = c_1\, e^x \cos 2x + c_2\, e^x \sin 2x + (2x^3 - 3x^2 + \tfrac{3}{2}x)\, e^{3x}.$$

EXAMPLE (b):   Solve the equation

$$(D^2 - 2D + 1)y = x\, e^x + 7x - 2. \tag{7}$$

Here the immediate use of the exponential shift would do no good, because removing the e^x factor from the first term on the right would only insert a factor e^{-x} in the second and third terms on the right. The terms $7x - 2$ on the right give us no trouble as they stand. Therefore we break (7) into two problems, obtaining a particular solution for each of the equations

$$(D^2 - 2D + 1)y_1 = x\, e^x \tag{8}$$

and

$$(D^2 - 2D + 1)y_2 = 7x - 2. \tag{9}$$

On (8) we use the exponential shift, passing from

$$e^{-x}(D - 1)^2 y_1 = x$$

to

$$D^2(e^{-x}y_1) = x. \tag{10}$$

A particular solution of (10) is easily obtained:

$$e^{-x}y_1 = \tfrac{1}{6}x^3,$$

so

$$y_1 = \tfrac{1}{6}x^3\, e^x. \tag{11}$$

Equation (9) is treated as in Chapter 7. Put

$$y_2 = Ax + B.$$

Then $Dy_2 = A$, and from (9) it is easily found that $A = 7$, $B = 12$. Thus a particular solution of (9) is

$$y_2 = 7x + 12. \tag{12}$$

Using (11), (12), and the roots of the auxiliary equation for (7), the general solution of (7) can now be written. It is

$$y = (c_1 + c_2 x)\, e^x + \tfrac{1}{6}x^3\, e^x + 7x + 12.$$

EXAMPLE (c):   Solve the equation

$$D^2(D + 4)^2 y = 96\, e^{-4x}. \tag{13}$$

At once we have $m = 0, 0, -4, -4$ and $m' = -4$. We seek first a particular solution. Therefore we integrate each member of (13) twice before using the exponential shift. From (13) it follows that

$$(D + 4)^2 y_p = 6\, e^{-4x}, \tag{14}$$

the constants of integration being disregarded because only a particular solution is sought. Equation (14) yields

$$e^{4x}(D + 4)^2 y_p = 6,$$

$$D^2(y_p\, e^{4x}) = 6,$$

$$y_p\, e^{4x} = 3x^2,$$

$$y_p = 3x^2\, e^{-4x}.$$

Thus the general solution of (13) is seen to be

$$y = c_1 + c_2 x + (c_3 + c_4 x + 3x^2)\, e^{-4x}.$$

The exponential shift is particularly helpful when applied in connection with terms for which the values of m' (using the notations of Chapter 7) are repetitions of values of m.

Exercises

In exercises 1 through 12 use the exponential shift to find a particular solution.

1. $(D - 3)^2 y = e^{3x}$.   ANS. $y = \frac{1}{2} x^2 e^{3x}$.
2. $(D - 1)^2 y = e^x$.   ANS. $y = \frac{1}{2} x^2 e^x$.
3. $(D + 2)^2 y = 12x e^{-2x}$.   ANS. $y = 2x^3 e^{-2x}$.
4. $(D + 1)^2 y = 3x e^{-x}$.   ANS. $y = \frac{1}{2} x^3 e^{-x}$.
5. $(D - 2)^3 y = 6x e^{2x}$.   ANS. $y = \frac{1}{4} x^4 e^{2x}$.
6. $(D + 4)^3 y = 8x e^{-4x}$.   ANS. $y = \frac{1}{3} x^4 e^{-4x}$.
7. $(D + 3)^3 y = 15x^2 e^{-3x}$.   ANS. $y = \frac{1}{4} x^5 e^{-3x}$.
8. $(D - 4)^3 y = 15x^2 e^{4x}$.   ANS. $y = \frac{1}{4} x^5 e^{4x}$.
9. $D^2(D - 2)^2 y = 16 e^{2x}$.   ANS. $y = 2x^2 e^{2x}$.
10. $D^2(D + 3)^2 y = 9 e^{-3x}$.   ANS. $y = \frac{1}{2} x^2 e^{-3x}$.
11. $(D^2 - D - 2) y = 18x e^{-x}$.   ANS. $y = -(3x^2 + 2x) e^{-x}$.
12. $(D^2 - D - 2) y = 36x e^{2x}$.   ANS. $y = e^{2x}(6x^2 - 4x)$.

In exercises 13 through 15, find a particular solution, using the exponential shift in part of your work, as in Example (b) above.

13. $(D - 2)^2 y = 20 - 3x e^{2x}$.   ANS. $y = 5 - \frac{1}{2} x^3 e^{2x}$.
14. $(D - 2)^2 y = 4 - 8x + 6x e^{2x}$.   ANS. $y = x^3 e^{2x} - 2x - 1$.

15. $y'' - 9y = 9(2x - 3 + 4xe^{3x})$.   ANS. $y = 3 - 2x + (3x^2 - x)e^{3x}$.
16. $y'' + 4y' + 4y = 4x - 6e^{-2x} + 3e^x$.   ANS. $y = \frac{1}{3}e^x - 3x^2 e^{-2x} + x - 1$.
17. $(D + 1)^2 y = e^{-x} + 3x$.   ANS. $y = \frac{1}{2}x^2 e^{-x} + 3x - 6$.
18. $(D^2 - 4)y = 16x e^{-2x} + 8x + 4$.   ANS. $y = -(2x + 1)(x e^{-2x} + 1)$.

In exercises 19 through 28 find the general solution.

19. $y'' - 4y = 8x e^{2x}$.   ANS. $y = c_1 e^{-2x} + (c_2 - \frac{1}{2}x + x^2)e^{2x}$.
20. $y'' - 9y = -72x e^{-3x}$.
21. $D(D + 1)^2 y = e^{-x}$.   ANS. $y = c_1 + (c_2 + c_3 x - \frac{1}{2}x^2)e^{-x}$.
22. $D^2(D - 2)^2 y = 2e^{2x}$.
23. $y'' + 2y' + y = 48 e^{-x} \cos 4x$.   ANS. $y = (c_1 + c_2 x - 3 \cos 4x)e^{-x}$.
24. $y'' + 4y' + 4y = 18 e^{-2x} \cos 3x$.
25. $(D - 1)^2 y = e^x \sec^2 x \tan x$.   ANS. $y = e^x(c_1 + c_2 x + \frac{1}{2}\tan x)$.
26. $(D^2 + 4D + 4)y = -x^{-2} e^{-2x}$.   ANS. $y = e^{-2x}(c_1 + c_2 x + \ln|x|)$.
27. $(D - a)^2 y = e^{ax} f''(x)$.   ANS. $y = e^{ax}[c_1 + c_2 x + f(x)]$.
28. $(D^2 + 7D + 12)y = e^{-3x} \sec^2 x(1 + 2 \tan x)$.
          ANS. $y = c_1 e^{-4x} + e^{-3x}(c_2 + \tan x)$.

48. The operator $1/f(D)$

In seeking a particular solution of

$$f(D)y = R(x), \tag{1}$$

it is natural to write

$$y = \frac{1}{f(D)} R(x) \tag{2}$$

and to try to define an operator $1/f(D)$ so that the function y of (2) will have meaning and will satisfy equation (1).

Instead of building a theory of such inverse differential operators, we shall adopt the following method of attack. Purely formal (unjustified) manipulations of the symbols will be performed, thus leading to a tentative evaluation of

$$\frac{1}{f(D)} R(x).$$

After all, the only thing that we require of our evaluation is that

$$f(D) \cdot \frac{1}{f(D)} R(x) = R(x). \tag{3}$$

Hence the burden of proof will be placed on a direct verification of the condition (3) in each instance.

49. Evaluation of $[1/f(D)] e^{ax}$

We proved (Section 32) with slightly different notation that

$$f(D) e^{ax} = e^{ax}f(a) \tag{1}$$

and

$$(D - a)^n(x^n e^{ax}) = n!\, e^{ax}. \tag{2}$$

Equation (1) suggests

$$\frac{1}{f(D)} e^{ax} = \frac{e^{ax}}{f(a)}, \qquad f(a) \neq 0. \tag{3}$$

Now from (1) it follows that

$$f(D)\frac{e^{ax}}{f(a)} = \frac{f(a)\, e^{ax}}{f(a)} = e^{ax}.$$

Hence (3) is verified.

Now suppose that $f(a) = 0$. Then $f(D)$ contains the factor $(D - a)$. Suppose that the factor occurs precisely n times in $f(D)$; that is, let

$$f(D) = \phi(D)(D - a)^n, \qquad \phi(a) \neq 0.$$

With the aid of (2) we obtain

$$\phi(D)(D - a)^n(x^n e^{ax}) = \phi(D)n!\, e^{ax},$$

from which, by (1), it follows that

$$\phi(D)(D - a)^n(x^n e^{ax}) = n!\phi(a)\, e^{ax}. \tag{4}$$

Therefore we write

$$\frac{1}{\phi(D)(D - a)^n} e^{ax} = \frac{x^n e^{ax}}{n!\phi(a)}, \qquad \phi(a) \neq 0, \tag{5}$$

which is easily verified. Indeed,

$$\phi(D)(D - a)^n \frac{x^n e^{ax}}{n!\phi(a)} = \frac{n!\phi(a)\, e^{ax}}{n!\phi(a)} = e^{ax}.$$

Note that formula (3) is included in formula (5) as the special case, $n = 0$. See also exercise 35, page 154.

EXAMPLE (a): Solve the equation

$$(D^2 + 1)y = e^{2x}. \tag{6}$$

Here the roots of the auxiliary equation are $m = \pm i$. Further,

$$f(D) = (D^2 + 1)$$

and

$$f(2) \neq 0.$$

Hence, using (3),

$$y_p = \frac{1}{D^2 + 1} e^{2x} = \frac{e^{2x}}{2^2 + 1} = \frac{1}{5} e^{2x},$$

so the solution of (6) is

$$y = c_1 \cos x + c_2 \sin x + \tfrac{1}{5} e^{2x}.$$

EXAMPLE (b):   Solve the equation

$$D^2(D - 1)^3(D + 1)y = e^x. \tag{7}$$

Here $m = 0, 0, 1, 1, 1, -1$. To get a particular solution of (7), we use formula (5) with $a = 1, n = 3, \phi(D) = D^2(D + 1)$. Then

$$\phi(1) = 1^2 \cdot 2$$

and a particular solution of (7) is given by

$$y_p = \frac{1}{(D - 1)^3 D^2(D + 1)} e^x = \frac{x^3 e^x}{3!1^2 \cdot 2} = \frac{1}{12} x^3 e^x.$$

Then the general solution of (7) is

$$y = c_1 + c_2 x + c_3 e^{-x} + (c_4 + c_5 x + c_6 x^2 + \tfrac{1}{12} x^3) e^x.$$

50. Evaluation of $(D^2 + a^2)^{-1} \sin ax$ and $(D^2 + a^2)^{-1} \cos ax$

No special device is needed for the evaluation of

$$(D^2 + a^2)^{-1} \sin bx$$

when $b \neq a$. In fact, it is easy to show that

$$\frac{1}{D^2 + a^2} \sin bx = \frac{\sin bx}{a^2 - b^2}, \qquad b \neq a;$$

and a similar result holds for the expression $(D^2 + a^2)^{-1} \cos bx$, when $b \neq a$.
    Consider next the evaluation of

$$\frac{1}{D^2 + a^2} \sin ax. \tag{1}$$

The formulas of the preceding section can be put to good use here, since

$$\sin ax = \frac{e^{aix} - e^{-aix}}{2i}.$$

Then

$$\frac{1}{D^2 + a^2} \sin ax = \frac{1}{2i} \frac{1}{(D - ai)(D + ai)} (e^{aix} - e^{-aix})$$

$$= \frac{1}{2i} \left(\frac{x\, e^{aix}}{1\,!2ai} - \frac{x\, e^{-aix}}{1\,!(-2ai)} \right)$$

$$= -\frac{x}{2a} \frac{e^{aix} + e^{-aix}}{2},$$

so

$$\frac{1}{D^2 + a^2} \sin ax = -\frac{x}{2a} \cos ax. \qquad (2)$$

The verification of (2) is left as an exercise.

Another useful result,

$$\frac{1}{D^2 + a^2} \cos ax = \frac{x}{2a} \sin ax, \qquad (3)$$

can be obtained in the same way.

Exercises

1. Verify formula (2) of Section 50.
2. Obtain and verify formula (3) of Section 50.

In exercises 3 through 34, find the general solution.

3. $(D^2 - 1)y = e^{2x}$.   ANS. $y = c_1 e^x + c_2 e^{-x} + \frac{1}{3} e^{2x}$.
4. $(D^2 - 1)y = e^x$.   ANS. $y = c_1 e^x + c_2 e^{-x} + \frac{1}{2} x e^x$.
5. $(D^2 + 1)y = \sin x$.   ANS. $y = c_1 \cos x + c_2 \sin x - \frac{1}{2} x \cos x$.
6. $(D^2 + 4)y = \cos 2x$.   ANS. $y = c_1 \cos 2x + c_2 \sin 2x + \frac{1}{4} x \sin 2x$.
7. $(D^2 + 9)y = e^{2x}$.   8. $(D^2 + 4)y = e^{3x}$.
9. $(4D^2 + 1)y = e^{-2x}$.   10. $D(D - 2)y = e^{-x}$.
11. $D(D - 2)^2 y = e^{2x}$.   12. $D(D + 3)^2 y = e^{-3x}$.
13. $(D^2 + 4)y = \cos 3x$.   14. $(D^2 + 9)y = \cos 3x$.
15. $(D^2 + 4)y = \sin 2x$.   16. $(D^2 + 36)y = \sin 6x$.
17. $(D^2 + 9)y = \sin 3x$.   18. $(D^2 + 36)y = \cos 6x$.
19. $(D^2 + 3D - 4)y = 12 e^{2x}$.   ANS. $y = c_1 e^{-4x} + c_2 e^x + 2 e^{2x}$.
20. $(D^2 + 3D - 4)y = 21 e^{3x}$.   ANS. $y = c_1 e^x + c_2 e^{-4x} + \frac{3}{2} e^{3x}$.
21. $(D^2 + 3D - 4)y = 15 e^x$.   ANS. $y = c_1 e^{-4x} + (c_2 + 3x)e^x$.

22. $(D^2 + 3D - 4)y = 20 e^{-4x}$.   ANS. $y = c_1 e^x + (c_2 - 4x)e^{-4x}$.

23. $(D^2 - 3D + 2)y = e^x + e^{2x}$.   ANS. $y = (c_1 - x)e^x + (c_2 + x)e^{2x}$.

24. $(4D^2 - 1)y = e^{x/2} + 12 e^x$.   ANS. $y = c_1 e^{-x/2} + (c_2 + \frac{1}{4}x)e^{x/2} + \frac{1}{3} e^x$.

25. $D^2(D - 2)^3 y = 48 e^{2x}$.   ANS. $y = c_1 + c_2 x + (c_3 + c_4 x + c_5 x^2 + 2x^3)e^{2x}$.

26. $(D^4 - 18D^2 + 81)y = 36 e^{3x}$. ANS. $y = (c_1 + c_2 x + \frac{1}{2}x^2)e^{3x} + (c_3 + c_4 x)e^{-3x}$.

27. $(D^2 + 16)y = 14 \cos 3x$.   ANS. $y = c_1 \cos 4x + c_2 \sin 4x + 2 \cos 3x$.

28. $(4D^2 + 1)y = 35 \sin 3x$.   ANS. $y = c_1 \cos \frac{1}{2}x + c_2 \sin \frac{1}{2}x - \sin 3x$.

29. $y'' + 16y = 24 \sin 4x$.   ANS. $y = (c_1 - 3x) \cos 4x + c_2 \sin 4x$.

30. $y'' + 16y = 48 \cos 4x$.   ANS. $y = c_1 \cos 4x + (c_2 + 6x) \sin 4x$.

31. $y'' + y = 12 \cos 2x - \sin x$.   ANS. $y = (c_1 + \frac{1}{2}x) \cos x + c_2 \sin x - 4 \cos 2x$.

32. $y'' + y = \sin 3x + 4 \cos x$.   ANS. $y = c_1 \cos x + (c_2 + 2x) \sin x - \frac{1}{8} \sin 3x$.

33. $(D^2 - 2D + 5)y = e^x \cos 2x$. *Hint:* use the exponential shift followed by formula (3) of Section 50.   ANS. $y = e^x(c_1 \cos 2x + c_2 \sin 2x) + \frac{1}{4} x e^x \sin 2x$.

34. $(D^2 + 2D + 5)y = e^{-x} \sin 2x$.

ANS. $y = e^{-x}(c_1 \cos 2x + c_2 \sin 2x) - \frac{1}{4} x e^{-x} \cos 2x$.

35. Prove that if $f(x) = (x - a)^n \phi(x)$, then $f^{(n)}(a) = n! \phi(a)$. Then use equation (5) of Section 49 above to prove that

$$\frac{1}{f(D)} e^{ax} = \frac{x^n e^{ax}}{f^{(n)}(a)},$$

where n is the smallest nonnegative integer for which $f^{(n)}(a) \neq 0$.*

In exercises 36 through 39 use the formula of exercise 35 above.

36. Exercise 21.   **37.** Exercise 22.

38. Exercise 25.   **39.** Exercise 26.

In exercises 40 to 45 verify the formulas† stated.

40. $\dfrac{1}{f(D)} \sin ax = \dfrac{f(-D) \sin ax}{f(ai)f(-ai)}; f(ai)f(-ai) \neq 0$.

41. $\dfrac{1}{f(D)} \cos ax = \dfrac{f(-D) \cos ax}{f(ai)f(-ai)}; f(ai)f(-ai) \neq 0$.

42. $\dfrac{1}{f(D)} \sinh ax = \dfrac{f(-D) \sinh ax}{f(a)f(-a)}; f(a)f(-a) \neq 0$.

43. $\dfrac{1}{f(D)} \cosh ax = \dfrac{f(-D) \cosh ax}{f(a)f(-a)}; f(a)f(-a) \neq 0$.

44. $\dfrac{1}{(D^2 + a^2)^n} \sin ax = \dfrac{x^n}{(2a)^n n!} \sin (ax - \frac{1}{2}n\pi)$.

45. $\dfrac{1}{(D^2 + a^2)^n} \cos ax = \dfrac{x^n}{(2a)^n n!} \cos (ax - \frac{1}{2}n\pi)$.

---

* See C. A. Hutchinson: Another note on linear operators. *Amer. Math. Mon.*, **46**:161 (1939).

† The formulas of exercises 40 through 43 were obtained by C. A. Hutchinson: An operational formula. *Amer. Math. Mon.*, **40**:482–483 (1933); those of exercises 44 and 45 were given by C. A. Hutchinson, Note on an operational formula, *Amer. Math. Mon.*, **44**:371–372 (1937).

In exercises 46 to 49 use the formulas of exercises 40 and 41 above.

46. Exercise 4, page 125.   **47.** Exercise 11, page 125.
48. Exercise 12, page 125.   **49.** Exercise 32, page 125.

Applications

51. Vibration of a spring

Consider a steel spring attached to a support and hanging downward. Within certain elastic limits the spring will obey Hooke's law: if the spring is stretched or compressed, its change in length will be proportional to the force exerted upon it and, when that force is removed, the spring will return to its original position with its length and other physical properties unchanged. There is, therefore, associated with each spring a numerical constant, the ratio of the force exerted to the displacement produced by that force. If a force of magnitude Q pounds (lb) stretches the spring c feet (ft), the relation

$$Q = kc \tag{1}$$

defines the spring constant k in units of pounds per foot (lb/ft).

Let a body B weighing w lb be attached to the lower end of a spring (Figure 13) and brought to the point of equilibrium where it can remain at rest. Once the weight B is moved from the point of equilibrium E in Figure 14, the motion of B will be determined by a differential equation and associated initial conditions.

Let t be time measured in seconds after some initial moment when the motion begins. Let x, in feet, be distance measured positive downward (negative upward) from the point of equilibrium, as in Figure 14. We assume that the motion of B takes place entirely in a vertical line, so the velocity and acceleration are given by the first and second derivatives of x with respect to t.

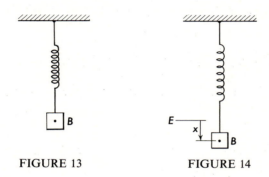

FIGURE 13   FIGURE 14

In addition to the force proportional to displacement (Hooke's law), there will in general be a retarding force caused by resistance of the medium in which the motion takes place or by friction. We are interested here only in such retarding forces as can be well approximated by a term proportional to the velocity because we restrict our study to problems involving linear differential equations. Such a retarding force will contribute to the total force acting on B a term $bx'(t)$, in which b is a constant to be determined experimentally for the medium in which the motion takes place. Some common retarding forces, such as one proportional to the cube of the velocity, lead to nonlinear differential equations.

The weight of the spring is usually negligible compared to the weight B, so we use for the mass of our system the weight of B divided by g, the constant acceleration of gravity. If no forces other than those described above act upon the weight, the displacement x must satisfy the equation

$$\frac{w}{g}x''(t) + bx'(t) + kx(t) = 0. \tag{2}$$

Suppose that an additional vertical force, due to the motion of the support or to presence of a magnetic field, and so on, is imposed upon the system. The new, impressed force, will depend upon time and we may use $F(t)$ to denote the acceleration that it alone would impart to the weight B. Then the impressed force is $(w/g)F(t)$ and equation (2) is replaced by

$$\frac{w}{g}x''(t) + bx'(t) + kx(t) = \frac{w}{g}F(t). \tag{3}$$

At time zero, let the weight be displaced by an amount x_0 from the equilibrium point and let the weight be given an initial velocity v_0. Either or both of x_0 and v_0 may be zero in specific instances. The problem of determining the position of the weight at any time t becomes that of solving the initial value problem consisting of the differential equation

$$\frac{w}{g}x''(t) + bx'(t) + kx(t) = \frac{w}{g}F(t), \qquad \text{for } t > 0, \tag{4}$$

and the initial conditions

$$x(0) = x_0, \qquad x'(0) = v_0. \tag{5}$$

It is convenient to rewrite equation (4) in the form

$$x''(t) + 2\gamma x'(t) + \beta^2 x(t) = F(t), \tag{6}$$

in which we have put

$$\frac{bg}{w} = 2\gamma, \qquad \frac{kg}{w} = \beta^2.$$

We may choose $\beta > 0$ and we know $\gamma \geqq 0$. Note that $\gamma = 0$ corresponds to a negligible retarding force.

A number of special cases of the initial value problem contained in equations (5) and (6) will now be studied.

52. Undamped vibrations

If $\gamma = 0$ in the problem of Section 51, the differential equation becomes

$$x''(t) + \beta^2 x(t) = F(t), \tag{1}$$

a second-order linear equation with constant coefficients in which $\beta^2 = kg/w$. The complementary function associated with the homogeneous equation $x''(t) + \beta^2 x(t) = 0$ is

$$x_c = c_1 \sin \beta t + c_2 \cos \beta t,$$

and the general solution of equation (1) will be of the form

$$x = c_1 \sin \beta t + c_2 \cos \beta t + x_p, \tag{2}$$

where x_p is any particular solution of the nonhomogeneous equation.

We now look at a number of examples of the motion described by equation (2) for different functions $F(t)$ in equation (1).

EXAMPLE (a):   Solve the spring problem with no damping but with $F(t) = A \sin \omega t$, where $\beta \neq \omega$. The case $\beta = \omega$ leads to resonance, which will be discussed in the next section.
   The differential equation of motion is

$$\frac{w}{g} x''(t) + kx(t) = \frac{w}{g} A \sin \omega t$$

and may be written

$$x''(t) + \beta^2 x(t) = A \sin \omega t, \tag{3}$$

with the introduction of $\beta^2 = kg/w$. We shall assume initial conditions

$$x(0) = x_0, \qquad x'(0) = v_0. \tag{4}$$

   A particular solution of equation (3) will be of the form

$$x_p = E \sin \omega t,$$

and we may obtain E by direct substitution into equation (3). We have

$$-E\omega^2 \sin \omega t + \beta^2 E \sin \omega t = A \sin \omega t,$$

an equation that is satisfied for all t only if we choose

$$E = \frac{A}{\beta^2 - \omega^2}.$$

   The general solution of (3) now becomes

$$x(t) = c_1 \sin \beta t + c_2 \cos \beta t + \frac{A}{\beta^2 - \omega^2} \sin \omega t \tag{5}$$

with derivative

$$x'(t) = c_1 \beta \cos \beta t - c_2 \beta \sin \beta t + \frac{A\omega}{\beta^2 - \omega^2} \cos \omega t.$$

The initial conditions (4) now require

$$x_0 = c_2 \qquad \text{and} \qquad v_0 = c_1 \beta + \frac{A\omega}{\beta^2 - \omega^2}$$

and force us to choose

$$c_1 = \frac{v_0}{\beta} - \frac{A\omega}{\beta(\beta^2 - \omega^2)} \qquad \text{and} \qquad c_2 = x_0.$$

From (5) it follows at once that

$$x(t) = \frac{v_0}{\beta}\sin \beta t + x_0 \cos \beta t - \frac{A\omega}{\beta(\beta^2 - \omega^2)}\sin \beta t + \frac{A}{\beta^2 - \omega^2}\sin \omega t. \quad (6)$$

The x of (6) has two parts. The first two terms represent the natural simple harmonic component of the motion, a motion that would be present if A were zero. The last two terms in (6) are caused by the presence of the external force $(w/g)A \sin \omega t$.

EXAMPLE (b): A spring is such that it would be stretched 6 inches (in.) by a 12-lb weight. Let the weight be attached to the spring and pulled down 4 in. below the equilibrium point. If the weight is started with an upward velocity of 2 ft/sec, describe the motion. No damping or impressed force is present.

  We know that the acceleration of gravity enters our work in the expression for the mass. We wish to use the value $g = 32$ feet per second per second (ft/sec^2) and we must use consistent units, so we put all lengths into feet.

  First we determine the spring constant k from the fact that the 12-lb weight stretches the spring 6 in., $\frac{1}{2}$ ft. Thus $12 = \frac{1}{2}k$ so that $k = 24$ lb/ft.

  The differential equation of the motion is therefore

$$\tfrac{12}{32}x''(t) + 24x(t) = 0. \quad (7)$$

At time zero the weight is 4 in. ($\frac{1}{3}$ ft) below the equilibrium point, so $x(0) = \frac{1}{3}$. The initial velocity is negative (upward), so $x'(0) = -2$. Thus our problem is that of solving

$$x''(t) + 64x(t) = 0; \qquad x(0) = \tfrac{1}{3}, x'(0) = -2. \quad (8)$$

  The general solution of equation (8) is

$$x(t) = c_1 \sin 8t + c_2 \cos 8t,$$

from which

$$x'(t) = 8c_1 \cos 8t - 8c_2 \sin 8t.$$

The initial conditions now require that

$$\tfrac{1}{3} = c_2 \qquad \text{and} \qquad -2 = 8c_1,$$

so that finally

$$x(t) = -\tfrac{1}{4}\sin 8t + \tfrac{1}{3}\cos 8t. \quad (9)$$

  A detailed study of the motion is straightforward once (9) has been obtained. The amplitude of the motion is

$$\sqrt{(\tfrac{1}{3})^2 + (\tfrac{1}{4})^2} = \tfrac{5}{12};$$

that is, the weight oscillates between points 5 in. above and below E. The period is $\frac{1}{4}\pi$ sec.

53. Resonance

In Example (a) of the previous section we postponed the study of the special case, $\beta = \omega$. In that case, the differential equation to be solved is

$$x''(t) + \beta^2 x(t) = A \sin \beta t, \tag{1}$$

where we had let $\beta^2 = kg/w$.

The complementary function associated with the homogeneous equation $x''(t) + \beta^2 x(t) = 0$ will be the same as it was before, but the previous particular solution x_p will not exist because $\beta = \omega$.

The method of undetermined coefficients may be applied here to seek a particular solution of the form

$$x_p = Pt \sin \beta t + Qt \cos \beta t, \tag{2}$$

where P and Q are constants to be determined. Direct substitution of the x_p of (2) into equation (1) yields

$$2P\beta \cos \beta t - 2Q\beta \sin \beta t = A \sin \beta t,$$

an equation that can be satisfied for all t only if $P = 0$ and $Q = -A/2\beta$. Thus

$$x_p = \frac{-At}{2\beta} \cos \beta t, \tag{3}$$

and the general solution of (1) is

$$x(t) = c_1 \sin \beta t + c_2 \cos \beta t - \frac{At}{2\beta} \cos \beta t, \tag{4}$$

from which we obtain

$$x'(t) = c_1 \beta \cos \beta t - c_2 \beta \sin \beta t + \frac{At}{2} \sin \beta t - \frac{A}{2\beta} \cos \beta t.$$

The initial conditions $x(0) = x_0$ and $x'(0) = v_0$ now force us to take

$$c_2 = x_0 \quad \text{and} \quad c_1 = \frac{v_0}{\beta} + \frac{A}{2\beta^2}.$$

The final solution may now be written

$$x(t) = x_0 \cos \beta t + \frac{v_0}{\beta} \sin \beta t + \frac{A}{2\beta^2}(\sin \beta t - \beta t \cos \beta t). \tag{5}$$

That (5) satisfies the initial value problem is readily verified.

In the solution (5) the terms proportional to cos βt and sin βt are bounded, but the term with $\beta t \cos \beta t$ can be made as large as we wish by proper choice of t. This building up of large amplitudes in the vibration is called *resonance*.

Exercises

1. A spring is such that a 5-lb weight stretches it 6 in. The 5-lb weight is attached, the spring reaches equilibrium, then the weight is pulled down 3 in. below the equilibrium point and started off with an upward velocity of 6 ft/sec. Find an equation giving the position of the weight at all subsequent times.
   ANS. $x = \frac{1}{4}(\cos 8t - 3 \sin 8t)$.

2. A spring is stretched 1.5 in. by a 2-lb weight. Let the weight be pushed up 3 in. above E and then released. Describe the motion. ANS. $x = -\frac{1}{4} \cos 16t$.

3. For the spring and weight of exercise 2, let the weight be pulled down 4 in. below E and given a downward initial velocity of 8 ft/sec. Describe the motion.
   ANS. $x = \frac{1}{3} \cos 16t + \frac{1}{2} \sin 16t$.

4. Show that the answer to exercise 3 can be written $x = 0.60 \sin(16t + \phi)$ where $\phi = \arctan \frac{2}{3}$.

5. A spring is such that a 4-lb weight stretches it 6 in. An impressed force $\frac{1}{2} \cos 8t$ is acting on the spring. If the 4-lb weight is started from the equilibrium point with an imparted upward velocity of 4 ft/sec, determine the position of the weight as a function of time. ANS. $x = \frac{1}{4}(t - 2) \sin 8t$.

6. A spring is such that it is stretched 6 in. by a 12-lb weight. The 12-lb weight is pulled down 3 in. below the equilibrium point and then released. If there is an impressed force of magnitude $9 \sin 4t$ lb, describe the motion. Assume that the impressed force acts downward for very small t.
   ANS. $x = \frac{1}{4} \cos 8t - \frac{1}{4} \sin 8t + \frac{1}{2} \sin 4t$.

7. Show that the answer to exercise 6 can be written
$$x = \frac{1}{4}\sqrt{2} \cos(8t + \pi/4) + \frac{1}{2} \sin 4t.$$

8. A spring is such that a 2-lb weight stretches it $\frac{1}{2}$ ft. An impressed force $\frac{1}{4} \sin 8t$ is acting upon the spring. If the 2-lb weight is released from a point 3 in. below the equilibrium point, determine the equation of motion.
   ANS. $x = \frac{1}{4}(1 - t) \cos 8t + \frac{1}{32} \sin 8t$ (ft).

9. For the motion of exercise 8, find the first four times at which stops occur and find the position at each stop. ANS. $t = \pi/8, \pi/4, 1, 3\pi/8$ (sec) and $x = -0.15, +0.05, +0.03, +0.04$ (ft), respectively.

10. Determine the approximate position to be expected, if nothing such as breakage interferes, at the time of the 65th stop, when $t = 8\pi$ (sec), in exercise 8.
    ANS. $x = -6.0$ (ft).

11. A spring is such that a 16-lb weight stretches it 1.5 in. The weight is pulled down to a point 4 in. below the equilibrium point and given an initial downward velocity of 4 ft/sec. An impressed force of $360 \cos 4t$ lb is applied. Find the position and velocity of the weight at time $t = \pi/8$ sec.
    ANS. At $t = \pi/8$ (sec), $x = -\frac{8}{3}$ (ft), $v = -8$ (ft/sec).

12. A spring is stretched 3 in. by a 5-lb weight. Let the weight be started from E with an upward velocity of 12 ft/sec. Describe the motion.   ANS. $x = -1.06 \sin 11.3t$

13. For the spring and weight of exercise 12, let the weight be pulled down 4 in. below E and then given an upward velocity of 8 ft/sec. Describe the motion.
    ANS. $x = 0.33 \cos 11.3t - 0.71 \sin 11.3t$.

14. Find the amplitude of the motion in exercise 13.   ANS.   0.78 ft.

15. A 20-lb weight stretches a certain spring 10 in. Let the spring first be compressed 4 in., and then the 20-lb weight attached and given an initial downward velocity of 8 ft/sec. Find how far the weight would drop.   ANS.   35 in.

16. A spring is such that an 8-lb weight would stretch it 6 in. Let a 4-lb weight be attached to the spring, which is then pushed up 2 in. above its equilibrium point and released. Describe the motion.   ANS. $x = -\frac{1}{6} \cos 11.3t$.

17. If the 4-lb weight of exercise 16 starts at the same point, 2 in. above E, but with an upward velocity of 15 ft/sec, when will the weight reach its lowest point?
    ANS.   At $t =$ approximately 0.4 sec.

18. A spring is such that it is stretched 4 in. by a 10-lb weight. Suppose the 10-lb weight to be pulled down 5 in. below E and then given a downward velocity of 15 ft/sec. Describe the motion.
    ANS. $x = 0.42 \cos 9.8t + 1.53 \sin 9.8t$
    $$= 1.59 \cos (9.8t - \phi), \text{ where } \phi = \text{arc tan } 3.64.$$

19. A spring is such that it is stretched 4 in. by an 8-lb weight. Suppose the weight to be pulled down 6 in. below E and then given an upward velocity of 8 ft/sec. Describe the motion.   ANS. $x = 0.50 \cos 9.8t - 0.82 \sin 9.8t$.

20. Show that the answer to exercise 19 can be written $x = 0.96 \cos (9.8t + \phi)$ where $\phi = \text{arc tan } 1.64$.

21. A spring is such that a 4 lb weight stretches it 6 in. The 4-lb weight is attached to the vertical spring and reaches its equilibrium point. The weight is then ($t = 0$) drawn downward 3 in. and released. There is a simple harmonic exterior force equal to $\sin 8t$ impressed upon the whole system. Find the time for each of the first four stops following $t = 0$. Put the stops in chronological order.
    ANS. $t = \pi/8, \frac{1}{2}, \pi/4, 3\pi/8$ (sec).

22. A spring is stretched 1.5 in. by a 4-lb weight. Let the weight be pulled down 3 in. below equilibrium and released. If there is an impressed force $8 \sin 16t$ acting upon the spring, describe the motion.   ANS. $x = \frac{1}{4}(1 - 8t) \cos 16t + \frac{1}{8} \sin 16t$.

23. For the motion of exercise 22, find the first four times at which stops occur and find the position at each stop.   ANS. $t = \frac{1}{8}, \pi/16, \pi/8, 3\pi/16$ (sec) and
    $$x = +0.11, +0.14, -0.54, +0.93 \text{ (ft), respectively.}$$

54. Damped vibrations

In the general linear spring problem of Section 51, we were confronted with

$$x''(t) + 2\gamma x'(t) + \beta^2 x(t) = F(t); \qquad x(0) = x_0, x'(0) = v_0, \qquad (1)$$

in which $2\gamma = bg/w$ and $\beta^2 = kg/w$, $\beta > 0$. The auxiliary equation $m^2 + 2\gamma m + \beta^2 = 0$ has roots $-\gamma \pm \sqrt{\gamma^2 - \beta^2}$ and we see that the nature of the complementary function depends upon whether $\beta > \gamma$, $\beta = \gamma$, or $\beta < \gamma$.

If $\beta > \gamma$, $\beta^2 - \gamma^2 > 0$, so let us put

$$\beta^2 - \gamma^2 = \delta^2. \tag{2}$$

Then the general solution of (1) will be

$$x(t) = e^{-\gamma t}(c_1 \cos \delta t + c_2 \sin \delta t) + \psi_1(t), \tag{3}$$

in which $\psi_1(t)$ is any particular solution of equation (1). The presence of the function $e^{-\gamma t}$, called a damping factor, will cause the natural part of the solution, that is, the part independent of the external force $(w/g)F(t)$, to approach zero as $t \to \infty$.

If in (1) we have $\beta = \gamma$, the two roots of the auxiliary equation are equal and the general solution becomes

$$x(t) = e^{-\gamma t}(c_1 + c_2 t) + \psi_2(t), \tag{4}$$

in which $\psi_2(t)$ is a particular solution of (1). Again the natural component has the damping factor $e^{-\gamma t}$ in it.

If in (1) we have $\beta < \gamma$ and $\gamma^2 - \beta^2 > 0$, then we can set

$$\gamma^2 - \beta^2 = \sigma^2, \qquad \sigma > 0. \tag{5}$$

Since $\sigma < \gamma$, the two roots of the auxiliary equation are both real and negative, and we have

$$x(t) = c_1 e^{(-\gamma + \sigma)t} + c_2 e^{(-\gamma - \sigma)t} + \psi_3(t). \tag{6}$$

Again $\psi_3(t)$ is a particular solution of (1), and we see that the damping factor $e^{-\gamma t}$ causes the natural component of (6) to approach zero as $t \to \infty$.

Suppose for the moment that we have $F(t) \equiv 0$, so the natural component of the motion is all that is under consideration. If $\beta > \gamma$, equation (3) holds and the motion is a *damped oscillatory* one. If $\beta = \gamma$, equation (4) holds and the motion is not oscillatory; it is called *critically damped* motion. If $\beta < \gamma$, (6) holds and the motion is said to be *overdamped*; the parameter γ is larger than it needs to be to remove the oscillations. Figure 15 shows a representative graph of each type of motion mentioned in this paragraph, a damped oscillatory motion, a critically damped motion, and an overdamped motion.

EXAMPLE: Solve the problem of Example (b), Section 52, with an added damping force of magnitude $0.6|v|$. Such a damping force can be realized by immersing the weight B in a thick liquid.

The initial value problem to be solved is

$$\tfrac{12}{32}x''(t) + 0.6x'(t) + 24x(t) = 0; \qquad x(0) = \tfrac{1}{3}, x'(0) = -2. \tag{7}$$

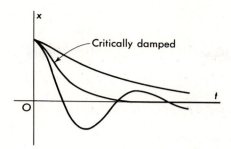

FIGURE 15

The auxiliary equation of (7) may be written

$$m^2 + 1.6m + 64 = 0,$$

an equation that has roots $-0.8 \pm \sqrt{63.36}\,i$. Therefore, the general solution of (7) is

$$x(t) = e^{-0.8t}(c_1 \cos 8.0t + c_2 \sin 8.0t)$$

and

$$x'(t) = e^{-0.8t}[(-8c_1 - 0.8c_2) \sin 8.0t + (8c_2 - 0.8c_1) \cos 8.0t].$$

The initial conditions in (7) now give us

$$\tfrac{1}{3} = c_1 \qquad \text{and} \qquad -2 = 8c_2 - 0.8c_1,$$

so that $c_1 = 0.33$ and $c_2 = -0.22$.

Therefore the desired solution is

$$x(t) = \exp(-0.8t)(0.33 \cos 8.0t - 0.22 \sin 8.0t), \tag{8}$$

a portion of its graph being shown in Figure 16.

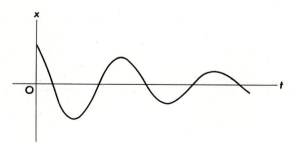

FIGURE 16

Exercises

1. A certain straight-line motion is determined by the differential equation

$$\frac{d^2x}{dt^2} + 2\gamma\frac{dx}{dt} + 169x = 0$$

and the conditions that when $t = 0$, $x = 0$, and $v = 8$ ft/sec.

(a) Find the value of γ that leads to critical damping, determine x in terms of t, and draw the graph for $0 \leq t \leq 0.2$.   ANS. $\gamma = 13(1/\text{sec})$, $x = 8t\,e^{-13t}$.

(b) Use $\gamma = 12$. Find x in terms of t and draw the graph.

ANS. $x = 1.6\,e^{-12t}\sin 5t$.

(c) Use $\gamma = 14$. Find x in terms of t and draw the graph.

ANS. $x = 0.77(e^{-8.8t} - e^{-19.2t})$.

2. A spring is such that a 2-lb weight stretches it $\frac{1}{2}$ ft. An impressed force $\frac{1}{4}\sin 8t$ and a damping force of magnitude $|v|$ are both acting on the spring. The weight starts $\frac{1}{4}$ ft below the equilibrium point with an imparted upward velocity of 3 ft/sec. Find a formula for the position of the weight at time t.

ANS. $x = \frac{3}{32}e^{-8t}(3 - 8t) - \frac{1}{32}\cos 8t$.

3. A spring is such that a 4-lb weight stretches it 0.64 ft. The 4-lb weight is pushed up $\frac{1}{3}$ ft above the point of equilibrium and then started with a downward velocity of 5 ft/sec. The motion takes place in a medium which furnishes a damping force of magnitude $\frac{1}{4}|v|$ at all times. Find the equation describing the position of the weight at time t.   ANS. $x = \frac{1}{3}e^{-t}(2\sin 7t - \cos 7t)$.

4. A spring is such that a 4-lb weight stretches it 0.32 ft. The weight is attached to the spring and moves in a medium which furnishes a damping force of magnitude $\frac{3}{2}|v|$. The weight is drawn down $\frac{1}{2}$ ft below the equilibrium point and given an initial upward velocity of 4 ft/sec. Find the position of the weight thereafter.

ANS. $x = \frac{1}{8}e^{-6t}(4\cos 8t - \sin 8t)$.

5. A spring is such that a 4-lb weight stretches the spring 0.4 ft. The 4-lb weight is attached to the spring (suspended from a fixed support) and the system is allowed to reach equilibrium. Then the weight is started from equilibrium position with an imparted upward velocity of 2 ft/sec. Assume that the motion takes place in a medium that furnishes a retarding force of magnitude numerically equal to the speed, in feet per second, of the moving weight. Determine the position of the weight as a function of time.   ANS. $x = -\frac{1}{4}e^{-4t}\sin 8t$.

6. A spring is stretched 6 in. by a 3-lb weight. The 3-lb weight is attached to the spring and then started from equilibrium with an imparted upward velocity of 12 ft/sec. Air resistance furnishes a retarding force equal in magnitude to $0.03|v|$. Find the equation of motion.   ANS. $x = -1.5\,e^{-0.16t}\sin 8t$.

7. A spring is such that a 2-lb weight stretches it 6 in. There is a damping force present, with magnitude the same as the magnitude of the velocity. An impressed force $(2\sin 8t)$ is acting on the spring. If, at $t = 0$, the weight is released from a point 3 in. below the equilibrium point, find its position for $t > 0$.

ANS. $x = (\frac{1}{2} + 4t)\,e^{-8t} - \frac{1}{4}\cos 8t$.

8. A spring is stretched 10 in. by a 4-lb weight. The weight is started 6 in. below the equilibrium point with an upward velocity of 8 ft/sec. If a resisting medium furnishes

a retarding force of magnitude $\frac{1}{4}|v|$, describe the motion.

ANS. $x = e^{-t}[0.50 \cos 6.1t - 1.23 \sin 6.1t]$.

9. For exercise 8, find the times of the first three stops and the position (to the nearest inch) of the weight at each stop.   ANS. $t_1 = 0.3$ sec, $x_1 = -12$ in.; $t_2 = 0.8$ sec, $x_2 = +6$ in.; $t_3 = 1.3$ sec, $x_3 = -4$ in.

10. A spring is stretched 4 in. by a 2-lb weight. The 2-lb weight is started from the equilibrium point with a downward velocity of 12 ft/sec. If air resistance furnishes a retarding force of magnitude 0.02 of the velocity, describe the motion.

ANS. $x = 1.22 \, e^{-0.16t} \sin 9.8t$.

11. For exercise 10, find how long it takes the damping factor to drop to one-tenth its initial value.   ANS.   14.4 sec.

12. For exercise 10, find the position of the weight at: (a) the first stop; (b) the second stop.   ANS.   (a) $x = 1.2$ ft; (b) $x = -1.1$ ft.

13. Let the motion of exercise 8, page 162, be retarded by a damping force of magnitude $0.6|v|$. Find the equation of motion.

ANS. $x = 0.30 \, e^{-4.8t} \cos 6.4t + 0.22 \, e^{-4.8t} \sin 6.4t - 0.05 \cos 8t$ (ft).

14. Show that whenever $t > 1$ (sec), the solution of exercise 13 can be replaced (to the nearest 0.01 ft) by $x = -0.05 \cos 8t$.

15. Let the motion of exercise 8, page 162, be retarded by a damping force of magnitude $|v|$. Find the equation of motion and also determine its form (to the nearest 0.01 ft) for $t > 1$ (sec).

ANS. $x = \frac{9}{32}(8t + 1) e^{-8t} - \frac{1}{32} \cos 8t$ (ft); for $t > 1$, $x = -\frac{1}{32} \cos 8t$.

16. Let the motion of exercise 8, page 162, be retarded by a damping force of magnitude $\frac{5}{3}|v|$. Find the equation of motion.

ANS. $x = 0.30 \, e^{-(8/3)t} - 0.03 \, e^{-24t} - 0.02 \cos 8t$.

17. Alter exercise 6, page 162, by inserting a damping force of magnitude one-half that of the velocity and then determine x.

ANS. $x = \exp\left(-\frac{2}{3}t\right)(0.30 \cos 8.0t - 0.22 \sin 8.0t) - 0.05 \cos 4t + 0.49 \sin 4t$.

18. A spring is stretched 6 in. by a 4 lb weight. Let the weight be pulled down 6 in. below equilibrium and given an initial upward velocity of 7 ft/sec. Assuming a damping force twice the magnitude of the velocity, describe the motion and sketch the graph at intervals of 0.05 sec for $0 \leq t \leq 0.3$ (sec).   ANS. $x = \frac{1}{2} e^{-8t}(1 - 6t)$.

19. An object weighing w lb is dropped from a height h ft above the earth. At time t (sec) after the object is dropped, let its distance from the starting point be x (ft), measured positive downward. Assuming air resistance to be negligible, show that x must satisfy the equation

$$\frac{w}{g} \frac{d^2x}{dt^2} = w$$

as long as $x < h$. Find x.   ANS. $x = \frac{1}{2}gt^2$.

20. Let the weight of exercise 19 be given an initial velocity v_0. Let v be the velocity at time t. Determine v and x.   ANS. $v = gt + v_0, x = \frac{1}{2}gt^2 + v_0 t$.

21. From the results in exercise 20, find a relation that does not contain t explicitly.

ANS. $v^2 = v_0^2 + 2gx$.

22. If air resistance furnishes an additional force proportional to the velocity in the motion studied in exercises 19 and 20, show that the equation of motion becomes

$$\frac{w}{g} \frac{d^2x}{dt^2} + b\frac{dx}{dt} = w. \tag{A}$$

Solve equation (A) given the conditions $t = 0$, $x = 0$, and $v = v_0$. Use $a = bg/w$.

ANS.　$x = a^{-1}gt + a^{-2}(av_0 - g)(1 - e^{-at})$.

23. To compare the results of exercises 20 and 22 when $a = bg/w$ is small, use the power series for e^{-at} in the answer for exercise 22 and discard all terms involving a^n for $n \geq 3$.　　ANS.　$x = \frac{1}{2}gt^2 + v_0 t - \frac{1}{6}at^2(3v_0 + gt) + \frac{1}{24}a^2 t^3(4v_0 + gt)$.

24. The equation of motion of the vertical fall of a man with a parachute may be roughly approximated by equation (A) of exercise 22. Suppose a 180-lb man drops from a great height and attains a velocity of 20 miles per hour (mph) after a long time. Determine the implied coefficient b of equation (A).　　ANS.　6.1 (lb)(sec)/ft.

25. A particle is moving along the x-axis according to the law

$$\frac{d^2x}{dt^2} + 6\frac{dx}{dt} + 25x = 0.$$

If the particle started at $x = 0$ with an initial velocity of 12 ft/sec to the left, determine: (a) x in terms of t; (b) the times at which stops occur; and (c) the ratio between the numerical values of x at successive stops.

　　　　　　　　　　(a) $x = -3 e^{-3t} \sin 4t$.

ANS.　(b) $t = 0.23 + \frac{1}{4}n\pi, n = 0, 1, 2, 3, \ldots$.

　　　　　　　　　　(c) 0.095.

55. The simple pendulum

A rod of length C ft is suspended by one end so it can swing freely in a vertical plane. Let a weight B (the bob) of w lb be attached to the free end of the rod, and let the weight of the rod be negligible compared to the weight of the bob.

Let θ (radians) be the angular displacement from the vertical, as shown in Figure 17, of the rod at time t (sec). The tangential component of the force

FIGURE 17

w (lb) is $w \sin \theta$ and it tends to decrease θ. Then, neglecting the weight of the rod and using $S = C\theta$ as a measure of arc length from the vertical position, we may conclude that

$$\frac{w}{g} \frac{d^2 S}{dt^2} = -w \sin \theta. \tag{1}$$

Since $S = C\theta$ and C is constant, (1) becomes

$$\frac{d^2 \theta}{dt^2} + \frac{g}{C} \sin \theta = 0. \tag{2}$$

The solution of equation (2) is not elementary; it involves an elliptic integral. If θ is small, however, $\sin \theta$ and θ are nearly equal and (2) is closely approximated by the much simpler equation

$$\frac{d^2 \theta}{dt^2} + \beta^2 \theta = 0; \qquad \beta^2 = \frac{g}{C}. \tag{3}$$

The solution of (3) with pertinent initial conditions gives usable results whenever those conditions are such that θ remains small, say $|\theta| < 0.3$ (radians).

Exercises

1. A clock has a 6-in. pendulum. The clock ticks once for each time that the pendulum completes a swing, returning to its original position. How many times does the clock tick in 30 sec?   ANS.   38 times.
2. A 6-in. pendulum is released from rest at an angle one-tenth of a radian from the vertical. Using $g = 32$ (ft/sec²), describe the motion.
                           ANS. $\theta = 0.1 \cos 8t$ (radians).
3. For the pendulum of exercise 2, find the maximum angular speed and its first time of occurrence.   ANS.   0.8 (radians/sec) at 0.2 sec.
4. A 6-in. pendulum is started with a velocity of 1 radian/sec, toward the vertical, from a position one-tenth radian from the vertical. Describe the motion.
             ANS. $\theta = \frac{1}{10} \cos 8t - \frac{1}{8} \sin 8t$ (radians).
5. For exercise 4, find to the nearest degree the maximum angular displacement from the vertical.   ANS.   9°.
6. Interpret as a pendulum problem and solve:

$$\frac{d^2 \theta}{dt^2} + \beta^2 \theta = 0 : \beta^2 = \frac{g}{C}, \qquad \text{when } t = 0, \theta = \theta_0, \omega = \frac{d\theta}{dt} = \omega_0.$$

         ANS. $\theta = \theta_0 \cos \beta t + \beta^{-1} \omega_0 \sin \beta t$ (radians).
7. Find the maximum angular displacement from the vertical for the pendulum of exercise 6.   ANS. $\theta_{\max} = (\theta_0^2 + \beta^{-2} \omega_0^2)^{1/2}$.

11

The Laplace Transform

56. The transform concept

The reader is already familiar with some operators that transform functions into functions. An outstanding example is the differential operator D, which transforms each function of a large class (those possessing a derivative) into another function.

We have already found that the operator D is useful in the treatment of linear differential equations with constant coefficients. In this chapter we study another transformation (a mapping of functions onto functions) which has played an increasingly important role in both pure and applied mathematics in the past few decades. The operator L, to be introduced in Section 57, is particularly effective in the study of initial value problems involving linear differential equations with constant coefficients.

One class of transformations, which are called integral transforms, may be defined by

$$T\{F(t)\} = \int_{-\infty}^{\infty} K(s, t)F(t)\, dt = f(s). \tag{1}$$

Given a function $K(s, t)$, called the kernel of the transformation, equation (1) associates with each $F(t)$ of the class of functions for which the above integral exists a function $f(s)$ defined by (1). Generalizations and abstractions of (1), as well as studies of special cases, are to be found in profusion in mathematical literature.

Various particular choices of $K(s, t)$ in (1) have led to special transforms, each with its own properties to make it useful in specific circumstances. The transform defined by choosing

$$K(s, t) = 0, \qquad \text{for } t < 0,$$
$$= e^{-st}, \qquad \text{for } t \geq 0,$$

is the one to which this chapter is devoted.

57. Definition of the Laplace transform

Let $F(t)$ be any function such that the integrations encountered may be legitimately performed on $F(t)$. The *Laplace transform* of $F(t)$ is denoted by $L\{F(t)\}$ and is defined by

$$L\{F(t)\} = \int_0^\infty e^{-st} F(t)\, dt. \tag{1}$$

The integral in (1) is a function of the parameter s; call that function $f(s)$. We may write

$$L\{F(t)\} = \int_0^\infty e^{-st} F(t)\, dt = f(s). \tag{2}$$

It is customary to refer to $f(s)$, as well as to the symbol $L\{F(t)\}$, as the transform, or the Laplace transform, of $F(t)$.

We may also look upon (2) as a definition of a Laplace operator L, which transforms each function $F(t)$ of a certain set of functions into some function $f(s)$.

It is easy to show that if the integral in (2) does converge, it will do so for all s greater than* some fixed value s_0. That is, equation (2) will define $f(s)$ for $s > s_0$. In extreme cases the integral may converge for all finite s.

It is important that the operator L, like the differential operator D, is a linear operator. If $F_1(t)$ and $F_2(t)$ have Laplace transforms and if c_1 and c_2 are any constants,

$$L\{c_1 F_1(t) + c_2 F_2(t)\} = c_1 L\{F_1(t)\} + c_2 L\{F_2(t)\}. \tag{3}$$

---

* If s is not to be restricted to real values, the convergence takes place for all s with real part greater than some fixed value.

Using elementary properties of definite integrals, the student can easily show the validity of equation (3).

We shall hereafter employ the relation (3) without restating the fact that the operator L is a linear one.

58. Transforms of elementary functions

The transforms of certain exponential and trigonometric functions and of polynomials will now be obtained. These results enter our work frequently.

EXAMPLE (a): Find $L\{e^{kt}\}$.
We proceed as follows:

$$L\{e^{kt}\} = \int_0^\infty e^{-st} \cdot e^{kt} \, dt = \int_0^\infty e^{-(s-k)t} \, dt.$$

For $s \leq k$, the exponent on e is positive or zero and the integral diverges. For $s > k$, the integral converges.
Indeed, for $s > k$,

$$L\{e^{kt}\} = \int_0^\infty e^{-(s-k)t} \, dt$$

$$= \left[\frac{-e^{-(s-k)t}}{s-k} \right]_0^\infty$$

$$= 0 + \frac{1}{s-k}.$$

Thus we find that

$$L\{e^{kt}\} = \frac{1}{s-k}, \qquad s > k. \tag{1}$$

Note the special case $k = 0$:

$$L\{1\} = \frac{1}{s}, \qquad s > 0. \tag{2}$$

EXAMPLE (b): Obtain $L\{\sin kt\}$.
From elementary calculus we obtain

$$\int e^{ax} \sin mx \, dx = \frac{e^{ax}(a \sin mx - m \cos mx)}{a^2 + m^2} + C.$$

Since

$$L\{\sin kt\} = \int_0^\infty e^{-st} \sin kt \, dt,$$

it follows that

$$L\{\sin kt\} = \left[\frac{e^{-st}(-s \sin kt - k \cos kt)}{s^2 + k^2} \right]_0^\infty. \tag{3}$$

For positive s, $e^{-st} \to 0$ as $t \to \infty$. Furthermore, $\sin kt$ and $\cos kt$ are bounded as $t \to \infty$. Therefore (3) yields

$$L\{\sin kt\} = 0 - \frac{1(0 - k)}{s^2 + k^2},$$

or

$$L\{\sin kt\} = \frac{k}{s^2 + k^2}, \qquad s > 0. \tag{4}$$

The result

$$L\{\cos kt\} = \frac{s}{s^2 + k^2}, \qquad s > 0, \tag{5}$$

can be obtained in a similar manner.

EXAMPLE (c): Obtain $L\{t^n\}$ for n a positive integer.
   By definition

$$L\{t^n\} = \int_0^\infty e^{-st} t^n \, dt.$$

Let us attack the integral using integration by parts with the choice exhibited in the table.

t^n	$e^{-st} \, dt$
$nt^{n-1} \, dt$	$-\dfrac{1}{s} e^{-st}$

We thus obtain

$$\int_0^\infty e^{-st} t^n \, dt = \left[\frac{-t^n e^{-st}}{s} \right]_0^\infty + \frac{n}{s} \int_0^\infty e^{-st} t^{n-1} \, dt. \tag{6}$$

For $s > 0$ and $n > 0$, the first term on the right in (6) is zero, and we are left with

$$\int_0^\infty e^{-st} t^n \, dt = \frac{n}{s} \int_0^\infty e^{-st} t^{n-1} \, dt, \qquad s > 0,$$

or

$$L\{t^n\} = \frac{n}{s} L\{t^{n-1}\}, \qquad s > 0. \tag{7}$$

From (7) we may conclude that, for $n > 1$,

$$L\{t^{n-1}\} = \frac{n-1}{s} L\{t^{n-2}\}$$

so

$$L\{t^n\} = \frac{n(n-1)}{s^2} L\{t^{n-2}\}. \tag{8}$$

Iteration of this process yields

$$L\{t^n\} = \frac{n(n-1)(n-2)\cdots 2\cdot 1}{s^n} L\{t^0\}.$$

From Example (a) above, we have

$$L\{t^0\} = L\{1\} = s^{-1}.$$

Hence, for n a positive integer,

$$L\{t^n\} = \frac{n!}{s^{n+1}}, \qquad s > 0. \tag{9}$$

The Laplace transform of $F(t)$ will exist even if the object function $F(t)$ is discontinuous, provided the integral in the definition of $L\{F(t)\}$ exists. Little will be done at this time with specific discontinuous $F(t)$, because more efficient methods for obtaining such transforms are to be developed later.

EXAMPLE (d):   Find the Laplace transform of $H(t)$ where

$$H(t) = t, \qquad 0 < t < 4,$$
$$= 5, \qquad t > 4.$$

Note that the fact that $H(t)$ is not defined at $t = 0$ and $t = 4$ has no bearing whatever on the existence, or the value, of $L\{H(t)\}$. We turn to the definition of $L\{H(t)\}$ to obtain

$$L\{H(t)\} = \int_0^\infty e^{-st} H(t)\, dt$$

$$= \int_0^4 e^{-st} t\, dt + \int_4^\infty e^{-st} 5\, dt.$$

Using integration by parts on the next-to-last integral above, we soon arrive, for $s > 0$, at

$$L\{H(t)\} = \left[-\frac{t}{s} e^{-st} - \frac{1}{s^2} e^{-st} \right]_0^4 + \left[-\frac{5}{s} e^{-st} \right]_4^\infty.$$

Thus

$$L\{H(t)\} = -\frac{4 e^{-4s}}{s} - \frac{e^{-4s}}{s^2} + 0 + \frac{1}{s^2} - 0 + \frac{5 e^{-4s}}{s}$$

$$= \frac{1}{s^2} + \frac{e^{-4s}}{s} - \frac{e^{-4s}}{s^2}.$$

Exercises

1. Show that $L\{\cos kt\} = \dfrac{s}{s^2 + k^2}$; for $s > 0$.

2. Euler's formula $e^{ikt} = \cos kt + i \sin kt$ can be used to obtain the formula $\cos kt = \frac{1}{2}(e^{ikt} + e^{-ikt})$. Show that the result of exercise 1 can now be obtained with a formal application of the Laplace transform.

3. Obtain the transform for $\sin kt$ by an argument similar to the one suggested in exercise 2.

4. Obtain $L\{t^2 + 4t - 5\}$.   ANS. $\dfrac{2}{s^3} + \dfrac{4}{s^2} - \dfrac{5}{s}, s > 0.$

5. Obtain $L\{t^3 - t^2 + 4t\}$.   ANS. $\dfrac{6}{s^4} - \dfrac{2}{s^3} + \dfrac{4}{s^2}, s > 0.$

6. Evaluate $L\{e^{-2t} + 4 e^{-3t}\}$.   ANS. $\dfrac{5s + 11}{(s + 2)(s + 3)}, s > -2.$

7. Evaluate $L\{3 e^{4t} - e^{-2t}\}$.   ANS. $\dfrac{2s + 10}{(s - 4)(s + 2)}, s > 4.$

8. Show that $L\{\cosh kt\} = \dfrac{s}{s^2 - k^2}$; for $s > |k|.$

9. Show that $L\{\sinh kt\} = \dfrac{k}{s^2 - k^2}$; for $s > |k|.$

10. Use the trigonometric identity $\cos^2 A = \frac{1}{2}(1 + \cos 2A)$ and equation (5), Section 58, to evaluate $L\{\cos^2 kt\}$.   ANS. $\dfrac{s^2 + 2k^2}{s(s^2 + 4k^2)}, s > 0.$

11. Parallel the method suggested in exercise 8 to obtain $L\{\sin^2 kt\}$.

            ANS. $\dfrac{2k^2}{s(s^2 + 4k^2)}, s > 0.$

12. Obtain $L\{\sin^2 kt\}$ directly from the answer to exercise 10.

13. Evaluate $L\{\sin kt \cos kt\}$ with the aid of a trigonometric identity.

ANS. $\dfrac{k}{s^2 + 4k^2}, s > 0.$

14. Evaluate $L\{e^{-at} - e^{-bt}\}.$

ANS. $\dfrac{b - a}{(s + a)(s + b)}, s > \max(-a, -b).$

15. Find $L\{\psi(t)\}$ where

$$\psi(t) = 4, \qquad 0 < t < 1,$$
$$= 3, \qquad t > 1.$$

ANS. $\dfrac{1}{s}(4 - e^{-s}), s > 0.$

16. Find $L\{\phi(t)\}$ where

$$\phi(t) = 1, \qquad 0 < t < 2,$$
$$= t, \qquad t > 2.$$

ANS. $\dfrac{1}{s} + \dfrac{e^{-2s}}{s} + \dfrac{e^{-2s}}{s^2}, s > 0.$

17. Find $L\{A(t)\}$ where

$$A(t) = 0, \qquad 0 < t < 1,$$
$$= t, \qquad 1 < t < 2,$$
$$= 0, \qquad t > 2.$$

ANS. $\left(\dfrac{1}{s^2} + \dfrac{1}{s}\right) e^{-s} - \left(\dfrac{1}{s^2} + \dfrac{2}{s}\right) e^{-2s}, s > 0.$

18. Find $L\{B(t)\}$ where

$$B(t) = \sin 2t, \qquad 0 < t < \pi,$$
$$= 0, \qquad t > \pi.$$

ANS. $\dfrac{2(1 - e^{-\pi s})}{s^2 + 4}, s > 0.$

59. Sectionally continuous functions

It should be apparent that, if we are to find problems for which the Laplace transform method is useful, we must learn a good deal more about the transforms of more complicated functions than those we considered in the previous sections.

Our approach will be to prove a number of useful properties of the Laplace transform and then consider initial value problems in which we can make use of those properties.

In Section 58 we began this study by actually determining the transforms of some simple functions. However, it soon becomes tiresome to test each $F(t)$ we encounter to determine whether the integral

$$\int_0^\infty e^{-st}F(t)\,dt \tag{1}$$

exists for some range of values of s. We therefore seek a fairly large class of functions for which we can prove once and for all that the integral (1) exists.

One of our avowed interests in the Laplace transform is in its usefulness as a tool in solving problems in more or less elementary applications, particularly initial value problems in differential equations. Therefore we do not hesitate to restrict our study to functions $F(t)$ that are continuous or even differentiable, except possibly at a discrete set of points, in the semi-infinite range $t \geq 0$.

For such functions, the existence of the integral (1) can be endangered only at points of discontinuity of $F(t)$ or by divergence due to behavior of the integrand as $t \to \infty$.

In elementary calculus we found that finite discontinuities, or finite jumps, of the integrand did not interfere with the existence of the integral. We therefore introduce a term to describe functions that are continuous except for such jumps.

DEFINITION: *The function $F(t)$ is said to be sectionally continuous over the closed interval $a \leq t \leq b$ if that interval can be divided into a finite number of subintervals $c \leq t \leq d$ such that in each subinterval:*

(a) *$F(t)$ is continuous in the open interval $c < t < d$,*
(b) *$F(t)$ approaches a limit as t approaches each endpoint from within the interval; that is, $\lim\limits_{t \to c^+} F(t)$ and $\lim\limits_{t \to d^-} F(t)$ exist.*

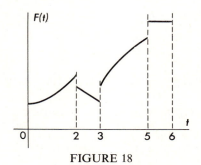

FIGURE 18

Figure 18 shows the graph of a function $F(t)$ that is sectionally continuous over the interval

$$0 \leq t \leq 6.$$

The student should realize that there is no implication that $F(t)$ must be sectionally continuous for $L\{F(t)\}$ to exist. Indeed, we shall meet several counterexamples to any such notion. The concept of sectionally continuous functions will, in Section 61, play a role in a set of conditions sufficient for the existence of the transform.

60. Functions of exponential order

If the integral of $e^{-st}F(t)$ between the limits 0 and t_0 exists for every finite positive t_0, the only remaining threat to the existence of the transform

$$\int_0^\infty e^{-st}F(t) \, dt \tag{1}$$

is the behavior of the integrand as $t \to \infty$.

We know that

$$\int_0^\infty e^{-ct} \, dt \tag{2}$$

converges for $c > 0$. This arouses our interest in functions $F(t)$ that are, for large t ($t \geq t_0$), essentially bounded by some exponential e^{bt} so that the integrand in (1) will behave like the integrand in (2) for s large enough.

DEFINITION: *The function $F(t)$ is said to be of exponential order as $t \to \infty$ if constants M and b and a fixed t-value t_0 exist such that*

$$|F(t)| < M \, e^{bt}, \qquad for \ t \geq t_0. \tag{3}$$

If b is to be emphasized, we say that $F(t)$ is of the order of e^{bt} as $t \to \infty$. We also write

$$F(t) = O(e^{bt}), \qquad t \to \infty, \tag{4}$$

to mean that $F(t)$ is of exponential order, the exponential being e^{bt}, as $t \to \infty$. That is, (4) is another way of expressing (3).

The integral in (1) may be split into parts as follows:

$$\int_0^\infty e^{-st}F(t) \, dt = \int_0^{t_0} e^{-st}F(t) \, dt + \int_{t_0}^\infty e^{-st}F(t) \, dt. \tag{5}$$

If $F(t)$ is of exponential order, $F(t) = O(e^{bt})$, the last integral in equation (5) exists because from the inequality (3) it follows that for $s > b$,

$$\int_{t_0}^{\infty} |e^{-st}F(t)| \, dt < M \int_{t_0}^{\infty} e^{-st} \cdot e^{bt} \, dt = \frac{M \exp[-t_0(s-b)]}{s-b}. \tag{6}$$

For $s > b$, the last member of (6) approaches zero as $t_0 \to \infty$. Therefore the last integral in (5) is absolutely convergent* for $s > b$. We have proved the following result.

THEOREM 8:   *If the integral of $e^{-st}F(t)$ between the limits 0 and t_0 exists for every finite positive t_0, and if $F(t)$ is of exponential order, $F(t) = O(e^{bt})$ as $t \to \infty$, the Laplace transform*

$$L\{F(t)\} = \int_0^{\infty} e^{-st}F(t) \, dt = f(s) \tag{7}$$

exists for $s > b$.

We know that a function that is sectionally continuous over an interval is integrable over that interval. This leads us to the following useful special case of Theorem 8.

THEOREM 9:   *If $F(t)$ is sectionally continuous over every finite interval in the range $t \geqq 0$, and if $F(t)$ is of exponential order, $F(t) = O(e^{bt})$ as $t \to \infty$, the Laplace transform $L\{F(t)\}$ exists for $s > b$.*

Functions of exponential order play a dominant role throughout our work. It is therefore wise to develop proficiency in determining whether or not a specified function is of exponential order.

Surely if a constant b exists such that

$$\lim_{t \to \infty} [e^{-bt}|F(t)|] \tag{8}$$

exists, the function $F(t)$ is of exponential order, indeed of the order of e^{bt}. To see this, let the value of the limit (8) be $K \neq 0$. Then, for t large enough, $|e^{-bt}F(t)|$ can be made as close to K as is desired, so certainly

$$|e^{-bt}F(t)| < 2K.$$

Therefore, for t sufficiently large,

$$|F(t)| < M e^{bt}, \tag{9}$$

with $M = 2K$. If the limit in (8) is zero, we may write (9) with $M = 1$.

---

* If complex s is to be used, the integral converges for $\mathrm{Re}(s) > b$.

On the other hand if, for every fixed c,

$$\lim_{t \to \infty} [e^{-ct}|F(t)|] = \infty, \tag{10}$$

the function $F(t)$ is not of exponential order. For, assume that b exists such that

$$|F(t)| < M e^{bt}, \qquad t \geq t_0; \tag{11}$$

then the choice $c = 2b$ would yield, by (11),

$$|e^{-2bt}F(t)| < M e^{-bt},$$

so $e^{-2bt}F(t) \to 0$ as $t \to \infty$, which disagrees with (10).

EXAMPLE (a):   Show that t^3 is of exponential order as $t \to \infty$.
  We consider, with b as yet unspecified,

$$\lim_{t \to \infty} (e^{-bt}t^3) = \lim_{t \to \infty} \frac{t^3}{e^{bt}}. \tag{12}$$

If $b > 0$, the limit in (12) is of a type treated in calculus. In fact,

$$\lim_{t \to \infty} \frac{t^3}{e^{bt}} = \lim_{t \to \infty} \frac{3t^2}{b\, e^{bt}} = \lim_{t \to \infty} \frac{6t}{b^2\, e^{bt}} = \lim_{t \to \infty} \frac{6}{b^3\, e^{bt}} = 0.$$

Therefore t^3 is of exponential order,

$$t^3 = O(e^{bt}), \qquad t \to \infty,$$

for any fixed positive b.

EXAMPLE (b):   Show that $\exp(t^2)$ is not of exponential order as $t \to \infty$.
  Consider

$$\lim_{t \to \infty} \frac{\exp(t^2)}{\exp(bt)}. \tag{13}$$

If $b \leq 0$, the limit in (13) is infinite. If $b > 0$,

$$\lim_{t \to \infty} \frac{\exp(t^2)}{\exp(bt)} = \lim_{t \to \infty} \exp[t(t - b)] = \infty.$$

Thus, no matter what fixed b we use, the limit in (13) is infinite and $\exp(t^2)$ cannot be of exponential order.
  The exercises at the end of the next section give additional opportunities for practice in determining whether or not a function is of exponential order.

61. Functions of class A

For brevity we shall hereafter use the term "a function of class A" for any function that

(a) is sectionally continuous over every finite interval in the range $t \geqq 0$ and
(b) is of exponential order as $t \to \infty$.

We may then reword Theorem 9 as follows.

THEOREM 10:   *If $F(t)$ is a function of class A, $L\{F(t)\}$ exists.*

It is important to realize that Theorem 10 states only that, for $L\{F(t)\}$ to exist, it is sufficient that $F(t)$ be of class A. The condition is not necessary. A classic example showing that functions other than those of class A do have Laplace transforms is

$$F(t) = t^{-1/2}.$$

This function is not sectionally continuous in every finite interval in the range $t \geqq 0$, because $F(t) \to \infty$ as $t \to 0^+$. But $t^{-1/2}$ is integrable from 0 to any positive t_0. Also $t^{-1/2} \to 0$ as $t \to \infty$, so $t^{-1/2}$ is of exponential order, with $M = 1$ and $b = 0$ in the inequality (3), page 178. Hence, by Theorem 8, page 179, $L\{t^{-1/2}\}$ exists.

Indeed, for $s > 0$,

$$L\{t^{-1/2}\} = \int_0^\infty e^{-st} t^{-1/2} \, dt,$$

in which the change of variable $st = y^2$ leads to

$$L\{t^{-1/2}\} = 2s^{-1/2} \int_0^\infty \exp(-y^2) \, dy, \qquad s > 0.$$

In elementary calculus we found that $\int_0^\infty \exp(-y^2) \, dy = \frac{1}{2}\sqrt{\pi}$.

Therefore

$$L\{t^{-1/2}\} = 2s^{-1/2} \cdot \tfrac{1}{2}\sqrt{\pi}$$

$$= \left(\frac{\pi}{s}\right)^{1/2}, \qquad s > 0, \tag{1}$$

even though $t^{-1/2} \to \infty$ as $t \to 0^+$. Additional examples are easily constructed and we shall meet some of them later in the book.

If $F(t)$ is of class A, $F(t)$ is bounded over the range $0 \leq t \leq t_0$,

$$|F(t)| < M_1, \qquad 0 \leq t \leq t_0. \tag{2}$$

But $F(t)$ is also of exponential order,

$$|F(t)| < M_2 e^{bt}, \qquad t \geq t_0. \tag{3}$$

If we choose M as the larger of M_1 and M_2 and c as the larger of b and zero, we may write

$$|F(t)| < M e^{ct}, \qquad t \geq 0. \tag{4}$$

Therefore, for any function $F(t)$ of class A,

$$\left| \int_0^\infty e^{-st} F(t)\, dt \right| < M \int_0^\infty e^{-st} \cdot e^{ct}\, dt = \frac{M}{s - c}, \qquad s > c. \tag{5}$$

Since the right member of (5) approaches zero as $s \to \infty$, we have proved the following useful result.

THEOREM 11:   *If $F(t)$ is of class A and if $L\{F(t)\} = f(s)$,*

$$\lim_{s \to \infty} f(s) = 0.$$

From (5) we may also conclude the stronger result that the transform $f(s)$ of a function $F(t)$ of class A must be such that $sf(s)$ is bounded as $s \to \infty$.

Exercises

1. Prove that if $F_1(t)$ and $F_2(t)$ are each of exponential order as $t \to \infty$, then $F_1(t) \cdot F_2(t)$ and $F_1(t) + F_2(t)$ are also of exponential order as $t \to \infty$.
2. Prove that if $F_1(t)$ and $F_2(t)$ are of class A (see page 181), then $F_1(t) + F_2(t)$ and $F_1(t) \cdot F_2(t)$ are also of class A.
3. Show that t^x is of exponential order as $t \to \infty$ for all real x.

In exercises 4 through 17, show that the given function is of class A. In these exercises, n denotes a nonnegative integer, k any real number.

4. $\sin kt$.   5. $\cos kt$.
6. $\cosh kt$.   7. $\sinh kt$.
8. t^n.   9. $t^n e^{kt}$.
10. $t^n \sin kt$.   11. $t^n \cos kt$.
12. $t^n \sinh kt$.   13. $t^n \cosh kt$.

14. $\dfrac{\sin kt}{t}$.   15. $\dfrac{1 - \exp(-t)}{t}$.

16. $\dfrac{1 - \cos kt}{t}$.   17. $\dfrac{\cos t - \cosh t}{t}$.

62. Transforms of derivatives

Any function of class A (see page 181) has a Laplace transform, but the derivative of such a function may or may not be of class A. For the function

$$F_1(t) = \sin[\exp(t)]$$

with derivative

$$F_1'(t) = \exp(t)\cos[\exp(t)],$$

both F_1 and F_1' are of exponential order as $t \to \infty$. Here F_1 is bounded so it is of the order of $\exp(0 \cdot t)$; F_1' is of the order of $\exp(t)$. On the other hand, the function

$$F_2(t) = \sin[\exp(t^2)]$$

with derivative

$$F_2'(t) = 2t\exp(t^2)\cos[\exp(t^2)]$$

is such that F_2 is of the order of $\exp(0 \cdot t)$, but F_2' is not of exponential order. From Example (b), page 180,

$$\lim_{t \to \infty} \frac{\exp(t^2)}{\exp(bt)} = \infty$$

for any real b. Since the factors $2t\cos[\exp(t^2)]$ do not even approach zero as $t \to \infty$, the product $F_2'\exp(-ct)$ cannot be bounded as $t \to \infty$ no matter how large a fixed c is chosen.

Therefore, in studying the transforms of derivatives, we shall stipulate that the derivatives themselves be of class A.

If $F(t)$ is continuous for $t \geq 0$ and of exponential order as $t \to \infty$, and if $F'(t)$ is of class A, the integral in

$$L\{F'(t)\} = \int_0^\infty e^{-st}F'(t)\,dt \tag{1}$$

may be simplified by integration by parts with the choice exhibited in the table.

e^{-st}	$F'(t)\,dt$
$-s\,e^{-st}\,dt$	$F(t)$

We thus obtain, for s greater than some fixed s_0,

$$\int_0^\infty e^{-st}F'(t)\,dt = \left[e^{-st}F(t)\right]_0^\infty + s\int_0^\infty e^{-st}F(t)\,dt,$$

or

$$L\{F'(t)\} = -F(0) + sL\{F(t)\}.\tag{2}$$

THEOREM 12: *If $F(t)$ is continuous for $t \geqq 0$ and of exponential order as $t \to \infty$, and if $F'(t)$ is of class A (see page 181), it follows from $L\{F(t)\} = f(s)$ that*

$$L\{F'(t)\} = sf(s) - F(0).\tag{3}$$

In treating a differential equation of order n, we seek solutions for which the highest-ordered derivative present is reasonably well behaved, say sectionally continuous. The integral of a sectionally continuous function is continuous. Hence, we lose nothing by requiring continuity for all derivatives of order lower than n. The requirement that the various derivatives be of exponential order is forced upon us by our desire to use the Laplace transform as a tool. For our purposes, iteration of Theorem 12 to obtain transforms of higher derivatives makes sense,

From (3) we obtain, if F, F', F'' are suitably restricted,

$$L\{F''(t)\} = sL\{F'(t)\} - F'(0),$$

or

$$L\{F''(t)\} = s^2 f(s) - sF(0) - F'(0),\tag{4}$$

and the process can be repeated as many times as we wish.

THEOREM 13: *If $F(t)$, $F'(t), \ldots, F^{(n-1)}(t)$ are continuous for $t \geqq 0$ and of exponential order as $t \to \infty$, and if $F^{(n)}(t)$ is of class A, then from*

$$L\{F(t)\} = f(s)$$

it follows that

$$L\{F^{(n)}(t)\} = s^n f(s) - \sum_{k=0}^{n-1} s^{n-1-k} F^{(k)}(0).\tag{5}$$

Thus

$$L\{F^{(3)}(t)\} = s^3 f(s) - s^2 F(0) - sF'(0) - F''(0),$$

$$L\{F^{(4)}(t)\} = s^4 f(s) - s^3 F(0) - s^2 F'(0) - sF''(0) - F^{(3)}(0), \text{ etc.}$$

Theorem 13 is basic in employing the Laplace transform to solve linear differential equations with constant coefficients. The theorem permits us to transform such differential equations into algebraic ones.

The restriction that $F(t)$ be continuous can be relaxed, but discontinuities in $F(t)$ bring in additional terms in the transform of $F'(t)$. As an example, consider an $F(t)$ that is continuous for $t \geqq 0$ except for a finite jump at

$t = t_1$, as in Figure 19. If $F(t)$ is also of exponential order as $t \to \infty$, and if $F'(t)$ is of class A, we may write

$$L\{F'(t)\} = \int_0^\infty e^{-st} F'(t)\, dt$$

$$= \int_0^{t_1} e^{-st} F'(t)\, dt + \int_{t_1}^\infty e^{-st} F'(t)\, dt.$$

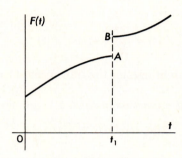

FIGURE 19

Then, integration by parts applied to the last two integrals yields

$$L\{F'(t)\} = \left[e^{-st} F(t) \right]_0^{t_1} + s \int_0^{t_1} e^{-st} F(t)\, dt + \left[e^{-st} F(t) \right]_{t_1}^\infty + s \int_{t_1}^\infty e^{-st} F(t)\, dt$$

$$= s \int_0^\infty e^{-st} F(t)\, dt + \exp(-st_1) F(t_1^-) - F(0) + 0 - \exp(-st_1) F(t_1^+)$$

$$= sL\{F(t)\} - F(0) - \exp(-st_1)[F(t_1^+) - F(t_1^-)].$$

In Figure 19 the directed distance AB is of length $[F(t_1^+) - F(t_1^-)]$.

THEOREM 14:   *If $F(t)$ is of exponential order as $t \to \infty$ and $F(t)$ is continuous for $t \geq 0$ except for a finite jump at $t = t_1$, and if $F'(t)$ is of class A, then from*

$$L\{F(t)\} = f(s)$$

it follows that

$$L\{F'(t)\} = sf(s) - F(0) - \exp(-st_1)[F(t_1^+) - F(t_1^-)]. \tag{6}$$

If $F(t)$ has more than one finite discontinuity, additional terms, similar to the last term in (6), enter the formula for $L\{F'(t)\}$.

63. Derivatives of transforms

For functions of class A, the theorems of advanced calculus show that it is legitimate to differentiate the Laplace transform integral. That is, if $F(t)$ is of class A, from

$$f(s) = \int_0^\infty e^{-st} F(t)\, dt \tag{1}$$

it follows that

$$f'(s) = \int_0^\infty (-t) e^{-st} F(t)\, dt. \tag{2}$$

The integral on the right in (2) is the transform of the function $(-t)F(t)$.

THEOREM 15:   *If $F(t)$ is a function of class A, it follows from*

$$L\{F(t)\} = f(s)$$

that

$$f'(s) = L\{-tF(t)\}. \tag{3}$$

When $F(t)$ is of class A, $(-t)^k F(t)$ is also of class A for any positive integer k.

THEOREM 16:   *If $F(t)$ is of class A, it follows from $L\{F(t)\} = f(s)$ that for any positive integer n*

$$\frac{d^n}{ds^n} f(s) = L\{(-t)^n F(t)\}. \tag{4}$$

These theorems are useful in several ways. One immediate application is to add to our list of transforms with very little labor. We know that

$$\frac{k}{s^2 + k^2} = L\{\sin kt\}, \tag{5}$$

and therefore, by Theorem 15,

$$\frac{-2ks}{(s^2 + k^2)^2} = L\{-t \sin kt\}.$$

Thus we obtain

$$\frac{s}{(s^2 + k^2)^2} = L\left\{\frac{t}{2k} \sin kt\right\}. \tag{6}$$

From the known formula

$$\frac{s}{s^2 + k^2} = L\{\cos kt\}$$

we obtain, by differentiation with respect to s,

$$\frac{k^2 - s^2}{(s^2 + k^2)^2} = L\{-t \cos kt\}. \tag{7}$$

Let us add to each side of (7) the corresponding member of

$$\frac{1}{s^2 + k^2} = L\left\{\frac{1}{k} \sin kt\right\}$$

to get

$$\frac{s^2 + k^2 + k^2 - s^2}{(s^2 + k^2)^2} = L\left\{\frac{1}{k} \sin kt - t \cos kt\right\},$$

from which it follows that

$$\frac{1}{(s^2 + k^2)^2} = L\left\{\frac{1}{2k^3}(\sin kt - kt \cos kt)\right\}. \tag{8}$$

64. The gamma function

For obtaining the Laplace transform of nonintegral powers of t, we need a function not usually discussed in elementary mathematics.

The gamma function $\Gamma(x)$ is defined by

$$\Gamma(x) = \int_0^\infty e^{-\beta}\beta^{x-1} \, d\beta, \qquad x > 0. \tag{1}$$

Substitution of $(x + 1)$ for x in (1) gives

$$\Gamma(x + 1) = \int_0^\infty e^{-\beta}\beta^x \, d\beta. \tag{2}$$

An integration by parts, integrating $e^{-\beta} \, d\beta$ and differentiating β^x, yields

$$\Gamma(x + 1) = \left[-e^{-\beta}\beta^x\right]_0^\infty + x \int_0^\infty e^{-\beta}\beta^{x-1} \, d\beta. \tag{3}$$

Because $x > 0$, $\beta^x \to 0$ as $\beta \to 0$, and, because x is fixed, $e^{-\beta}\beta^x \to 0$ as $\beta \to \infty$. Thus

$$\Gamma(x + 1) = x \int_0^\infty e^{-\beta}\beta^{x-1} \, d\beta = x\Gamma(x). \tag{4}$$

THEOREM 17:　　*For $x > 0$, $\Gamma(x + 1) = x\Gamma(x)$.*

Suppose that n is a positive integer. Iteration of Theorem 17 gives us

$$\Gamma(n + 1) = n\Gamma(n)$$
$$= n(n - 1)\Gamma(n - 1)$$
$$\vdots$$
$$= n(n - 1)(n - 2)\cdots 2 \cdot 1 \cdot \Gamma(1)$$
$$= n!\,\Gamma(1).$$

But, by definition,

$$\Gamma(1) = \int_0^\infty e^{-\beta}\beta^0\,d\beta = \left[-e^{-\beta}\right]_0^\infty = 1.$$

THEOREM 18:   *For positive integral n, $\Gamma(n + 1) = n!$.*

In the integral for $\Gamma(x + 1)$ in (2), let us put $\beta = st$ with $s > 0$ and t as the new variable of integration. This yields, since $t \to 0$ as $\beta \to 0$ and $t \to \infty$ as $\beta \to \infty$,

$$\Gamma(x + 1) = \int_0^\infty e^{-st}s^x t^x s\,dt = s^{x+1}\int_0^\infty e^{-st}t^x\,dt, \tag{5}$$

which is valid for $x + 1 > 0$. We thus obtain

$$\frac{\Gamma(x + 1)}{s^{x+1}} = \int_0^\infty e^{-st}t^x\,dt, \qquad s > 0, x > -1,$$

which in our Laplace transform notation says that

$$L\{t^x\} = \frac{\Gamma(x + 1)}{s^{x+1}}, \qquad s > 0, x > -1. \tag{6}$$

If in (6) we put $x = -\tfrac{1}{2}$, we get

$$L\{t^{-1/2}\} = \frac{\Gamma(\tfrac{1}{2})}{s^{1/2}}.$$

But we already know that $L\{t^{-1/2}\} = (\pi/s)^{1/2}$. Hence

$$\Gamma(\tfrac{1}{2}) = \sqrt{\pi}. \tag{7}$$

65. Periodic functions

Suppose that the function $F(t)$ is periodic with period ω:

$$F(t + \omega) = F(t). \tag{1}$$

The function is completely determined by (1) once the nature of $F(t)$ throughout one period, $0 \leq t < \omega$, is given. If $F(t)$ has a transform,

$$L\{F(t)\} = \int_0^\infty e^{-st} F(t)\, dt, \tag{2}$$

the integral can be written as a sum of integrals,

$$L\{F(t)\} = \sum_{n=0}^\infty \int_{n\omega}^{(n+1)\omega} e^{-st} F(t)\, dt. \tag{3}$$

Let us put $t = n\omega + \beta$. Then (3) becomes

$$L\{F(t)\} = \sum_{n=0}^\infty \int_0^\omega \exp(-sn\omega - s\beta) F(\beta + n\omega)\, d\beta.$$

But $F(\beta + n\omega) = F(\beta)$, by iteration of (1). Hence

$$L\{F(t)\} = \sum_{n=0}^\infty \exp(-sn\omega) \int_0^\omega \exp(-s\beta) F(\beta)\, d\beta. \tag{4}$$

The integral on the right in (4) is independent of n and we can sum the series on the right;

$$\sum_{n=0}^\infty \exp(-sn\omega) = \sum_{n=0}^\infty [\exp(-s\omega)]^n = \frac{1}{1 - e^{-s\omega}}.$$

THEOREM 19: *If $F(t)$ has a Laplace transform and if $F(t + \omega) = F(t)$,*

$$L\{F(t)\} = \frac{\displaystyle\int_0^\omega e^{-s\beta} F(\beta)\, d\beta}{1 - e^{-s\omega}}. \tag{5}$$

Next suppose that a function $H(t)$ has a period $2c$ and that we demand that $H(t)$ be zero throughout the right half of each period. That is,

$$H(t + 2c) = H(t), \tag{6}$$

$$H(t) = g(t), \qquad 0 \leq t < c, \tag{7}$$

$$= 0, \qquad c \leq t < 2c.$$

Then we say that $H(t)$ is a half-wave rectification of $g(t)$. Using (5) we may conclude that for the $H(t)$ defined by (6) and (7),

$$L\{H(t)\} = \frac{\displaystyle\int_0^c \exp(-s\beta) g(\beta)\, d\beta}{1 - \exp(-2cs)}. \tag{8}$$

EXAMPLE (a): Find the transform of the function $\psi(t, c)$ shown in Figure 20 and defined by

$$\psi(t, c) = 1, \qquad 0 < t < c, \tag{9}$$

$$= 0, \qquad c < t < 2c;$$

$$\psi(t + 2c, c) = \psi(t, c). \tag{10}$$

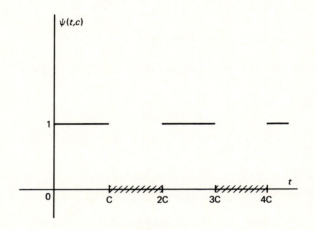

FIGURE 20

We may use equation (8) and the fact that

$$\int_0^c \exp(-s\beta)\, d\beta = \frac{1 - \exp(-sc)}{s}$$

to conclude that

$$L\{\psi(t, c)\} = \frac{1}{s} \cdot \frac{1 - \exp(-sc)}{1 - \exp(-2sc)} = \frac{1}{s} \cdot \frac{1}{1 + \exp(-sc)}. \tag{11}$$

EXAMPLE (b): Find the transform of the square-wave function $Q(t, c)$ shown in Figure 21 and defined by

$$Q(t, c) = 1, \qquad 0 < t < c, \tag{12}$$

$$= -1, \qquad c < t < 2c;$$

$$Q(t + 2c, c) = Q(t, c). \tag{13}$$

This transform can be obtained by using Theorem 19, but also

$$Q(t, c) = 2\psi(t, c) - 1; \tag{14}$$

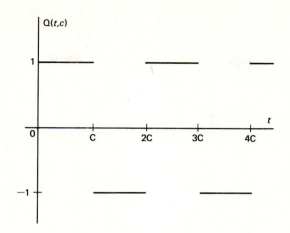

FIGURE 21

hence, from (11),

$$L\{Q(t, c)\} = \frac{1}{s}\left[\frac{2}{1 + \exp(-sc)} - 1\right] = \frac{1}{s} \cdot \frac{1 - \exp(-sc)}{1 + \exp(-sc)}. \tag{15}$$

By multiplying numerator and denominator of the last fraction above by $\exp(\frac{1}{2}sc)$, we may put (15) in the form

$$L\{Q(t, c)\} = \frac{1}{s}\tanh\frac{cs}{2}. \tag{16}$$

Exercises

1. Show that $L\{t^{1/2}\} = \dfrac{1}{2s}\left(\dfrac{\pi}{s}\right)^{1/2}, \quad s > 0.$

2. Show that $L\{t^{5/2}\} = \dfrac{15}{8s^3}\left(\dfrac{\pi}{s}\right)^{1/2}, \quad s > 0.$

3. Use equation (4), page 184, to derive $L\{\sin kt\}$.
4. Use equation (4), page 184, to derive $L\{\cos kt\}$.
5. Check the known transforms of $\sin kt$ and $\cos kt$ against one another by using Theorem 12, page 184.
6. If n is a positive integer, obtain $L\{t^n e^{kt}\}$ from the known $L\{e^{kt}\}$ by using Theorem 16, page 186.   ANS. $\dfrac{n!}{(s - k)^{n+1}}, \quad s > k.$

7. Find $L\{t^2 \sin kt\}$.   ANS. $\dfrac{2k(3s^2 - k^2)}{(s^2 + k^2)^3}, \quad s > 0.$

8. Find $L\{t^2 \cos kt\}$.
ANS. $\dfrac{2s(s^2 - 3k^2)}{(s^2 + k^2)^3}$, $s > 0$.

9. For the function

$$F(t) = t + 1, \qquad 0 \leq t \leq 2,$$
$$= 3, \qquad\qquad t > 2,$$

graph $F(t)$ and $F'(t)$. Find $L\{F(t)\}$. Find $L\{F'(t)\}$ in two ways.
ANS. $L\{F'(t)\} = s^{-1}(1 - e^{-2s})$, $s > 0$.

10. For the function

$$H(t) = t + 1, \qquad 0 \leq t \leq 2,$$
$$= 6, \qquad\qquad t > 2,$$

parallel exercise 9 above.

11. Define a triangular-wave function $T(t, c)$ by

$$T(t, c) = t, \qquad\qquad 0 \leq t \leq c,$$
$$= 2c - t, \qquad c < t < 2c;$$
$$T(t + 2c, c) = T(t, c).$$

Sketch $T(t, c)$ and find its Laplace transform.
ANS. $\dfrac{1}{s^2} \tanh \dfrac{cs}{2}$.

12. Show that the derivative of the function $T(t, c)$ of exercise 11 is, except at certain points, the function $Q(t, c)$ of Example (b), Section 65. Obtain $L\{T(t, c)\}$ from $L\{Q(t, c)\}$.

13. Find $L\{|\sin kt|\}$.
ANS. $\dfrac{k}{s^2 + k^2} \cdot \coth \pi \dfrac{s}{2k}$.

14. Find $L\{|\cos kt|\}$.

15. Define the function $G(t)$ by

$$G(t) = e^t, \qquad\qquad 0 \leq t < c,$$
$$G(t + c) = G(t), \qquad t \geq 0.$$

Sketch the graph of $G(t)$ and find its Laplace transform.

ANS. $\dfrac{-1}{s - 1} \cdot \dfrac{1 - \exp c(1 - s)}{1 - \exp(-cs)}$, $s > 1$.

16. Define the function $S(t)$ by

$$S(t) = 1 - t, \qquad 0 \leq t < 1,$$
$$S(t + 1) = S(t), \qquad t \geq 0.$$

Sketch the graph of $S(t)$ and find its Laplace transform.

17. Sketch a half-wave rectification of the function sin ωt, as described below, and find its transform.

$$F(t) = \sin \omega t, \qquad 0 \leq t \leq \frac{\pi}{\omega},$$

$$= 0, \qquad \frac{\pi}{\omega} < t < \frac{2\pi}{\omega};$$

$$F\left(t + \frac{2\pi}{\omega}\right) = F(t).$$

ANS. $\dfrac{\omega}{s^2 + \omega^2} \cdot \dfrac{1}{1 - \exp\left(-s\pi/\omega\right)}.$

18. Find $L\{F(t)\}$ where $F(t) = t$ for $0 < t < \omega$ and $F(t + \omega) = F(t)$.

ANS. $\dfrac{1}{s^2} - \dfrac{\omega}{s}\dfrac{\exp\left(-s\omega\right)}{1 - \exp\left(-s\omega\right)} = \dfrac{1}{s^2} + \dfrac{\omega}{2s}\left(1 - \coth\dfrac{\omega s}{2}\right).$

19. Prove that if $L\{F(t)\} = f(s)$ and if $F(t)/t$ is of class A,

$$L\{F(t)/t\} = \int_s^\infty f(\beta)\, d\beta.$$

Hint: Use Theorem 15, page 186.

Inverse Transforms

66. Definition of an inverse transform

Suppose that the function $F(t)$ is to be determined from a differential equation with initial conditions. The Laplace operator L is used to transform the original problem into a new problem from which the transform $f(s)$ is to be found. If the Laplace transformation is to be effective, the new problem must be simpler than the original problem. We first find $f(s)$ and then must obtain $F(t)$ from $f(s)$. It is therefore desirable to develop methods for finding the object function $F(t)$ when its transform $f(s)$ is known.

If

$$L\{F(t)\} = f(s), \tag{1}$$

we say that $F(t)$ is an *inverse Laplace transform*, or an inverse transform, of $f(s)$ and we write

$$F(t) = L^{-1}\{f(s)\}. \tag{2}$$

Since (1) means that

$$\int_0^\infty e^{-st}F(t)\,dt = f(s), \tag{3}$$

it follows at once that an inverse transform is not unique. For example, if $F_1(t)$ and $F_2(t)$ are identical except at a discrete set of points and differ at these points, the value of the integral in (3) is the same for the two functions; their transforms are identical.

Let us employ the term *null function* for any function $N(t)$ for which

$$\int_0^{t_0} N(t)\, dt = 0 \tag{4}$$

for every positive t_0. Lerch's theorem (not proved here) states that if $L\{F_1(t)\} = L\{F_2(t)\}$, then $F_1(t) - F_2(t) = N(t)$. That is, an inverse Laplace transform is unique except for the addition of an arbitrary null function.

The only continuous null function is the zero function. If an $f(s)$ has a continuous inverse $F(t)$, then $F(t)$ is the only continuous inverse of $f(s)$. If $f(s)$ has an inverse $F_1(t)$ continuous over a specified closed interval, every inverse that is also continuous over that interval is identical with $F_1(t)$ on that interval. Essentially, inverses of the same $f(s)$ differ at most at their points of discontinuity.

In applications, failure of uniqueness caused by addition of a null function is not vital, because the effect of that null function on physical properties of the solution is null. In the problems we treat, the inverse $F(t)$ is required either to be continuous for $t \geq 0$ or to be sectionally continuous with the values of $F(t)$ at the points of discontinuity specified by each problem. The $F(t)$ is then unique.

A crude, but sometimes effective, method for finding inverse Laplace transforms is to construct a table of transforms (page 231) and then to use it in reverse to find inverse transforms.

We know from exercise 1, page 175, that

$$L\{\cos kt\} = \frac{s}{s^2 + k^2}. \tag{5}$$

Therefore

$$L^{-1}\left\{\frac{s}{s^2 + k^2}\right\} = \cos kt. \tag{6}$$

We shall refine the above method, and actually make it quite powerful, by developing theorems by which a given $f(s)$ may be expanded into component parts whose inverses are known (found in the table). Other theorems will permit us to write $f(s)$ in alternative forms that yield the desired inverse. The most fundamental of such theorems is one that states that the inverse transformation is a linear operation.

THEOREM 20: *If c_1 and c_2 are constants,*

$$L^{-1}\{c_1 f_1(s) + c_2 f_2(s)\} = c_1 L^{-1}\{f_1(s)\} + c_2 L^{-1}\{f_2(s)\}.$$

Next let us prove a simple, but extremely useful, theorem on the manipulation of inverse transforms. From

$$f(s) = \int_0^\infty e^{-st} F(t) \, dt, \tag{7}$$

we obtain

$$f(s - a) = \int_0^\infty e^{-(s-a)t} F(t) \, dt$$

$$= \int_0^\infty e^{-st} [e^{at} F(t)] \, dt.$$

Thus, from $L^{-1}\{f(s)\} = F(t)$ it follows that

$$L^{-1}\{f(s - a)\} = e^{at} F(t),$$

or

$$L^{-1}\{f(s - a)\} = e^{at} L^{-1}\{f(s)\}. \tag{8}$$

Equation (8) may be rewritten with the exponential transferred to the other side of the equation. We thus obtain the following result.

THEOREM 21: $L^{-1}\{f(s)\} = e^{-at} L^{-1}\{f(s - a)\}.$

EXAMPLE (a):   Find $L^{-1}\left\{\dfrac{15}{s^2 + 4s + 13}\right\}.$

First complete the square in the denominator.

$$L^{-1}\left\{\frac{15}{s^2 + 4s + 13}\right\} = L^{-1}\left\{\frac{15}{(s + 2)^2 + 9}\right\}.$$

Since we know that $L^{-1}\left\{\dfrac{k}{s^2 + k^2}\right\} = \sin kt$, we proceed as follows:

$$L^{-1}\left\{\frac{15}{s^2 + 4s + 13}\right\} = 5L^{-1}\left\{\frac{3}{(s + 2)^2 + 9}\right\} = 5 e^{-2t} L^{-1}\left\{\frac{3}{s^2 + 9}\right\}$$

$$= 5 e^{-2t} \sin 3t,$$

in which we have used Theorem 21.

EXAMPLE (b):   Evaluate $L^{-1}\left\{\dfrac{s + 1}{s^2 + 6s + 25}\right\}.$

We write

$$L^{-1}\left\{\frac{s + 1}{s^2 + 6s + 25}\right\} = L^{-1}\left\{\frac{s + 1}{(s + 3)^2 + 16}\right\}.$$

Then

$$L^{-1}\left\{\frac{s+1}{s^2+6s+25}\right\} = e^{-3t}L^{-1}\left\{\frac{s-2}{s^2+16}\right\}$$

$$= e^{-3t}\left[L^{-1}\left\{\frac{s}{s^2+16}\right\} - \tfrac{1}{2}L^{-1}\left\{\frac{4}{s^2+16}\right\}\right]$$

$$= e^{-3t}(\cos 4t - \tfrac{1}{2}\sin 4t).$$

Exercises

In exercises 1 through 10 obtain $L^{-1}\{f(s)\}$ from the given $f(s)$.

1. $\dfrac{1}{s^2+2s+10}$.   ANS. $\tfrac{1}{3}e^{-t}\sin 3t$.

2. $\dfrac{1}{s^2-4s+8}$.   ANS. $\tfrac{1}{2}e^{2t}\sin 2t$.

3. $\dfrac{3s}{s^2+4s+13}$.   ANS. $e^{-2t}(3\cos 3t - 2\sin 3t)$.

✳4. $\dfrac{s}{s^2+6s+13}$.   ANS. $e^{-3t}(\cos 2t - \tfrac{3}{2}\sin 2t)$.

5. $\dfrac{1}{s^2+4s+4}$.   ANS. $t\,e^{-2t}$.

6. $\dfrac{s}{s^2+4s+4}$.   ANS. $e^{-2t}(1-2t)$.

7. $\dfrac{2s-3}{s^2-4s+8}$.   ANS. $e^{2t}(2\cos 2t + \tfrac{1}{2}\sin 2t)$.

8. $\dfrac{3s+1}{s^2+6s+13}$.   ANS. $e^{-3t}(3\cos 2t - 4\sin 2t)$.

✳9. $\dfrac{2s+3}{(s+4)^3}$.   ANS. $e^{-4t}(2t - \tfrac{5}{2}t^2)$.

10. $\dfrac{s^2}{(s-1)^4}$.   ANS. $e^t(t + t^2 + \tfrac{1}{6}t^3)$.

11. Show that for n a nonnegative integer

Read

$$L^{-1}\left\{\frac{1}{(s+a)^{n+1}}\right\} = \frac{t^n e^{-at}}{n!}.$$

12. Show that for $m > -1$,

Read Prove

$$L^{-1}\left\{\frac{1}{(s+a)^{m+1}}\right\} = \frac{t^m e^{-at}}{\Gamma(m+1)}.$$

13. Show that

$$L^{-1}\left\{\frac{1}{(s+a)^2+b^2}\right\}=\frac{1}{b}e^{-at}\sin bt.$$

14. Show that

$$L^{-1}\left\{\frac{s}{(s+a)^2+b^2}\right\}=\frac{1}{b}e^{-at}(b\cos bt-a\sin bt).$$

15. For $a > 0$, show that, from $L^{-1}\{f(s)\} = F(t)$, it follows that

$$L^{-1}\{f(as)\}=\frac{1}{a}F\left(\frac{t}{a}\right).$$

16. For $a > 0$, show that, from $L^{-1}\{f(s)\} = F(t)$, it follows that

$$L^{-1}\{f(as+b)\}=\frac{1}{a}\exp\left(-\frac{bt}{a}\right)F\left(\frac{t}{a}\right).$$

67. Partial fractions

In using the Laplace transform to solve differential equations, we often need to obtain the inverse transform of a rational fraction

$$\frac{N(s)}{D(s)}.\tag{1}$$

The numerator and denominator in (1) are polynomials in s and the degree of $D(s)$ is larger than the degree of $N(s)$. The fraction (1) has the partial fractions expansion used in calculus. Because of the linearity of the inverse operator L^{-1}, the partial fractions expansion of (1) permits us to replace a complicated problem in obtaining an inverse transform with a set of simpler problems.

EXAMPLE (a):   Obtain $L^{-1}\left\{\dfrac{s^2-6}{s^3+4s^2+3s}\right\}.$

Since the denominator is a product of distinct linear factors, we know that constants A, B, C exist such that

$$\frac{s^2-6}{s^3+4s^2+3s}=\frac{s^2-6}{s(s+1)(s+3)}=\frac{A}{s}+\frac{B}{s+1}+\frac{C}{s+3}.$$

Multiplying each term by the lowest common denominator, we obtain the identity

$$s^2-6=A(s+1)(s+3)+Bs(s+3)+Cs(s+1),\tag{2}$$

from which we need to determine A, B, and C. Using the values $s = 0$, -1, -3 successively in (2), we get

$$s = 0: \qquad -6 = A(1)(3),$$
$$s = -1: \qquad -5 = B(-1)(2),$$
$$s = -3: \qquad 3 = C(-3)(-2),$$

from which $A = -2$, $B = \frac{5}{2}$, $C = \frac{1}{2}$. Therefore

$$\frac{s^2 - 6}{s^3 + 4s^2 + 3s} = \frac{-2}{s} + \frac{\frac{5}{2}}{s + 1} + \frac{\frac{1}{2}}{s + 3}.$$

Since $L^{-1}\left\{\frac{1}{s}\right\} = 1$ and $L^{-1}\left\{\frac{1}{s + a}\right\} = e^{-at}$, we get the desired result,

$$L^{-1}\left\{\frac{s^2 - 6}{s^3 + 4s^2 + 3s}\right\} = -2 + \frac{5}{2}e^{-t} + \frac{1}{2}e^{-3t}.$$

EXAMPLE (b):   Obtain $L^{-1}\left\{\dfrac{5s^3 - 6s - 3}{s^3(s + 1)^2}\right\}$.

Since the denominator contains repeated linear factors, we must assume partial fractions of the form shown:

$$\frac{5s^3 - 6s - 3}{s^3(s + 1)^2} = \frac{A_1}{s} + \frac{A_2}{s^2} + \frac{A_3}{s^3} + \frac{B_1}{s + 1} + \frac{B_2}{(s + 1)^2}. \qquad (3)$$

Corresponding to a denominator factor $(x - \gamma)^r$, we must in general assume r partial fractions of the form

$$\frac{A_1}{x - \gamma} + \frac{A_2}{(x - \gamma)^2} + \cdots + \frac{A_r}{(x - \gamma)^r}.$$

From (3) we get

$$5s^3 - 6s - 3 = A_1 s^2(s + 1)^2 + A_2 s(s + 1)^2$$
$$+ A_3(s + 1)^2 + B_1 s^3(s + 1) + B_2 s^3, \qquad (4)$$

which must be an identity in s. To get the necessary five equations for the determination of A_1, A_2, A_3, B_1, B_2, two elementary methods are popular. Specific values of s can be used in (4), or the coefficients of like powers of s in the two members of (4) may be equated. We employ whatever combination of these methods yields simple equations to be solved for $A_1, A_2, \ldots, B_2$. From (4) we obtain

$$s = 0: \qquad -3 = A_3(1),$$
$$s = -1: \qquad -2 = B_2(-1),$$

coeff. of s^4: $0 = A_1 + B_1,$

coeff. of s^3: $5 = 2A_1 + A_2 + B_1 + B_2,$

coeff. of s: $-6 = A_2 + 2A_3.$

The above equations yield $A_1 = 3$, $A_2 = 0$, $A_3 = -3$, $B_1 = -3$, $B_2 = 2$. Therefore we find that

$$L^{-1}\left\{\frac{5s^3 - 6s - 3}{s^3(s+1)^2}\right\} = L^{-1}\left\{\frac{3}{s} - \frac{3}{s^3} - \frac{3}{s+1} + \frac{2}{(s+1)^2}\right\}$$

$$= 3 - \tfrac{3}{2}t^2 - 3e^{-t} + 2te^{-t}.$$

EXAMPLE (c):   Obtain $L^{-1}\left\{\dfrac{16}{s(s^2+4)^2}\right\}$.

Since quadratic factors require the corresponding partial fractions to have linear numerators, we start with an expansion of the form

$$\frac{16}{s(s^2+4)^2} = \frac{A}{s} + \frac{B_1 s + C_1}{s^2+4} + \frac{B_2 s + C_2}{(s^2+4)^2}.$$

From the identity

$$16 = A(s^2+4)^2 + (B_1 s + C_1)s(s^2+4) + (B_2 s + C_2)s,$$

it is not difficult to find the values $A = 1$, $B_1 = -1$, $B_2 = -4$, $C_1 = 0$, $C_2 = 0$. We thus obtain

$$L^{-1}\left\{\frac{16}{s(s^2+4)^2}\right\} = L^{-1}\left\{\frac{1}{s} - \frac{s}{s^2+4} - \frac{4s}{(s^2+4)^2}\right\}$$

$$= 1 - \cos 2t - t\sin 2t.$$

Exercises

In exercises 1 through 10, find an inverse transform of the given $f(s)$.

1. $\dfrac{1}{s^2 + as}$.   ANS. $\dfrac{1}{a}(1 - e^{-at})$.

2. $\dfrac{s+2}{s^2 - 6s + 8}$.   ANS. $3e^{4t} - 2e^{2t}$.

3. $\dfrac{2s^2 + 5s - 4}{s^3 + s^2 - 2s}$.   ANS. $2 + e^t - e^{-2t}$.

4. $\dfrac{2s^2 + 1}{s(s+1)^2}$.   ANS. $1 + e^{-t} - 3te^{-t}$.

5. $\dfrac{4s+4}{s^2(s-2)}$.   ANS. $3e^{2t} - 3 - 2t$.

6. $\dfrac{1}{s^3(s^2 + 1)}$.   ANS. $\frac{1}{2}t^2 - 1 + \cos t$.

7. $\dfrac{5s - 2}{s^2(s + 2)(s - 1)}$.   ANS. $t - 2 + e^t + e^{-2t}$.

8. $\dfrac{1}{(s^2 + a^2)(s^2 + b^2)}$. $a^2 \neq b^2, ab \neq 0$.   ANS. $\dfrac{b \sin at - a \sin bt}{ab(b^2 - a^2)}$.

9. $\dfrac{s}{(s^2 + a^2)(s^2 + b^2)}$, $a^2 \neq b^2, ab \neq 0$.   ANS. $\dfrac{\cos at - \cos bt}{b^2 - a^2}$.

10. $\dfrac{s^2}{(s^2 + a^2)(s^2 + b^2)}$, $a^2 \neq b^2, ab \neq 0$.   ANS. $\dfrac{a \sin at - b \sin bt}{a^2 - b^2}$.

68. Initial value problems

Because of Theorem 13, page 184, the Laplace operator will transform a linear differential equation with constant coefficients into an algebraic equation in the transformed function. If upon solving this algebraic equation for the transformed function we are able to obtain the inverse transform, we may have a solution of the original differential equation. Several examples will now be treated in detail so that we can get some feeling for the advantages and disadvantages of the transform method. One fact is apparent from the nature of the transform of derivatives: this method is most readily applied if the appropriate initial conditions are given along with the differential equation. If they are not, the algebra is more complicated.

EXAMPLE (a):   Solve the initial value problem

$$y''(t) + y(t) = 0; \qquad y(0) = 0, y'(0) = 1. \tag{1}$$

Applying the Laplace transform to both sides of the differential equation gives

$$L\{y'' + y\} = 0,$$

and because of the linearity of the transform

$$L\{y''\} + L\{y\} = 0.$$

An application of Theorem 13 now yields

$$s^2 L\{y(t)\} - 1 + L\{y(t)\} = 0,$$

an equation that may easily be solved for $L\{y(t)\}$. We have

$$L\{y(t)\} = \frac{1}{s^2 + 1}. \tag{2}$$

We know that $\sin t$ is a function that satisfies (2), and it is a simple matter to verify that $\sin t$ is the solution of (1).

EXAMPLE (b): Solve the problem

$$y''(t) + \beta^2 y(t) = A \sin \omega t; \qquad y(0) = 1, y'(0) = 0. \tag{3}$$

Here A, β, ω are constants. Because $\beta = 0$ would make the problem one of elementary calculus and because a change in sign of β or ω would not alter the character of the problem, we may assume that β and ω are positive.
Let

$$L\{y(t)\} = u(s).$$

Then

$$L\{y'(t)\} = su(s) - 1,$$
$$L\{y''(t)\} = s^2 u(s) - s \cdot 1 - 0,$$

and application of the operator L transforms the problem (3) into

$$s^2 u(s) - s + \beta^2 u(s) = \frac{A\omega}{s^2 + \omega^2},$$

from which

$$u(s) = \frac{s}{s^2 + \beta^2} + \frac{A\omega}{(s^2 + \beta^2)(s^2 + \omega^2)}. \tag{4}$$

We need the inverse transform of the right member of (4). The form of that inverse depends upon whether β and ω are equal or unequal.
If $\omega \neq \beta$,

$$u(s) = \frac{s}{s^2 + \beta^2} + \frac{A\omega}{\beta^2 - \omega^2}\left(\frac{1}{s^2 + \omega^2} - \frac{1}{s^2 + \beta^2}\right)$$

$$= \frac{s}{s^2 + \beta^2} + \frac{A}{\beta(\beta^2 - \omega^2)}\left(\frac{\omega\beta}{s^2 + \omega^2} - \frac{\omega\beta}{s^2 + \beta^2}\right).$$

Now $y(t) = L^{-1}\{u(s)\}$ so, for $\omega \neq \beta$,

$$y(t) = \cos \beta t + \frac{A}{\beta(\beta^2 - \omega^2)}(\beta \sin \omega t - \omega \sin \beta t). \tag{5}$$

If $\omega = \beta$, the transform (4) becomes

$$u(s) = \frac{s}{s^2 + \beta^2} + \frac{A\beta}{(s^2 + \beta^2)^2}. \tag{6}$$

We know from equation (8) of Section 63 that

$$L^{-1}\left\{\frac{1}{(s^2 + \beta^2)^2}\right\} = \frac{1}{2\beta^3}(\sin \beta t - \beta t \cos \beta t).$$

Hence, for $\omega = \beta$,

$$y(t) = \cos \beta t + \frac{A}{2\beta^2}(\sin \beta t - \beta t \cos \beta t). \tag{7}$$

It is a simple matter to show that this function is indeed the solution of the given initial value problem.

Note that the initial conditions were satisfied automatically by this method when Theorem 13 was applied. We get not the general solution with arbitrary constants still to be determined but that particular solution which satisfies the desired initial conditions. The transform method also gives us some insight into the reason that the solution takes different forms according to whether ω and β are equal or unequal.

EXAMPLE (c):   Solve the problem

$$x''(t) + 2x'(t) + x(t) = 3t\, e^{-t}; \qquad x(0) = 4, \quad x'(0) = 2. \tag{8}$$

Let $L\{x(t)\} = y(s)$. Then the operator L converts (8) into

$$s^2 y(s) - 4s - 2 + 2[sy(s) - 4] + y(s) = \frac{3}{(s + 1)^2},$$

or

$$y(s) = \frac{4s + 10}{(s + 1)^2} + \frac{3}{(s + 1)^4}. \tag{9}$$

We may write

$$y(s) = \frac{4(s + 1) + 6}{(s + 1)^2} + \frac{3}{(s + 1)^4}$$

or

$$y(s) = \frac{4}{s + 1} + \frac{6}{(s + 1)^2} + \frac{3}{(s + 1)^4}.$$

Employing the inverse transform, we obtain

$$x(t) = (4 + 6t + \tfrac{1}{2}t^3)e^{-t}. \tag{10}$$

Again the knowledge of initial conditions contributed to the efficiency of our method. In obtaining and in using equation (9), those terms that came from the initial values $x(0)$ and $x'(0)$ were not combined with the term that came from the transform of the right member of the differential equation.

To combine such terms rarely simplifies and frequently complicates the task of obtaining the inverse transform.

From the solution (10) the student should obtain the derivatives

$$x'(t) = (2 - 6t + \tfrac{3}{2}t^2 - \tfrac{1}{2}t^3)e^{-t},$$

$$x''(t) = (-8 + 9t - 3t^2 + \tfrac{1}{2}t^3)e^{-t},$$

and thus verify that the x of (10) satisfies both the differential equation and the initial conditions of the problem (8). Such verification not only checks our work but also removes any need to justify temporary assumptions about the right to use the Laplace transform theorems on the function $x(t)$ during the time that the function is still unknown.

EXAMPLE (d): Solve the problem

$$w''(x) + 2w'(x) + w(x) = x; \qquad w(0) = -3, w(1) = -1. \tag{11}$$

In this example the boundary conditions are not both of the initial condition type. Using x, rather than t, as independent variable, let

$$L\{w(x)\} = g(s). \tag{12}$$

We know $w(0) = -3$, but we also need $w'(0)$ in order to write the transform of $w''(x)$. Hence we put

$$w'(0) = B \tag{13}$$

and hope to determine B later by using the condition that $w(1) = -1$.

The transformed problem is

$$s^2 g(s) - s(-3) - B + 2[sg(s) - (-3)] + g(s) = \frac{1}{s^2}$$

from which

$$g(s) = \frac{-3(s + 1) + B - 3}{(s + 1)^2} + \frac{1}{s^2(s + 1)^2}. \tag{14}$$

But, by the usual partial fractions expansion,

$$\frac{1}{s^2(s + 1)^2} = -\frac{2}{s} + \frac{1}{s^2} + \frac{2}{s + 1} + \frac{1}{(s + 1)^2},$$

so

$$g(s) = \frac{1}{s^2} - \frac{2}{s} - \frac{1}{s + 1} + \frac{B - 2}{(s + 1)^2}, \tag{15}$$

from which we obtain

$$w(x) = x - 2 - e^{-x} + (B - 2)x\, e^{-x}. \tag{16}$$

We have yet to impose the condition that $w(1) = -1$. From (16) with $x = 1$, we get

$$-1 = 1 - 2 - e^{-1} + (B - 2)e^{-1},$$

so $B = 3$.

Thus our final result is

$$w(x) = x - 2 - e^{-x} + xe^{-x}. \tag{17}$$

The problem in Example (d) may be solved efficiently by the methods of Chapter 7. See also exercises 23 through 44 below.

Exercises

In exercises 1 through 22, solve the problem by the Laplace transform method. Verify that your solution satisfies the differential equation and the initial conditions.

1. $y' = e^t; y(0) = 2.$ ANS. $y = e^t + 1.$
2. $y' = 2e^t; y(0) = -1.$ ANS. $y = 2e^t - 3.$
3. $y' + y = e^{2t}; y(0) = 0.$ ANS. $y = \frac{1}{3}e^{2t} - \frac{1}{3}e^{-t}.$
4. $y' - y = e^{-t}; y(0) = 1.$ ANS. $y = \frac{3}{2}e^t - \frac{1}{2}e^{-t}.$
5. $y'' + a^2 y = 0; y(0) = 1, y'(0) = 0.$ ANS. $y = \cos at.$
6. $y'' + a^2 y = 0; y(0) = 0, y'(0) = a.$ ANS. $y = \sin at.$
7. $y'' - 3y' + 2y = e^{3t}; y(0) = y'(0) = 0.$ ANS. $y = \frac{1}{2}e^t - e^{2t} + \frac{1}{2}e^{3t}.$
8. $y'' + y = e^{-t}; y(0) = y'(0) = 0.$ ANS. $y = \frac{1}{2}(\sin t - \cos t + e^{-t}).$
9. $y'' - 2y' = -4; y(0) = 0, y'(0) = 4.$ ANS. $y = e^{2t} + 2t - 1.$
10. $y'' + y' - 2y = -4; y(0) = 2, y'(0) = 3.$ ANS. $y = 2 + e^t - e^{-2t}.$
11. $x''(t) - 4x'(t) + 4x(t) = 4e^{2t}; x(0) = -1, x'(0) = -4.$
     ANS. $x(t) = e^{2t}(2t^2 - 2t - 1).$
12. $x''(t) + x(t) = 6\sin 2t; x(0) = 3, x'(0) = 1.$ ANS. $x(t) = 5\sin t + 3\cos t - 2\sin 2t.$
13. $y''(t) - y(t) = 4\cos t; y(0) = 0, y'(0) = 1.$ ANS. $y(t) = 2\cosh t + \sinh t - 2\cos t.$
14. $y''(t) - 6y'(t) + 9y(t) = 6t^2 e^{3t}; y(0) = y'(0) = 0.$ ANS. $y(t) = \frac{1}{2}t^4 e^{3t}.$
15. $x''(t) + 4x(t) = t + 4; x(0) = 1, x'(0) = 0.$ ANS. $x(t) = 1 + \frac{1}{4}t - \frac{1}{8}\sin 2t.$
16. $x''(t) - 2x'(t) = 6 - 4t; x(0) = 2, x'(0) = 0.$ ANS. $x(t) = t^2 - 2t + 1 + e^{2t}.$
17. $x''(t) + x(t) = 4e^t; x(0) = 1, x'(0) = 3.$ ANS. $x(t) = \sin t - \cos t + 2e^t.$
18. $x''(t) + x'(t) - 2x(t) = 6; x(0) = 1, x'(t) = 1.$ ANS. $x(t) = e^{-2t} + 3e^t - 3.$
19. $y''(x) + 9y(x) = 40e^x; y(0) = 5, y'(0) = -2.$ ANS. $y(x) = 4e^x + \cos 3x - 2\sin 3x.$
20. $y''(x) + y(x) = 4e^x; y(0) = 0, y'(0) = 0.$ ANS. $y(x) = 2(e^x - \cos x - \sin x).$
21. $x''(t) + 3x'(t) + 2x(t) = 4t^2; x(0) = 0, x'(0) = 0.$
     ANS. $x(t) = 2t^2 - 6t + 7 - 8e^{-t} + e^{-2t}.$
22. $x''(t) - 4x'(t) + 4x(t) = 4\cos 2t; x(0) = 2, x'(0) = 5.$
     ANS. $x(t) = 2e^{2t}(1 + t) - \frac{1}{2}\sin 2t.$

In exercises 23 through 42, use the Laplace transform method with the realization that these exercises were not constructed with the Laplace transform technique in mind. Compare your work with that done in solving the same problems by the methods of Chapter 7. See pages 125–126.

23. Exercise 1.   24. Exercise 2.

25. Exercise 3.

27. Exercise 14.

29. Exercise 21.

31. Exercise 23.

33. Exercise 37.

35. Exercise 39.

37. Exercise 41.

39. Exercise 43.

41. Exercise 45.

43. Solve the problem

26. Exercise 11.

28. Exercise 20.

30. Exercise 22.

32. Exercise 36.

34. Exercise 38.

36. Exercise 40.

38. Exercise 42.

40. Exercise 44.

42. Exercise 46.

$$x''(t) - 4x'(t) + 4x(t) = e^{2t}; \qquad x'(0) = 0, x(1) = 0.$$

ANS. $x(t) = \frac{1}{2}(1 - t)^2 e^{2t}.$

44. Solve the problem

$$x''(t) + 4x(t) = -8t^2; \qquad x(0) = 3, x(\tfrac{1}{4}\pi) = 0.$$

ANS. $x(t) = 2\cos 2t + (\tfrac{1}{8}\pi^2 - 1)\sin 2t + 1 - 2t^2.$

69. A step function

Applications frequently deal with situations that change abruptly at specified times. We need a notation for a function that will suppress a given term up to a certain value of t and insert the term for all larger t. The function we are about to introduce leads us to a powerful tool for constructing inverse transforms.

Let us define function $\alpha(t)$ by

$$\alpha(t) = 0, \qquad t < 0, \tag{1}$$

$$= 1, \qquad t \geqq 0.$$

The graph of $\alpha(t)$ is shown in Figure 22.

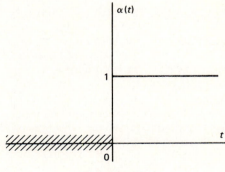

FIGURE 22

The definition (1) says that $\alpha(t)$ is zero when the argument is negative and $\alpha(t)$ is unity when the argument is positive or is zero. It follows that

$$\alpha(t - c) = 0, \qquad t < c, \tag{2}$$

$$= 1, \qquad t \geq c.$$

The α function permits easy designation of the result of translating the graph of $F(t)$. If the graph of

$$y = F(t), \qquad t \geq 0, \tag{3}$$

is as shown in Figure 23, the graph of

$$y = \alpha(t - c)F(t - c), \qquad t \geq c, \tag{4}$$

is that shown in Figure 24. Furthermore, if $F(x)$ is defined for $-c \leq x < 0$, then $F(t - c)$ is defined for $0 \leq t < c$ and the y of (4) is zero for $0 \leq t < c$ because of the negative argument in $\alpha(t - c)$. Notice that the values of $F(x)$ for negative x have no bearing on this result because each value is multiplied by zero (from the α); only the existence of F for negative arguments is needed.

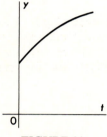

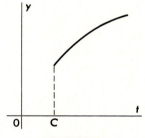

FIGURE 23   FIGURE 24

The Laplace transform of $\alpha(t - c)F(t - c)$ is related to that of $F(t)$. Consider

$$L\{\alpha(t - c)F(t - c)\} = \int_0^\infty e^{-st}\alpha(t - c)F(t - c)\,dt.$$

Since $\alpha(t - c) = 0$ for $0 \leq t < c$ and $\alpha(t - c) = 1$ for $t \geq c$, we get

$$L\{\alpha(t - c)F(t - c)\} = \int_c^\infty e^{-st}F(t - c)\,dt.$$

Now put $t - c = v$ in the integral to obtain

$$L\{\alpha(t - c)F(t - c)\} = \int_0^\infty e^{-s(c+v)}F(v)\,dv$$

$$= e^{-cs}\int_0^\infty e^{-sv}F(v)\,dv.$$

Since a definite integral is independent of the variable of integration,

$$\int_0^\infty e^{-sv} F(v)\, dv = \int_0^\infty e^{-st} F(t)\, dt = L\{F(t)\} = f(s).$$

Therefore we have shown that

$$L\{\alpha(t - c) F(t - c)\} = e^{-cs} L\{F(t)\} = e^{-cs} f(s). \tag{5}$$

THEOREM 22:   *If* $L^{-1}\{f(s)\} = F(t)$, *if* $c \geq 0$, *and if* $F(t)$ *be assigned values* (*no matter what ones*) *for* $-c \leq t < 0$,

$$L^{-1}\{e^{-cs} f(s)\} = F(t - c)\alpha(t - c). \tag{6}$$

EXAMPLE (a):   Find $L\{y(t)\}$ where (Figure 25)

$$y(t) = t^2, \qquad 0 < t < 2,$$
$$= 6, \qquad\quad t > 2.$$

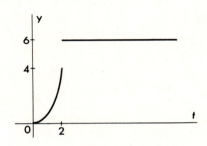

FIGURE 25

Here, direct use of the definition of a transform yields

$$L\{y(t)\} = \int_0^2 t^2 e^{-st}\, dt + \int_2^\infty 6 e^{-st}\, dt.$$

Although the above integrations are not difficult, we prefer to use the α function.

Since $\alpha(t - 2) = 0$ for $t < 2$ and $\alpha(t - 2) = 1$ for $t \geq 2$, we build the $y(t)$ in the following way. The crude trial

$$y_1 = t^2$$

works for $0 < t < 2$, but we wish to knock out the t^2 when $t > 2$. Hence we write

$$y_2 = t^2 - t^2 \alpha(t - 2).$$

This gives t^2 for $t < 2$ and zero for $t > 2$. Then we add the term $6\alpha(t - 2)$ and finally arrive at

$$y(t) = t^2 - t^2\alpha(t - 2) + 6\alpha(t - 2). \tag{7}$$

The y of (7) is the y of our example and, of course, it can be written at once after a little practice with the α function.

Unfortunately, the y of (7) is not yet in the best form for our purpose. The theorem we wish to use gives us

$$L\{F(t - c)\alpha(t - c)\} = e^{-cs}f(s).$$

Therefore we must have the coefficient of $\alpha(t - 2)$ expressed as a function of $(t - 2)$. Since

$$-t^2 + 6 = -(t^2 - 4t + 4) - 4(t - 2) + 2,$$

$$y(t) = t^2 - (t - 2)^2\alpha(t - 2) - 4(t - 2)\alpha(t - 2) + 2\alpha(t - 2), \tag{8}$$

from which it follows at once that

$$L\{y(t)\} = \frac{2}{s^3} - \frac{2\,e^{-2s}}{s^3} - \frac{4\,e^{-2s}}{s^2} + \frac{2\,e^{-2s}}{s}.$$

EXAMPLE (b):   Find and sketch a function $g(t)$ for which

$$g(t) = L^{-1}\left\{\frac{3}{s} - \frac{4\,e^{-s}}{s^2} + \frac{4\,e^{-3s}}{s^2}\right\}.$$

We know that $L^{-1}\{4/s^2\} = 4t$. By Theorem 22 we then get

$$L^{-1}\left\{\frac{4\,e^{-s}}{s^2}\right\} = 4(t - 1)\alpha(t - 1)$$

and

$$L^{-1}\left\{\frac{4\,e^{-3s}}{s^2}\right\} = 4(t - 3)\alpha(t - 3).$$

We may therefore write

$$g(t) = 3 - 4(t - 1)\alpha(t - 1) + 4(t - 3)\alpha(t - 3). \tag{9}$$

To write $g(t)$ without the α function, consider first the interval

$$0 \leqq t < 1$$

in which $\alpha(t - 1) = 0$ and $\alpha(t - 3) = 0$. We find

$$g(t) = 3, \qquad 0 \leqq t < 1. \tag{10}$$

For $1 \leqq t < 3$, $\alpha(t - 1) = 1$, and $\alpha(t - 3) = 0$. Hence

$$g(t) = 3 - 4(t - 1) = 7 - 4t, \qquad 1 \leqq t < 3. \tag{11}$$

For $t \geq 3$, $\alpha(t - 1) = 1$ and $\alpha(t - 3) = 1$, so

$$g(t) = 3 - 4(t - 1) + 4(t - 3) = -5, \qquad t \geq 3. \tag{12}$$

Equations (10), (11), and (12) are equivalent to equation (9). The graph of $g(t)$ is shown in Figure 26.

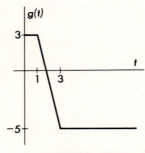

FIGURE 26

EXAMPLE (c): Solve the problem

$$x''(t) + 4x(t) = \psi(t); \qquad x(0) = 1, \ x'(0) = 0, \tag{13}$$

in which $\psi(t)$ is defined by

$$\psi(t) = 4t, \qquad 0 \leq t \leq 1, \tag{14}$$
$$= 4, \qquad\qquad t > 1.$$

We seek, of course, a solution valid in the range $t \geq 0$ in which the function $\psi(t)$ is defined.

In this problem another phase of the power of the Laplace transform method begins to emerge. The fact that the function $\psi(t)$ in the differential equation has discontinuous derivatives makes the use of the classical method of undetermined coefficients somewhat awkward, but such discontinuities do not interfere at all with the simplicity of the Laplace transform method.

In attacking this problem, let us put $L\{x(t)\} = h(s)$. We need to obtain $L\{\psi(t)\}$. In terms of the α function we may write, from (14),

$$\psi(t) = 4t - 4(t - 1)\alpha(t - 1), \qquad t \geq 0. \tag{15}$$

From (15) it follows that

$$L\{\psi(t)\} = \frac{4}{s^2} - \frac{4 e^{-s}}{s^2}.$$

Therefore the application of the operator L transforms problem (13) into

$$s^2 h(s) - s - 0 + 4h(s) = \frac{4}{s^2} - \frac{4e^{-s}}{s^2},$$

from which

$$h(s) = \frac{s}{s^2 + 4} + \frac{4}{s^2(s^2 + 4)} - \frac{4e^{-s}}{s^2(s^2 + 4)}. \tag{16}$$

Now

$$\frac{4}{s^2(s^2 + 4)} = \frac{1}{s^2} - \frac{1}{s^2 + 4},$$

so (16) becomes

$$h(s) = \frac{s}{s^2 + 4} + \frac{1}{s^2} - \frac{1}{s^2 + 4} - \left(\frac{1}{s^2} - \frac{1}{s^2 + 4}\right) e^{-s}. \tag{17}$$

Since $x(t) = L^{-1}\{h(s)\}$, we obtain the desired solution

$$x(t) = \cos 2t + t - \tfrac{1}{2}\sin 2t - [(t - 1) - \tfrac{1}{2}\sin 2(t - 1)]\alpha(t - 1). \tag{18}$$

It is easy to verify our solution. From (18) it follows that

$$x'(t) = -2\sin 2t + 1 - \cos 2t - [1 - \cos 2(t - 1)]\alpha(t - 1), \tag{19}$$

$$x''(t) = -4\cos 2t + 2\sin 2t - 2\sin 2(t - 1)\alpha(t - 1). \tag{20}$$

Therefore $x(0) = 1$ and $x'(0) = 0$, as desired. Also, from (18) and (20), we get

$$x''(t) + 4x(t) = 4t - 4(t - 1)\alpha(t - 1) = \psi(t), \qquad t \geq 0.$$

Exercises

In exercises 1 through 7 sketch the graph of the given function for $t \geq 0$.

1. $\alpha(t - c)$.
2. $\alpha(t - 1) + 2\alpha(t - 2) - 3\alpha(t - 4)$.
3. $(t - 3)\alpha(t - 3)$.
4. $\sin(t - \pi) \cdot \alpha(t - \pi)$.
5. $(t - 3)^2\alpha(t - 3)$.
6. $t^2 - (t - 1)^2\alpha(t - 1)$.
7. $t^2 - t^2\alpha(t - 2)$.

In exercises 8 through 15 express $F(t)$ in terms of the α function and find $L\{F(t)\}$.

8. $F(t) = 3, \qquad 0 < t < 1,$
   $\quad\;\; = t, \qquad\quad\; t > 1.$

   ANS. $\dfrac{3}{s} + e^{-s}\left(\dfrac{1}{s^2} - \dfrac{2}{s}\right).$

9. $F(t) = 4, \qquad 0 < t < 2,$
   $\quad\;\; = 2t - 1, \qquad t > 2.$

   ANS. $\dfrac{4}{s} + e^{-2s}\left(\dfrac{2}{s^2} - \dfrac{1}{s}\right).$

10. $F(t) = t^2, \qquad 0 < t < 2,$
$\quad = 3, \qquad\qquad t > 2.$

ANS. $\dfrac{2}{s^3} - e^{-2s}\left(\dfrac{1}{s} + \dfrac{4}{s^2} + \dfrac{2}{s^3}\right).$

11. $F(t) = t^2, \qquad 0 < t < 1,$
$\quad = 3, \qquad\qquad 1 < t < 2,$
$\quad = 0, \qquad\qquad t > 2.$

ANS. $\dfrac{2}{s^3} + e^{-s}\left(\dfrac{2}{s} - \dfrac{2}{s^2} - \dfrac{2}{s^3}\right) - \dfrac{3\,e^{-2s}}{s}.$

12. $F(t) = t^2, \qquad\quad 0 < t < 2,$
$\quad = t - 1, \qquad 2 < t < 3,$
$\quad = 7, \qquad\qquad t > 3.$

ANS. $\dfrac{2}{s^3} - e^{-2s}\left(\dfrac{3}{s} + \dfrac{3}{s^2} + \dfrac{2}{s^3}\right) + e^{-3s}\left(\dfrac{5}{s} - \dfrac{1}{s^2}\right).$

13. $F(t) = e^{-t}, \qquad 0 < t < 2,$
$\quad = 0, \qquad\qquad t > 2.$

ANS. $\dfrac{1 - \exp\,(-2s - 2)}{s + 1}.$

14. $F(t) = \sin 3t, \qquad 0 < t < \frac{1}{2}\pi,$
$\quad = 0, \qquad\qquad\quad t > \frac{1}{2}\pi.$

ANS. $\dfrac{3 + s\exp\,(-\frac{1}{2}\pi s)}{s^2 + 9}.$

15. $F(t) = \sin 3t, \qquad 0 < t < \pi,$
$\quad = 0, \qquad\qquad\quad t > \pi.$

ANS. $\dfrac{3(1 + e^{-\pi s})}{s^2 + 9}.$

16. Find and sketch an inverse Laplace transform of

$$\frac{5\,e^{-3s}}{s} - \frac{e^{-s}}{s}.$$

ANS. $F(t) = 5\alpha(t - 3) - \alpha(t - 1).$

17. Evaluate $L^{-1}\left\{\dfrac{e^{-4s}}{(s + 2)^3}\right\}.$

ANS. $\frac{1}{2}(t - 4)^2 \exp\,[-2(t - 4)]\alpha(t - 4).$

18. If $F(t)$ is to be continuous for $t \geq 0$ and

$$F(t) = L^{-1}\left\{\frac{e^{-3s}}{(s + 1)^3}\right\},$$

evaluate $F(2), F(5), F(7)$.   ANS. $F(2) = 0, F(5) = 2\,e^{-2}, F(7) = 8\,e^{-4}.$

19. If $F(t)$ is to be continuous for $t \geq 0$ and

$$F(t) = L^{-1}\left\{\frac{(1 - e^{-2s})(1 - 3\,e^{-2s})}{s^2}\right\},$$

evaluate $F(1), F(3), F(5)$.   ANS. $F(1) = 1, F(3) = -1, F(5) = -4.$

20. Prove that $\psi(t, c) = \displaystyle\sum_{n=0}^{\infty} (-1)^n \alpha(t - nc)$ is the same function as was used in Example (a), Section 65. Note that for any specific t, the series is finite; no question of convergence is involved.

21. Obtain the transform of the half-wave rectification $F(t)$ of $\sin t$ by writing

$$F(t) = \sin t\,\psi(t, \pi)$$

in terms of the ψ of exercise 20 above. Use the fact that

$$(-1)^n \sin t = \sin\,(t - n\pi).$$

Check your result with that of exercise 17, page 193.

In exercises 22 through 25, solve the problem using the Laplace transform. Verify that your solution satisfies the differential equation and the initial conditions.

22. $x''(t) + x(t) = F(t); x(0) = 0, x'(0) = 0$, in which

$$F(t) = 4, \qquad 0 \le t \le 2,$$
$$= t + 2, \qquad t > 2.$$

ANS. $x(t) = 4 - 4 \cos t + [(t - 2) - \sin (t - 2)]\alpha(t - 2)$.

23. $x''(t) + x(t) = H(t); x(0) = 1, x'(0) = 0$, in which

$$H(t) = 3, \qquad 0 \le t \le 4,$$
$$= 2t - 5, \qquad t > 4.$$

ANS. $x(t) = 3 - 2 \cos t + 2[t - 4 - \sin (t - 4)]\alpha(t - 4)$.

24. $x''(t) + x(t) = G(t); x(0) = 0, x'(0) = 1$, in which

$$G(t) = 1, \qquad 0 \le t < \pi/2,$$
$$= 0, \qquad t \ge \pi/2.$$

ANS. $x(t) = 1 - \cos t + \sin t - \alpha(t - \pi/2)(1 - \sin t)$.

25. $x''(t) + 4x(t) = M(t); x(0) = x'(0) = 0$, in which

$$M(t) = \sin t - \alpha(t - 2\pi) \sin (t - 2\pi).$$

ANS. $x(t) = \frac{1}{6}[1 - \alpha(t - 2\pi)](2 \sin t - \sin 2t)$.

26. Compute $y(\frac{1}{2}\pi)$ and $y(2 + \frac{1}{2}\pi)$ for the function $y(x)$ that satisfies the initial value problem

$$y''(x) + y(x) = (x - 2)\alpha(x - 2); \qquad y(0) = 0, y'(0) = 0.$$

ANS. $y(\frac{1}{2}\pi) = 0, y(2 + \frac{1}{2}\pi) = \frac{1}{2}\pi - 1$.

27. Compute $x(1)$ and $x(4)$ for the function $x(t)$ that satisfies the initial value problem

$$x''(t) + 2x'(t) + x(t) = 2 + (t - 3)\alpha(t - 3); \qquad x(0) = 2, x'(0) = 1.$$

ANS. $x(1) = 2 + e^{-1}, x(4) = 1 + 3 e^{-1} + 4 e^{-4}$.

70. A convolution theorem

We now seek a formula for the inverse transform of a product of transforms. Given

$$L^{-1}\{f(s)\} = F(t), \qquad L^{-1}\{g(s)\} = G(t), \tag{1}$$

in which $F(t)$ and $G(t)$ are assumed to be functions of class A, we shall obtain a formula for

$$L^{-1}\{f(s)g(s)\}. \tag{2}$$

Since $f(s)$ is the transform of $F(t)$, we may write

$$f(s) = \int_0^\infty e^{-st} F(t) \, dt. \tag{3}$$

Since $g(s)$ is the transform of $G(t)$,

$$g(s) = \int_0^\infty e^{-s\beta} G(\beta) \, d\beta, \tag{4}$$

in which, to avoid confusion, we have used β (rather than t) as the variable of integration in the definite integral.

By equation (4), we have

$$f(s)g(s) = \int_0^\infty e^{-s\beta} f(s) G(\beta) \, d\beta. \tag{5}$$

On the right in (5) we encounter the product $e^{-s\beta} f(s)$. By Theorem 22, page 208, we know that from

$$L^{-1}\{f(s)\} = F(t) \tag{6}$$

it follows that

$$L^{-1}\{e^{-s\beta} f(s)\} = F(t - \beta)\alpha(t - \beta), \tag{7}$$

in which α is the step function discussed in Section 69. Equation (7) means that

$$e^{-s\beta} f(s) = \int_0^\infty e^{-st} F(t - \beta)\alpha(t - \beta) \, dt. \tag{8}$$

With the aid of (8) we may put equation (5) in the form

$$f(s)g(s) = \int_0^\infty \int_0^\infty e^{-st} G(\beta) F(t - \beta)\alpha(t - \beta) \, dt \, d\beta. \tag{9}$$

Since $\alpha(t - \beta) = 0$ for $0 < t < \beta$ and $\alpha(t - \beta) = 1$ for $t \geq \beta$, equation (9) may be rewritten as

$$f(s)g(s) = \int_0^\infty \int_\beta^\infty e^{-st} G(\beta) F(t - \beta) \, dt \, d\beta. \tag{10}$$

In (10), the integration in the $t\beta$-plane covers the shaded region shown in Figure 27. The elements are summed from $t = \beta$ to $t = \infty$ and then from $\beta = 0$ to $\beta = \infty$.

In advanced calculus it is shown that, because $F(t)$ and $G(t)$ are functions of class A, it is legitimate to interchange the order of integration on the right in equation (10). From Figure 27 we see that, in the new order of integration, the elements are to be summed from $\beta = 0$ to $\beta = t$ and then from $t = 0$

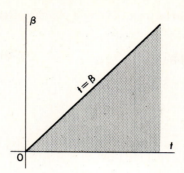

FIGURE 27

to $t = \infty$. We thus obtain

$$f(s)g(s) = \int_0^\infty \int_0^t e^{-st} G(\beta)F(t - \beta)\, d\beta\, dt,$$

or

$$f(s)g(s) = \int_0^\infty e^{-st} \left[\int_0^t G(\beta)F(t - \beta)\, d\beta \right] dt. \tag{11}$$

Since the right member of (11) is precisely the Laplace transform of

$$\int_0^t G(\beta)F(t - \beta)\, d\beta,$$

we have arrived at the desired result, which is called the convolution theorem for the Laplace transform.

THEOREM 23: *If $L^{-1}\{f(s)\} = F(t)$, if $L^{-1}\{g(s)\} = G(t)$, and if $F(t)$ and $G(t)$ are functions of class A (see page* 181), *then*

$$L^{-1}\{f(s)g(s)\} = \int_0^t G(\beta)F(t - \beta)\, d\beta. \tag{12}$$

It is easy to show that the right member of equation (12) is also a function of class A.

Of course F and G are interchangeable in (12) because f and g enter (12) symmetrically. We may replace (12) by

$$L^{-1}\{f(s)g(s)\} = \int_0^t F(\beta)G(t - \beta)\, d\beta, \tag{13}$$

a result which also follows from (12) by a change of variable of integration.

EXAMPLE (a):   Evaluate $L^{-1}\{f(s)/s\}$.

Let $L^{-1}\{f(s)\} = F(t)$. Since

$$L^{-1}\left\{\frac{1}{s}\right\} = 1,$$

we use Theorem 23 to conclude that

$$L^{-1}\left\{\frac{f(s)}{s}\right\} = \int_0^t F(\beta)\,d\beta.$$

EXAMPLE (b):   Solve the problem

$$x''(t) + k^2 x(t) = F(t); \qquad x(0) = A, x'(0) = B. \tag{14}$$

Here k, A, B are constants and $F(t)$ is a function whose Laplace transform exists. Let

$$L\{x(t)\} = u(s), \qquad L\{F(t)\} = f(s).$$

Then the Laplace operator transforms problem (14) into

$$s^2 u(s) - As - B + k^2 u(s) = f(s),$$

$$u(s) = \frac{As + B}{s^2 + k^2} + \frac{f(s)}{s^2 + k^2}. \tag{15}$$

To get the inverse transform of the last term in (15), we use the convolution theorem. Thus we arrive at

$$x(t) = A \cos kt + \frac{B}{k} \sin kt + \frac{1}{k}\int_0^t F(t - \beta) \sin k\beta\,d\beta,$$

or

$$x(t) = A \cos kt + \frac{B}{k} \sin kt + \frac{1}{k}\int_0^t F(\beta) \sin k(t - \beta)\,d\beta. \tag{16}$$

Verification of the solution (16) is simple. Once that check has been performed, the need for the assumption that $F(t)$ has a Laplace transform is removed. It does not matter what method we use to get a solution (with certain exceptions naturally imposed during college examinations) if the validity of the result can be verified from the result itself.

Exercises

In exercises 1 through 3 find the Laplace transform of the given convolution integral.

1. $\int_0^t (t - \beta) \sin 3\beta \, d\beta.$ ANS. $\dfrac{3}{s^2(s^2 + 9)}.$

2. $\int_0^t e^{-(t-\beta)} \sin \beta \, d\beta.$ ANS. $\dfrac{1}{(s + 1)(s^2 + 1)}.$

3. $\int_0^t (t - \beta)^3 e^{\beta} \, d\beta.$ ANS. $\dfrac{6}{s^4(s - 1)}.$

In exercises 4 through 7 find an inverse transform of the given $f(s)$ using the convolution theorem.

4. $\dfrac{1}{s(s^2 + k^2)}.$ ANS. $\dfrac{1}{k^2}(1 - \cos kt).$

5. $\dfrac{1}{s(s + 2)}.$ ANS. $\tfrac{1}{2}(1 - e^{-2t}).$

6. $\dfrac{4}{s^2(s - 2)}.$ ANS. $e^{2t} - 2t - 1.$

7. $\dfrac{1}{(s^2 + 1)^2}.$ ANS. $\tfrac{1}{2}(\sin t - t \cos t).$

8. Solve the problem

$$x''(t) + 2x'(t) + x(t) = F(t); \qquad x(0) = 0, x'(0) = 0.$$

$$\text{ANS.} \quad x(t) = \int_0^t \beta \, e^{-\beta} F(t - \beta) \, d\beta.$$

9. Solve the problem

$$y''(t) - k^2 y(t) = H(t); \qquad y(0) = 0, y'(0) = 0.$$

$$\text{ANS.} \quad y(t) = \frac{1}{k} \int_0^t H(t - \beta) \sinh k\beta \, d\beta.$$

10. Solve the problem

$$y''(t) + 4y'(t) + 13y(t) = F(t); \qquad y(0) = 0, y'(0) = 0.$$

11. Solve the problem

$$x''(t) + 6x'(t) + 9x(t) = F(t); \qquad x(0) = A, x'(0) = B.$$

$$\text{ANS.} \quad x(t) = e^{-3t}[A + (B + 3A)t] + \int_0^t \beta \, e^{-3\beta} F(t - \beta) \, d\beta.$$

71. Special integral equations

A differential equation may be loosely described as one that contains a derivative of a dependent variable; the equation contains a dependent variable under a derivative sign. An equation that contains a dependent variable under an integral sign is called an integral equation.

Because of the convolution theorem, the Laplace transform is an excellent tool for solving a very special class of integral equations. We know from Theorem 23 that if

$$L\{F(t)\} = f(s)$$

and

$$L\{G(t)\} = g(s),$$

then

$$L\left\{ \int_0^t F(\beta)G(t - \beta)\,d\beta \right\} = f(s)g(s). \tag{1}$$

The relation (1) suggests the use of the Laplace transform in solving equations that contain convolution integrals.

EXAMPLE (a):   Find $F(t)$ from the integral equation

$$F(t) = 4t - 3 \int_0^t F(\beta) \sin (t - \beta)\,d\beta. \tag{2}$$

The integral in (2) is in precisely the right form to permit the use of the convolution theorem. Let

$$L\{F(t)\} = f(s).$$

Then, because

$$L\{\sin t\} = \frac{1}{s^2 + 1},$$

application of Theorem 23 yields

$$L\left\{ \int_0^t F(\beta) \sin (t - \beta)\,d\beta \right\} = \frac{f(s)}{s^2 + 1}.$$

Therefore, the Laplace operator converts equation (2) into

$$f(s) = \frac{4}{s^2} - \frac{3f(s)}{s^2 + 1}. \tag{3}$$

We need to obtain $f(s)$ from (3) and then $F(t)$ from $f(s)$. From (3) we get

$$\left(1 + \frac{3}{s^2 + 1}\right) f(s) = \frac{4}{s^2},$$

or

$$f(s) = \frac{4(s^2 + 1)}{s^2(s^2 + 4)} = \frac{1}{s^2} + \frac{3}{s^2 + 4}.$$

Therefore

$$F(t) = L^{-1}\left\{\frac{1}{s^2} + \frac{3}{s^2 + 4}\right\},$$

or

$$F(t) = t + \tfrac{3}{2}\sin 2t. \tag{4}$$

That the $F(t)$ of (4) is a solution of equation (2) may be verified directly. Such a check is frequently tedious. We shall show that for the F of (4), the right-hand side of equation (2) reduces to the left-hand side of (2). Since

$$\text{RHS} = 4t - 3\int_0^t (\beta + \tfrac{3}{2}\sin 2\beta)\sin(t - \beta)\,d\beta,$$

we integrate by parts with the choice shown in the table.

$(\beta + \tfrac{3}{2}\sin 2\beta)$	$\sin(t - \beta)\,d\beta$
$(1 + 3\cos 2\beta)\,d\beta$	$\cos(t - \beta)$

It thus follows that

$$\text{RHS} = 4t - 3\left[(\beta + \tfrac{3}{2}\sin 2\beta)\cos(t - \beta)\right]_0^t$$

$$+ 3\int_0^t (1 + 3\cos 2\beta)\cos(t - \beta)\,d\beta,$$

from which

$$\text{RHS} = 4t - 3(t + \tfrac{3}{2}\sin 2t) + 3\int_0^t \cos(t - \beta)\,d\beta$$

$$+ 9\int_0^t \cos 2\beta \cos(t - \beta)\,d\beta,$$

or

$$\text{RHS} = t - \tfrac{9}{2}\sin 2t - 3\left[\sin(t - \beta)\right]_0^t + \tfrac{9}{2}\int_0^t [\cos(t + \beta) + \cos(t - 3\beta)]\,d\beta.$$

This leads us to the result

$$\text{RHS} = t - \tfrac{9}{2}\sin 2t + 3\sin t + \tfrac{9}{2}\left[\sin(t + \beta) - \tfrac{1}{3}\sin(t - 3\beta)\right]_0^t$$

$$= t - \tfrac{9}{2}\sin 2t + 3\sin t + \tfrac{9}{2}\sin 2t + \tfrac{3}{2}\sin 2t - \tfrac{9}{2}\sin t + \tfrac{3}{2}\sin t,$$

or

$$\text{RHS} = t + \tfrac{3}{2}\sin 2t = F(t) = \text{LHS},$$

as desired.

It is important to realize that the original equation

$$F(t) = 4t - 3\int_0^t F(\beta)\sin(t - \beta)\,d\beta \tag{2}$$

could equally well have been encountered in the equivalent form

$$F(t) = 4t - 3\int_0^t F(t - \beta)\sin\beta\,d\beta.$$

An essential ingredient for the success of the method being used is that the integral involved be in exactly the convolution integral form. We must have zero to the independent variable as limits of integration and an integrand that is the product of a function of the variable of integration by a function of the difference between the independent variable and the variable of integration. The fact that integrals of that form appear with significant frequency in physical problems keeps the topic of this section from being relegated to the role of a mathematical parlor game.

EXAMPLE (b):   Solve the equation

$$g(x) = \tfrac{1}{2}x^2 - \int_0^x (x - y)g(y)\,dy. \tag{5}$$

Again the integral involved is one of the convolution type with x playing the role of the independent variable. Let the Laplace transform of $g(x)$ be some as yet unknown function $h(z)$:

$$L\{g(x)\} = h(z). \tag{6}$$

Since $L\{\tfrac{1}{2}x^2\} = 1/z^3$ and $L\{x\} = 1/z^2$, we may apply the operator L throughout (5) and obtain

$$h(z) = \frac{1}{z^3} - \frac{h(z)}{z^2},$$

from which

$$\left(1 + \frac{1}{z^2}\right)h(z) = \frac{1}{z^3},$$

or

$$h(z) = \frac{1}{z(z^2 + 1)} = \frac{z^2 + 1 - z^2}{z(z^2 + 1)} = \frac{1}{z} - \frac{z}{z^2 + 1}.$$

Then

$$g(x) = L^{-1}\left\{\frac{1}{z} - \frac{z}{z^2 + 1}\right\}$$

or

$$g(x) = 1 - \cos x. \tag{7}$$

Verification of (7) is simple. For the right member of (5) we get

$$\text{RHS} = \tfrac{1}{2}x^2 - \int_0^x (x - y)(1 - \cos y)\, dy$$

$$= \tfrac{1}{2}x^2 - \left[(x - y)(y - \sin y)\right]_0^x - \int_0^x (y - \sin y)\, dy$$

$$= \tfrac{1}{2}x^2 - 0 - \left[\tfrac{1}{2}y^2 + \cos y\right]_0^x$$

$$= \tfrac{1}{2}x^2 - \tfrac{1}{2}x^2 - \cos x + 1 = 1 - \cos x = \text{LHS}.$$

Exercises

In exercises 1 through 4 solve the given equation and verify your solution.

1. $F(t) = 1 + 2\int_0^t F(t - \beta)\, e^{-2\beta}\, d\beta.$ 　　　　ANS. 　$F(t) = 1 + 2t.$

2. $F(t) = 1 + \int_0^t F(\beta) \sin(t - \beta)\, d\beta.$ 　　　　ANS. 　$F(t) = 1 + \tfrac{1}{2}t^2.$

3. $F(t) = t + \int_0^t F(t - \beta)\, e^{-\beta}\, d\beta.$ 　　　　ANS. 　$F(t) = t + \tfrac{1}{2}t^2.$

4. $F(t) = 4t^2 - \int_0^t F(t - \beta)\, e^{-\beta}\, d\beta.$ 　　ANS. 　$F(t) = -1 + 2t + 2t^2 + e^{-2t}.$

In exercises 5 through 8 solve the given equation. If sufficient time is available, verify your solution.

5. $F(t) = t^3 + \int_0^t F(\beta) \sin (t - \beta) \, d\beta.$ ANS. $F(t) = t^3 + \frac{1}{20} t^5.$

6. $F(t) = 8t^2 - 3 \int_0^t F(\beta) \sin (t - \beta) \, d\beta.$

7. $F(t) = t^2 - 2 \int_0^t F(t - \beta) \sinh 2\beta \, d\beta.$ ANS. $F(t) = t^2 - \frac{1}{3} t^4.$

8. $F(t) = 1 + 2 \int_0^t F(t - \beta) \cos \beta \, d\beta.$ ANS. $F(t) = 1 + 2te^t.$

In exercises 9 through 12 solve the given equation.

9. $H(t) = 9 e^{2t} - 2 \int_0^t H(t - \beta) \cos \beta \, d\beta.$

10. $H(y) = y^2 + \int_0^y H(x) \sin (y - x) \, dx.$ ANS. $H(y) = y^2 + \frac{1}{12} y^4.$

11. $g(x) = e^{-x} - 2 \int_0^x g(\beta) \cos (x - \beta) \, d\beta.$ ANS. $g(x) = e^{-x}(1 - x)^2.$

12. $y(t) = 6t + 4 \int_0^t (\beta - t)^2 y(\beta) \, d\beta.$ ANS. $y(t) = e^{2t} - e^{-t}(\cos\sqrt{3}t - \sqrt{3} \sin \sqrt{3}t).$

13. Solve the following equation for $F(t)$ with the condition that $F(0) = 4$:

$$F'(t) = t + \int_0^t F(t - \beta) \cos \beta \, d\beta.$$ ANS. $F(t) = 4 + \frac{5}{2} t^2 + \frac{1}{24} t^4.$

14. Solve the following equation for $F(t)$ with the condition that $F(0) = 0$:

$$F'(t) = \sin t + \int_0^t F(t - \beta) \cos \beta \, d\beta.$$ ANS. $F(t) = \frac{1}{2} t^2.$

15. Show that the equation of exercise 3 above can be put in the form

$$e^t F(t) = t e^t + \int_0^t e^\beta F(\beta) \, d\beta. \tag{A}$$

Differentiate each member of (A) with respect to t and thus replace the integral equation with a differential equation. Note that $F(0) = 0$. Find $F(t)$ by this method.

16. Solve the equation

$$\int_0^t F(t - \beta) e^{-\beta} \, d\beta = t$$

by two methods; use the convolution theorem and the basic idea introduced in exercise 15. Note that no differential equation need be solved in this instance.

72. Transform methods and the vibration of springs

All of the applications studied in Chapter 10 gave rise to linear differential equations with initial conditions. Those initial value problems were solved in that context by using the theory of linear differential equations developed in the earlier chapters. The same initial value problems may of course be solved by using Laplace transformations. We illustrate these techniques by reexamining some of the problems considered before.

EXAMPLE (a):   Solve the spring problem of Example (a), Section 52, with no damping but with $F(t) = A \sin \omega t$.
    As before, the problem to be solved is

$$x''(t) + \beta^2 x(t) = \sin \omega t, \tag{1}$$

with initial conditions

$$x(0) = x_0, \qquad x'(0) = v_0. \tag{2}$$

Let $L\{x(t)\} = u(s)$. Then (1) and (2) yield

$$s^2 u(s) - s x_0 - v_0 + \beta^2 u(s) = \frac{A\omega}{s^2 + \omega^2},$$

or

$$u(s) = \frac{s x_0 + v_0}{s^2 + \beta^2} + \frac{A\omega}{(s^2 + \beta^2)(s^2 + \omega^2)}. \tag{3}$$

The last term in (3) will lead to different inverse transforms according to whether $\omega = \beta$ or $\omega \neq \beta$. The case $\omega = \beta$ leads to resonance, which will be discussed in Example (d).
    If $\omega \neq \beta$, equation (3) yields

$$u(s) = \frac{s x_0 + v_0}{s^2 + \beta^2} + \frac{A\omega}{\omega^2 - \beta^2}\left(\frac{1}{s^2 + \beta^2} - \frac{1}{s^2 + \omega^2}\right). \tag{4}$$

From (4) it follows at once that

$$x(t) = x_0 \cos \beta t + v_0 \beta^{-1} \sin \beta t + \frac{A\omega}{\beta(\omega^2 - \beta^2)} \sin \beta t - \frac{A}{\omega^2 - \beta^2} \sin \omega t. \tag{5}$$

That the x of (5) is a solution of the problem (1) and (2) is easily verified. A study of (5) is simple and leads at once to conclusions such as that $x(t)$ is bounded, and so on. The first two terms on the right in (5) yield the natural harmonic component of the motion, the last two terms form the forced component.

This is the same solution that was found for the same problem in Section 52 using a very different approach.

EXAMPLE (b):   Solve the spring problem of Example (b), Section 52, using Laplace transformations.
The initial value problem is

$$x''(t) + 64x(t) = 0; \qquad x(0) = \tfrac{1}{3}, \qquad x'(0) = -2. \tag{6}$$

We let $L\{x(t)\} = u(s)$ and conclude at once that

$$s^2 u(s) - \tfrac{1}{3}s + 2 + 64u(s) = 0,$$

from which

$$u(s) = \frac{\tfrac{1}{3}s - 2}{s^2 + 64}.$$

Then

$$x(t) = \tfrac{1}{3}\cos 8t - \tfrac{1}{4}\sin 8t. \tag{7}$$

EXAMPLE (c):   A spring, with spring constant 0.75 lb/ft, lies on a long smooth (frictionless) table. A 6-lb weight is attached to the spring and is at rest (velocity zero) at the equilibrium position. A 1.5-lb force is applied to the support along the line of action of the spring for 4 sec and is then removed. Discuss the motion.
We must solve the problem

$$\tfrac{6}{32}x''(t) + \tfrac{3}{4}x(t) = H(t); \qquad x(0) = 0, x'(0) = 0, \tag{8}$$

in which

$$H(t) = 1.5, \qquad 0 < t < 4,$$
$$= 0, \qquad\qquad t > 4.$$

Now $H(t) = 1.5[1 - \alpha(t - 4)]$ in terms of the α function of Section 69. Therefore we rewrite our problem (8) in the form.

$$x''(t) + 4x(t) = 8[1 - \alpha(t - 4)]; \qquad x(0) = 0, x'(0) = 0. \tag{9}$$

Let $L\{x(t)\} = u(s)$. Then (9) yields

$$s^2 u(s) + 4u(s) = \frac{8}{s}(1 - e^{-4s}),$$

or

$$u(s) = \frac{8(1 - e^{-4s})}{s(s^2 + 4)}$$

$$= 2\left(\frac{1}{s} - \frac{s}{s^2 + 4}\right)(1 - e^{-4s}).$$

The desired solution is

$$x(t) = 2(1 - \cos 2t) - 2[1 - \cos 2(t - 4)]\alpha(t - 4). \tag{10}$$

Of course, the solution (10) can be broken down into the two relations

$$\text{for } 0 \leq t \leq 4, \qquad x(t) = 2(1 - \cos 2t), \tag{11}$$

$$\text{for } t > 4, \qquad x(t) = 2[\cos 2(t - 4) - \cos 2t], \tag{12}$$

if those forms seem simpler to use.

Verification of the solution (10), or (11) and (12), is direct. The student should show that

$$\lim_{t \to 4^-} x(t) = \lim_{t \to 4^+} x(t) = 2(1 - \cos 8) = 2.29$$

and

$$\lim_{t \to 4^-} x'(t) = \lim_{t \to 4^+} x'(t) = 4 \sin 8 = 3.96.$$

From (10) or (11) we see that in the range $0 < t < 4$, the maximum deviation of the weight from the starting point is $x = 4$ ft and occurs at $t = \frac{1}{2}\pi = 1.57$ sec. At $t = 4$, $x = 2.29$ ft, as shown above. For $t > 4$, equation (12) takes over and thereafter the motion is simple harmonic with a maximum x of 3.03 ft. Indeed, for $t > 4$,

$$\max |x(t)| = 2\sqrt{(1 - \cos 8)^2 + \sin^2 8}$$

$$= 2\sqrt{2}\sqrt{1 - \cos 8}$$

$$= 2\sqrt{2.2910} = 3.03.$$

Example (c) is one type of problem for which the Laplace transform technique is particularly useful. Such problems can be solved by the older classical methods, but with much less simplicity and dispatch.

EXAMPLE (d): Solve the problem of undamped vibration of a spring of Example (a) above in the case $\omega = \beta$.

Our problem is to solve

$$x''(t) + \beta^2 x(t) = A \sin \beta t; \qquad x(0) = x_0, x'(0) = v_0, \tag{13}$$

with the aid of

$$u(s) = \frac{sx_0 + v_0}{s^2 + \beta^2} + \frac{A\beta}{(s^2 + \beta^2)^2}. \tag{14}$$

We already know, from page 187, that

$$L^{-1}\left\{\frac{1}{(s^2 + \beta^2)^2}\right\} = \frac{1}{2\beta^3}(\sin \beta t - \beta t \cos \beta t).$$

Therefore (14) leads us to the solution

$$x(t) = x_0 \cos \beta t + \frac{v_0}{\beta}\sin \beta t + \frac{A}{2\beta^2}(\sin \beta t - \beta t \cos \beta t). \qquad (15)$$

Again this solution is the same as the solution obtained in equation (5), Section 53, and we have resonance occurring.

EXAMPLE (e): Solve the problem of the example of Section 54,

$$\tfrac{12}{32}x''(t) + 0.6x'(t) + 24x(t) = 0; \qquad x(0) = \tfrac{1}{3}, x'(0) = -2. \qquad (16)$$

Put $L\{x(t)\} = u(s)$. Then (16) yields

$$(s^2 + 1.6s + 64)u(s) = \tfrac{1}{3}(s - 4.4),$$

from which we obtain

$$x(t) = \tfrac{1}{3}L^{-1}\left\{\frac{s - 4.4}{(s + 0.8)^2 + 63.36}\right\}$$

$$= \tfrac{1}{3}\exp(-0.8t)L^{-1}\left\{\frac{s - 5.2}{s^2 + 63.36}\right\}.$$

Therefore the desired solution is

$$x(t) = \exp(-0.8t)(0.33\cos 8.0t - 0.22\sin 8.0t), \qquad (17)$$

a portion of its graph being shown in Figure 16.

Exercises

Each of the exercises of Chapter 10 is an appropriate exercise here. It would be instructive to solve a problem both with and without the Laplace transform and to compare the two methods.

73. The deflection of beams

As a further example of an application in which transform methods are useful, we consider a beam of length $2c$, as shown in Figure 28. Denote distance from one end of the beam by x, the deflection of the beam by y. If the beam is subjected to a vertical load $W(x)$, the deflection y must satisfy the equation

$$EI\frac{d^4y}{dx^4} = W(x), \qquad \text{for } 0 < x < 2c, \qquad (1)$$

in which E, the modulus of elasticity, and I, a moment of inertia, are known constants associated with the particular beam.

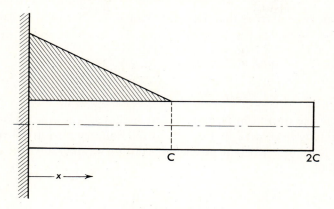

FIGURE 28

The slope of the curve of deflection is $y'(x)$, the bending moment is $EIy''(x)$, and the shearing force is $EIy'''(x)$. Common boundary conditions are of the following types:

(a) Beam imbedded in a support: $y = 0$ and $y' = 0$ at the point.
(b) Beam simply supported: $y = 0$ and $y'' = 0$ at the point.
(c) Beam free: $y'' = 0$ and $y''' = 0$ at the point.

Problems in the transverse displacement of a beam take the form of the differential equation (1) with boundary conditions at each end of the beam. Such problems can be solved by integration with the use of a little algebra. There are, however, two reasons for employing our transform method in such problems. Frequently the load function, or its derivative, is discontinuous. Beam problems also give us a chance to examine a useful device in which a problem over a finite range is solved with the aid of an associated problem over an infinite range.

EXAMPLE: Find the displacement y throughout the beam of Figure 28, in which the load is assumed to decrease uniformly from w_0 at $x = 0$ to zero at $x = c$ and to remain zero from $x = c$ to $x = 2c$. The weight of the beam is to be negligible. The beam is imbedded at $x = 0$ and free at $x = 2c$. We are to solve the problem

$$EI\frac{d^4y}{dx^4} = \frac{w_0}{c}[c - x + (x - c)\alpha(x - c)], \qquad \text{for } 0 < x < 2c; \qquad (2)$$

$$y(0) = 0, \qquad y'(0) = 0; \tag{3}$$

$$y''(2c) = 0, \qquad y'''(2c) = 0. \tag{4}$$

The student should verify that the right member of (2) is the stipulated load function

$$W(x) = \frac{w_0}{c}(c - x), \qquad \text{for } 0 \leqq x \leqq c, \tag{5}$$

$$= 0, \qquad \text{for } c < x \leqq 2c.$$

To apply the transform technique, with x playing the role for which we usually employ t, we need first to extend the range of x so it will run from 0 to ∞. That is, instead of the problem (2), (3), (4), we shall solve the problem consisting of

$$EI\frac{d^4y}{dx^4} = H(x), \qquad \text{for } 0 < x < \infty, \tag{6}$$

and the conditions (3) and (4). In (6) the function $H(x)$ is to be chosen by us except that $H(x)$ must agree with $W(x)$ over the range $0 < x < 2c$. The solution of the problem (6), (3), (4) will then be used only in the range $0 \leqq x \leqq 2c$. Of the various choices for $H(x)$, it seems simplest to use

$$H(x) = \frac{w_0}{c}[c - x + (x - c)\alpha(x - c)], \qquad \text{for } 0 < x < \infty. \tag{7}$$

That is, in practice we ordinarily retain the equation (2) and merely extend the range from $0 < x < 2c$ to $0 < x < \infty$. The student must, however, keep in mind that we cannot apply the Laplace operator to the function $W(x)$ of (5), since that function is not defined over the entire range $0 < x < \infty$. We shall solve (6) and conclude that the solution is valid for (2) on the range $0 \leqq x \leqq 2c$, over which (2) and (6) are identical.

Let

$$L\{EI\, y(x)\} = u(s);$$

$$u(s) = EI \int_0^\infty e^{-sx} y(x)\, dx. \tag{8}$$

To transform $EI y^{(4)}(x)$ we need to use the values of $EIy(x)$ and its first three derivatives at $x = 0$. From (3) we know that

$$EIy(0) = 0, \qquad EIy'(0) = 0.$$

Put

$$EIy''(0) = A, \qquad EIy'''(0) = B. \tag{9}$$

The constants A and B must be determined by using the conditions (4).

By our usual methods we obtain, for the $H(x)$ of (7),

$$L\{H(x)\} = \frac{w_0}{c} L\{c - x + (x - c)\alpha(x - c)\}$$

$$= \frac{w_0}{c}\left(\frac{c}{s} - \frac{1}{s^2} + \frac{e^{-cs}}{s^2}\right).$$

Thus the differential equation (6) is transformed into

$$s^4 u(s) - s^3 \cdot 0 - s^2 \cdot 0 - s \cdot A - B = \frac{w_0}{c}\left(\frac{c}{s} - \frac{1}{s^2} + \frac{e^{-cs}}{s^2}\right),$$

from which we get

$$u(s) = \frac{A}{s^3} + \frac{B}{s^4} + \frac{w_0}{c}\left(\frac{c}{s^5} - \frac{1}{s^6} + \frac{e^{-cs}}{s^6}\right). \tag{10}$$

Now $L^{-1}\{u(s)\} = EIy(x)$. Hence

$$EIy(x) = \tfrac{1}{2}Ax^2 + \tfrac{1}{6}Bx^3 + \frac{w_0}{120c}[5cx^4 - x^5 + (x - c)^5\alpha(x - c)]. \tag{11}$$

From (11) we obtain

$$EIy'(x) = Ax + \tfrac{1}{2}Bx^2 + \frac{w_0}{24c}[4cx^3 - x^4 + (x - 4)^4\alpha(x - c)], \tag{12}$$

$$EIy''(x) = A + Bx + \frac{w_0}{6c}[3cx^2 - x^3 + (x - c)^3\alpha(x - c)], \tag{13}$$

$$EIy'''(x) = B + \frac{w_0}{2c}[2cx - x^2 + (x - c)^2\alpha(x - c)]. \tag{14}$$

By differentiating both members of equations (14), we can see that the y of (11) is a solution of (6) over the infinite range and, more important, a solution of (2) over the range $0 < x < 2c$.

With the aid of equations (11) through (14) we can now determine A and B to make the y satisfy appropriate conditions at $x = 2c$, whether the beam be free, imbedded, or pin-supported there. In our example the beam is to be free at $x = 2c$; the solution is to satisfy the conditions

$$y''(2c) = 0, \qquad y'''(2c) = 0. \tag{4}$$

Using (13) and (14), and a little work, we find that (4) requires $A = \tfrac{1}{6}w_0c^2$, $B = -\tfrac{1}{2}w_0c$. We are thus led to the solution

$$EIy(x) = \tfrac{1}{12}w_0c^2x^2 - \tfrac{1}{12}w_0cx^3 + \frac{w_0}{120c}[5cx^4 - x^5 + (x - c)^5\alpha(x - c)], \tag{15}$$

for $0 \leq x \leq 2c$.

The student should verify by differentiations and appropriate substitutions that the y of (15) satisfies the original differential equation (2) and boundary conditions (3) and (4).

From (15) we can obtain whatever information we wish. For example, at $x = \frac{1}{2}c$ the bending moment is

$$EIy''(\tfrac{1}{2}c) = w_0 c^2[\tfrac{1}{6} - \tfrac{1}{4} + \tfrac{1}{6}(-\tfrac{1}{8} + \tfrac{3}{4} + 0)] = \tfrac{1}{48}w_0 c^2.$$

Exercises

In exercises 1 through 4, find the y that satisfies equation (1), page 226, with the given load function $W(x)$ and the given conditions at the ends of the beam. [See (a), (b), (c), page 227.] Verify your solutions.

1. $W(x)$ as in the example introduced on page 227, beam imbedded at both $x = 0$ and $x = 2c$.

   ANS. $EIy(x) = \tfrac{23}{480}w_0 c^2 x^2 - \tfrac{3}{40}w_0 c x^3$

   $$+ \frac{w_0}{120c}[5cx^4 - x^5 + (x - c)^5\alpha(x - c)].$$

2. $W(x) = 0,$ for $0 < x < \frac{1}{2}c$,

   $\qquad = w_0,$ for $\frac{1}{2}c < x < \frac{3}{2}c$,

   $\qquad = 0,$ for $\frac{3}{2}c < x < 2c$;

   beam imbedded at $x = 0$, free at $x = 2c$.
   ANS. $EIy(x) = \tfrac{1}{2}w_0 c^2 x^2 - \tfrac{1}{6}w_0 c x^3$

   $$+\tfrac{1}{24}w_0[(x - \tfrac{1}{2}c)^4\alpha(x - \tfrac{1}{2}c) - (x - \tfrac{3}{2}c)^4\alpha(x - \tfrac{3}{2}c)].$$

3. $W(x) = w_0[1 - \alpha(x - c)]$ (describe the load); beam to be imbedded at $x = 0$ and pin-supported (simply supported) at $x = 2c$.
   ANS. $EIy(x) = \tfrac{9}{64}w_0 c^2 x^2 - \tfrac{19}{128}w_0 c x^3 + \tfrac{1}{24}w_0[x^4 - (x - c)^4\alpha(x - c)].$

4. $W(x) = \dfrac{w_0}{c}(2c - x),$ for $0 < x < c$,

   $\qquad = w_0,$ for $c < x < 2c$;

   beam to be imbedded at $x = 0$ and free at $x = 2c$.
   ANS. $EIy(x) = \tfrac{13}{12}w_0 c^2 x^2 - \tfrac{5}{12}w_0 c x^3 + \tfrac{1}{24}w_0 x^4$

   $$+ \frac{w_0}{120c}[5cx^4 - x^5 + (x - c)^5\alpha(x - c)].$$

TABLE OF TRANSFORMS

Whenever n is used, it denotes a nonnegative integer. The range of validity may be determined from the appropriate text material. Many other transforms will be found in the examples and exercises.

$f(s) = L\{F(t)\}$	$F(t)$
$f(s - a)$	$e^{at} F(t)$
$f(as + b)$	$\dfrac{1}{a} \exp\left(-\dfrac{bt}{a}\right) F\left(\dfrac{t}{a}\right)$
$\dfrac{1}{s} e^{-cs}, \quad c > 0$	$\alpha(t - c) = 0, \quad 0 \leq t < c,$ $\qquad\qquad = 1, \quad t \geq c$
$e^{-cs} f(s), \quad c > 0$	$F(t - c)\alpha(t - c)$
$f_1(s) f_2(s)$	$\displaystyle\int_0^t F_1(\beta) F_2(t - \beta)\, d\beta$
$\dfrac{1}{s}$	1
$\dfrac{1}{s^{n+1}}$	$\dfrac{t^n}{n!}$
$\dfrac{1}{s^{x+1}}, \quad x > -1$	$\dfrac{t^x}{\Gamma(x + 1)}$
$s^{-1/2}$	$(\pi t)^{-1/2}$
$\dfrac{1}{s + a}$	e^{-at}
$\dfrac{1}{(s + a)^{n+1}}$	$\dfrac{t^n e^{-at}}{n!}$
$\dfrac{k}{s^2 + k^2}$	$\sin kt$
$\dfrac{s}{s^2 + k^2}$	$\cos kt$
$\dfrac{k}{s^2 - k^2}$	$\sinh kt$
$\dfrac{s}{s^2 - k^2}$	$\cosh kt$

$f(s) = L\{F(t)\}$	$F(t)$
$\dfrac{2k^3}{(s^2 + k^2)^2}$	$\sin kt - kt \cos kt$
$\dfrac{2ks}{(s^2 + k^2)^2}$	$t \sin kt$
$\ln\left(1 + \dfrac{1}{s}\right)$	$\dfrac{1 - e^{-t}}{t}$
$\ln\dfrac{s + k}{s - k}$	$\dfrac{2 \sinh kt}{t}$
$\ln\left(1 - \dfrac{k^2}{s^2}\right)$	$\dfrac{2}{t}(1 - \cosh kt)$
$\ln\left(1 + \dfrac{k^2}{s^2}\right)$	$\dfrac{2}{t}(1 - \cos kt)$
$\arctan\dfrac{k}{s}$	$\dfrac{\sin kt}{t}$

Linear Systems of Equations

74. Introduction

We shall see in the next chapter that certain problems in the studies of electrical circuits and arms races lead naturally to systems of linear differential equations with constant coefficients. Although the subject of systems of equations can be studied in a wider context involving coefficients that are not constant, we shall not do so in this book.

75. Elementary elimination calculus

Let us consider the problem of finding solutions for the system of equations

$$y'' - y + 5v' = x, \tag{1}$$
$$2y' - v'' + 4v = 2,$$

with independent variable x and dependent variables y and v. It is reasonable to attack the system (1) by eliminating one dependent variable to obtain a single equation in the other dependent variable. Using $D = d/dx$, write the system (1) in the form

$$(D^2 - 1)y + 5Dv = x, \tag{2}$$

$$2Dy - (D^2 - 4)v = 2.$$

Then, elimination of one dependent variable is straightforward. We may, for instance, operate upon the first equation with $2D$ and upon the second equation with $(D^2 - 1)$, and then subtract one from the other, obtaining

$$[10D^2 + (D^2 - 1)(D^2 - 4)]v = 2Dx - (D^2 - 1)2,$$

or

$$(D^4 + 5D^2 + 4)v = 4. \tag{3}$$

In a similar manner v may be eliminated; the resultant equation for y is

$$[(D^2 - 1)(D^2 - 4) + 10D^2]y = (D^2 - 4)x + 5D(2),$$

or

$$(D^4 + 5D^2 + 4)y = -4x. \tag{4}$$

From equations (3) and (4) it follows at once that

$$v = 1 + a_1 \cos x + a_2 \sin x + a_3 \cos 2x + a_4 \sin 2x \tag{5}$$

and

$$y = -x + b_1 \cos x + b_2 \sin x + b_3 \cos 2x + b_4 \sin 2x. \tag{6}$$

The a's and b's have yet to be chosen to make (5) and (6) satisfy the original equations, rather than just the equations (3) and (4), which resulted from the original ones after certain eliminations were performed.

Combining (by substitution) the v of (5) and the y of (6) with the first equation of the system (2) leads to the identity

$$x - 2b_1 \cos x - 2b_2 \sin x - 5b_3 \cos 2x - 5b_4 \sin 2x \tag{7}$$

$$- 5a_1 \sin x + 5a_2 \cos x - 10a_3 \sin 2x + 10a_4 \cos 2x \equiv x.$$

That (7) be an identity in x demands that

$$-2b_1 + 5a_2 = 0, \tag{8}$$

$$-2b_2 - 5a_1 = 0,$$

$$-5b_3 + 10a_4 = 0,$$

$$-5b_4 - 10a_3 = 0.$$

Relations between the a's and b's equivalent to the relations (8) follow from substitution of the v of (5) and the y of (6) into the second equation of the system (2).

We conclude that a set of solutions of the system (2) is

$$v = 1 + a_1 \cos x + a_2 \sin x + a_3 \cos 2x + a_4 \sin 2x, \tag{9}$$

$$y = -x + \tfrac{5}{2}a_2 \cos x - \tfrac{5}{2}a_1 \sin x + 2a_4 \cos 2x - 2a_3 \sin 2x,$$

in which a_1, a_2, a_3, a_4 are arbitrary constants.

The equations (3) and (4) for v and y can be written with the aid of determinants. From the system (2) above we may write at once

$$\begin{vmatrix} (D^2 - 1) & 5D \\ \\ 2D & -(D^2 - 4) \end{vmatrix} v = \begin{vmatrix} (D^2 - 1) & x \\ \\ 2D & 2 \end{vmatrix},$$

which reduces to equation (3) above, if care is used in the interpretation of the right-hand member. The determinant on the right is to be interpreted as

$$(D^2 - 1)(2) - 2D(x),$$

not as the differential operator $2(D^2 - 1) - x(2D)$.

Determinants are extremely useful in any treatment of the theory of systems of linear equations. For many simple systems that arise in practice, no such powerful tool is needed.

There are other techniques for treating a system such as (2). For example, we may first obtain (3) and from it v as in equation (5). Next we wish to find an equation giving y in terms of v; that is, we seek to eliminate from the system (2) those terms that involve derivatives of y. From (2) we obtain the two equations

$$(2D^2 - 2)y + 10Dv = 2x, \tag{10}$$

$$2D^2y - (D^3 - 4D)v = 0, \tag{11}$$

the latter by operating with D on each member of the second equation of the system (2). From (10) and (11) it follows at once that

$$2y - D^3v - 6Dv = -2x,$$

or

$$y = -x + \tfrac{1}{2}D^3v + 3Dv. \tag{12}$$

The v of equation (5) may now be used in (12) to compute the y, which was given in the solution (9). In this method of solution there is no need to obtain (4), (6), (7), or (8).

Exercises

Use elementary elimination calculus to solve the following systems of equations.

1. $u' = 4u - v,$
   $v' = -4u + 4v.$

ANS. $u = c_1 e^{2x} + c_2 e^{6x},$
$v = 2c_1 e^{2x} - 2c_2 e^{6x}.$

2. $w' = w - y - z,$
   $y' = y + 3z,$
   $z' = 3y + z.$

ANS. $w = c_1 e^x + 2c_2 e^{4x},$
$y = -3c_2 e^{4x} + c_3 e^{-2x},$
$z = -3c_2 e^{4x} - c_3 e^{-2x}.$

3. $y' = 2y + z,$
   $z' = -4y + 2z.$

ANS. $y = c_1 e^{2t} \cos 2t + b_2 e^{2t} \sin 2t,$
$z = -2c_1 e^{2t} \sin 2t + 2b_2 e^{2t} \cos 2t.$

4. $dx/dt = y,$
   $dy/dt = -4x + 4y.$

ANS. $x = c_1 e^{2t} + c_2t e^{2t},$
$y = 2c_1 e^{2t} + c_2(2t e^{2t} + e^{2t}).$

5. $v' - 2v + 2w' = 2 - 4 e^{2x},$
   $2v' - 3v + 3w' - w = 0.$

ANS. $v = a_1 e^x + a_2 e^{-2x} + 5 e^{2x} - 1,$
$w = \tfrac{1}{2}a_1 e^x - a_2 e^{-2x} - e^{2x} + 3.$

6. $(3D + 2)v + (D - 6)w = 5 e^x,$
   $(4D + 2)v + (D - 8)w = 5 e^x + 2x - 3.$

ANS. $v = a_1 \cos 2x + a_2 \sin 2x + 2 e^x - 3x + 5,$
$w = a_2 \cos 2x - a_1 \sin 2x + e^x - x.$

7. $(D^2 + 6)y + Dv = 0,$
   $(D + 2)y + (D - 2)v = 2.$

ANS. $y = c_1 e^{3x} + c_2 \cos 2x + c_3 \sin 2x,$
$v = -1 - 5c_1 e^{3x} + c_3 \cos 2x - c_2 \sin 2x.$

8. $D^2y - (2D - 1)v = 1,$
   $(2D + 1)y + (D^2 - 4)v = 0.$

ANS. $v = 1 + a_1 e^x + a_2 e^{-x} + a_3 \cos x + a_4 \sin x,$
$y = 4 + a_1 e^x - 3a_2 e^{-x} + (a_3 - 2a_4) \cos x + (2a_3 + a_4) \sin x.$

9. $(D^2 - 3D)y - (D - 2)z = 14x + 7,$
   $(D - 3)y + Dz = 1.$

ANS. $y = 2 + a_1 e^x + a_2 e^{3x} + a_3 e^{-2x},$
$z = 7 + 7x + 2a_1 e^x - \tfrac{5}{2}a_3 e^{-2x}.$

10. $(D^3 + D^2 - 1)u + (D^3 + 2D^2 + 3D + 1)v = 3 - x,$
   $(D - 1)u + (D + 1)v = 3 - x.$

ANS. $u = 2x + a_1 + a_2 \cos x + a_3 \sin x,$
$v = x + a_1 - a_3 \cos x + a_2 \sin x.$

11. $(D^2 + 1)y + 4(D - 1)v = 4 e^x,$
   $(D - 1)y + (D + 9)v = 0;$ when $x = 0, y = 5, y' = 0, v = \tfrac{1}{2}.$

ANS. $y = 2 e^x + 2 e^{-x} + e^{-2x}(\cos x + 2 \sin x),$
$v = \tfrac{1}{2}e^{-x} + e^{-2x} \sin x.$

12. $\dfrac{d^2x}{dt^2} + x - 2\dfrac{dy}{dt} = 2t,$

$2\dfrac{dx}{dt} - x + \dfrac{dy}{dt} - 2y = 7.$

ANS. $x = 2t - 2 + b_1 e^t + b_2 e^{-t} + b_3 e^{-2t},$
$y = -t - 1 + b_1 e^t - b_2 e^{-t} - \tfrac{5}{4}b_3 e^{-2t}.$

13. $2(D + 1)y + (D - 1)w = x + 1,$
   $(D + 3)y + (D + 1)w = 4x + 14.$

ANS. $y = x + 2 + c_1 e^{-x} \cos 2x + c_2 e^{-x} \sin 2x,$
$w = x + 6 + (c_2 - c_1) e^{-x} \cos 2x - (c_1 + c_2) e^{-x} \sin 2x.$

14. $(D + 1)y + (D - 4)v = 6 \cos x,$
   $(D - 1)y + (D^2 + 4)v = -6 \sin x.$

ANS. $y = 3 \cos x + 3 \sin x + a_1 + a_2 \cos 3x + a_3 \sin 3x,$
$v = \tfrac{1}{4}a_1 - \tfrac{1}{5}(a_2 - 3a_3) \cos 3x - \tfrac{1}{5}(3a_2 + a_3) \sin 3x.$

15. $2Du + (D - 1)v + (D + 2)w = 0,$
   $(D + 2)u + (2D - 3)v - (D - 6)w = 0,$
   $2Du - (D + 3)v - Dw = 0.$

ANS. $u = 3c_1 e^{2x} + 3c_2 e^{-x} + 3c_3 e^{x/2},$
$v = 4c_1 e^{2x} - 5c_2 e^{-x} + c_3 e^{x/2},$
$w = -4c_1 e^{2x} - 4c_2 e^{-x} - c_3 e^{x/2}.$

16. $D^2y + (D - 1)v = 0,$
   $(2D - 1)y + (D - 1)w = 0,$
   $(D + 3)y + (D - 4)v + 3w = 0.$

ANS. $y = a_1 + a_2 e^x + 3a_3 e^{4x},$
$v = b_2 e^x - a_2x e^x - 16a_3 e^{4x},$
$w = -a_1 + (b_2 - a_2) e^x - a_2x e^x - 7a_3 e^{4x}.$

76. First order systems with constant coefficients

In the following sections we shall show how matrix algebra can be used to reduce the problem of solving systems of differential equations to an algebraic routine. Before we do that it is important to realize that a system of linear equations of order higher than the first can be written in terms of a first-order system.

Consider for example the single equation

$$y'' + 2y' - y = e^x. \tag{1}$$

If we let $u = y'$ then equation (1) becomes

$$u' = y - 2u + e^x.$$

In other words the single second-order equation (1) has been replaced by the first-order system

$$y' = u, \tag{2}$$

$$u' = y - 2u + e^x.$$

In a similar manner the third-order equation

$$y''' + 2y'' - y' + 3y = x \tag{3}$$

can be written as a system of first order equations by choosing new variables

$$u = y', \quad \text{and} \quad v = u' = y''.$$

Then equation (3) becomes

$$v' = -2v + u - 3y + x,$$

and we can consider the first-order system

$$y' = u,$$
$$u' = v, \qquad\qquad (4)$$
$$v' = u - 2v - 3y + x,$$

as being equivalent to (3).

The system of second-order equations (1) in Section 75 can be replaced by a first-order system if we let $u = v'$ and $w = y'$ so that

$$u' = 4v + 2w - 2,$$
$$v' = u,$$
$$w' = -5u + y + x, \qquad\qquad (5)$$
$$y' = w.$$

Exercises

In exercises 1 through 5, replace the given equation by a system of first order equations.

1. $y'' - 6y' + 8y = x + 2.$ ANS. $\quad y' = u,$
$$u' = 6u - 8y + x + 2.$$

2. $y'' + 4y' + 4y = e^x.$ ANS. $\quad y' = u,$
$$u' = -4u - 4y + e^x.$$

3. $y'' + py' + qy = f(x).$ ANS. $\quad y' = u,$
$$u' = -pu - qy + f(x).$$

4. $y''' + py'' + qy' + ry = f(x).$ ANS. $\quad y' = u,$
$$u' = v,$$
$$v' = -pv - qu - ry + f(x).$$

5. $y^{(4)} - y = 0.$ ANS. $\quad y' = u,$
$$u' = v,$$
$$v' = w,$$
$$w' = y.$$

In exercises 6 through 9, replace the given system by an equivalent system of first order equations.

6. Exercise 5, Section 75.
7. Exercise 6, Section 75.
8. Exercise 7, Section 75.
9. Exercise 8, Section 75.

77. Solution of a first order system

Consider the first order system

$$\frac{dx}{dt} = y, \tag{1}$$

$$\frac{dy}{dt} = -2x + 3y.$$

We can rewrite this system in the form

$$Dx - y = 0, \tag{2}$$

$$2x + (D - 3)y = 0.$$

Operating on the first equation with the operator $D - 3$ and adding the two equations eliminates the variable y to give

$$(D^2 - 3D + 2)x = 0. \tag{3}$$

In a similar manner we can eliminate x from system (3) to obtain

$$(D^2 - 3D + 2)y = 0. \tag{4}$$

Thus we realize that the solutions of equations (1) are of the form

$$x = c_1 e^{2t} + c_2 e^t,$$

$$y = c_3 e^{2t} + c_4 e^t,$$

where there are some relations among the four constants c_1, c_2, c_3, c_4 that we can determine by substitution back into equations (1).

An alternative way of viewing the nature of the solutions of system (1) is to expect from the beginning the existence of solutions of the form

$$x = c_1 e^{mt}, \tag{5}$$

$$y = c_2 e^{mt},$$

where the constants c_1, c_2, and m must be determined by substitution into (1). If we do this we find

$$c_1 m e^{mt} = c_2 e^{mt},$$

and

$$c_2 m e^{mt} = -2c_1 e^{mt} + 3c_2 e^{mt},$$

or

$$-mc_1 + c_2 = 0, \tag{6}$$

$$-2c_1 + (3 - m)c_2 = 0.$$

The system (6) can have nontrivial solutions for c_1 and c_2 only if the determinant

$$\begin{vmatrix} -m & 1 \\ -2 & 3-m \end{vmatrix} \tag{7}$$

is zero. That is,

$$m^2 - 3m + 2 = (m-1)(m-2) = 0.$$

Moreover, for the choice $m = 1$, the system (6) yields the condition $c_2 = c_1$, and for $m = 2$ we would be forced to take $c_2 = 2c_1$. Thus there would be two distinct solutions of the form of equations (5), namely

$$\begin{aligned} x &= c_1\, e^t, \\ y &= c_1\, e^t, \end{aligned} \quad \text{and} \quad \begin{aligned} x &= c_1\, e^{2t}, \\ y &= 2c_1\, e^{2t}. \end{aligned} \tag{8}$$

A careful perusal of what we have done here should lead one to suspect that the elementary algebra problem of finding nontrivial solutions to the system (6) holds the entire key to the problem of solving the system of differential equations (1). The formalization of this procedure is best accomplished with the assistance of some vector and matrix notation. The next section summarizes the minimum amount of matrix algebra required for our purposes.

78. Some matrix algebra

We shall presume at this point that the student is familiar with the elementary calculus of vector functions. The basic ideas involved are deduced from the definition

$$\frac{d}{dt}\Big(f_1(t), f_2(t), \ldots, f_n(t)\Big) = \left(\frac{df_1}{dt}, \frac{df_2}{dt}, \ldots, \frac{df_n}{dt}\right) \tag{1}$$

and from the properties of vector addition, multiplication of vectors by numbers, and the scalar product of vectors. Most elementary calculus texts and courses provide all of the required algebra.

It is not always the case that students who have completed a course in elementary calculus are familiar with matrix algebra. We therefore include here a brief introduction suitable for our purposes.

A *matrix* is a rectangular array of numbers. For example, the following arrays are matrices:

$$A = \begin{pmatrix} 2 & 1 \\ 1 & 3 \end{pmatrix}, \qquad B = \begin{pmatrix} 3 & 2 \\ 1 & 1 \\ 2 & -1 \end{pmatrix}, \qquad C = \begin{pmatrix} 1 & 2 & 1 \\ 3 & 1 & 2 \end{pmatrix}.$$

Each of the numbers in a matrix is called an *element* of that matrix.

A matrix is said to be of *dimension* $n \times m$ (n by m) if it has n rows and m columns. Thus the general $n \times m$ matrix can be written

$$\begin{pmatrix} a_{11} & a_{12} \ldots a_{1m} \\ a_{21} & a_{22} \ldots a_{2m} \\ \vdots \\ a_{n1} & a_{n2} \ldots a_{nm} \end{pmatrix},$$

where a_{ij} indicates the number in the ith row and jth column.

Two matrices of the same dimension are said to be *equal* if the corresponding elements are equal.

Addition is defined only for two matrices of the same dimension, and is defined elementwise. For example, the sum of two 2×3 matrices is accomplished as follows:

$$\begin{pmatrix} a & b & c \\ d & e & f \end{pmatrix} + \begin{pmatrix} g & h & i \\ j & k & l \end{pmatrix} = \begin{pmatrix} a+g & b+h & c+i \\ d+j & e+k & f+l \end{pmatrix}. \tag{2}$$

Any matrix may be multiplied by a number by multiplying each of its elements by that number. For example,

$$k \begin{pmatrix} a & b \\ c & d \\ e & f \end{pmatrix} = \begin{pmatrix} ka & kb \\ kc & kd \\ ke & kf \end{pmatrix}. \tag{3}$$

The student should recognize that the algebra of matrices, which is dictated by the definitions in the preceding paragraphs, is essentially the same kind of algebra as the algebra of vectors. This comes from the fact that the operations of addition and multiplication by numbers are done elementwise both for matrix algebra and vector algebra. Indeed, one can regard the vector (a, b, c) as a 1×3 matrix and then interpret the usual vector algebra as a special case of the matrix algebra described above. The only caution to be observed is that the matrices

$$(a \quad b \quad c) \text{ and } \begin{pmatrix} a \\ b \\ c \end{pmatrix}$$

are very different from the matrix point of view, although we may wish to identify them as the same vector in some physical or geometrical context. We shall call the first vector a row vector and the second a column vector.

In what follows we shall often form the product of a row vector, a matrix of dimension $1 \times n$, with a column vector, a matrix of dimension $n \times 1$. The product is the familiar scalar product of elementary calculus

$$(a_1 \quad a_2 \quad \cdots \quad a_n) \cdot \begin{pmatrix} b_1 \\ b_2 \\ \vdots \\ b_n \end{pmatrix} = a_1 b_1 + a_2 b_2 + \cdots + a_n b_n, \tag{4}$$

with the insistence that the row vector always be written on the left and the column vector on the right.

The *product* of two matrices can now be defined in terms of scalar products of row and column vectors. An $n \times m$ matrix and a $p \times q$ matrix can be multiplied only if $m = p$; that is, the number of columns of the first matrix must equal the number of rows in the second matrix. The resulting product matrix has dimension $n \times q$. The definition is most easily described as follows:

$$A \cdot B = \begin{pmatrix} A_1 \cdot B^1 & A_1 \cdot B^2 & \cdots & A_1 \cdot B^q \\ A_2 \cdot B^1 & A_2 \cdot B^2 & \cdots & A_2 \cdot B^q \\ & \vdots & & \\ A_n \cdot B^1 & A_n \cdot B^2 & \cdots & A_n \cdot B^q \end{pmatrix}, \tag{5}$$

where A_i is the ith row vector of the matrix A and B^j is the jth column vector of the matrix B. Thus the element in the ith row and jth column of the product is the ordinary scalar product of the vectors A_i and B^j.

A few examples will help fix these definitions in our minds.

EXAMPLE (a):

$$\begin{pmatrix} 2 & 1 \\ -1 & 3 \end{pmatrix} + 2 \begin{pmatrix} -1 & 2 \\ 1 & 1 \end{pmatrix} = \begin{pmatrix} 2 & 1 \\ -1 & 3 \end{pmatrix} + \begin{pmatrix} -2 & 4 \\ 2 & 2 \end{pmatrix} = \begin{pmatrix} 0 & 5 \\ 1 & 5 \end{pmatrix}.$$

EXAMPLE (b):

$$\begin{pmatrix} t^2 + t \\ t^2 - t \end{pmatrix} = \begin{pmatrix} t^2 \\ t^2 \end{pmatrix} + \begin{pmatrix} t \\ -t \end{pmatrix} = \begin{pmatrix} 1 \\ 1 \end{pmatrix} t^2 + \begin{pmatrix} 1 \\ -1 \end{pmatrix} t.$$

EXAMPLE (c):

$$\begin{pmatrix} 2 & 1 \\ -1 & 3 \end{pmatrix}\begin{pmatrix} x \\ y \end{pmatrix} = \begin{pmatrix} 2x + y \\ -x + 3y \end{pmatrix}.$$

EXAMPLE (d):

$$\begin{pmatrix} a & b \\ c & d \end{pmatrix} \cdot \begin{pmatrix} p & q \\ r & s \end{pmatrix} = \begin{pmatrix} ap + br & aq + bs \\ cp + dr & cq + ds \end{pmatrix}.$$

EXAMPLE (e):

$$\begin{pmatrix} 2 & 1 \\ 1 & -1 \end{pmatrix} \cdot \begin{pmatrix} 3 & 1 \\ 4 & 1 \end{pmatrix} = \begin{pmatrix} 2\cdot 3 + 1\cdot 4 & 2\cdot 1 + 1\cdot 1 \\ 1\cdot 3 + (-1)\cdot 4 & 1\cdot 1 + (-1)\cdot 1 \end{pmatrix} = \begin{pmatrix} 10 & 3 \\ -1 & 0 \end{pmatrix}.$$

EXAMPLE (f):   The product

$$\begin{pmatrix} 1 & 2 \\ 1 & 3 \end{pmatrix} \cdot \begin{pmatrix} 1 & 1 \\ 2 & 1 \\ 3 & 1 \end{pmatrix}$$

is not defined because the number of columns in the first matrix is two and the number of rows in the second matrix is not two. On the other hand

$$\begin{pmatrix} 1 & 1 \\ 2 & 1 \\ 3 & 1 \end{pmatrix} \cdot \begin{pmatrix} 1 & 2 \\ 1 & 3 \end{pmatrix} = \begin{pmatrix} 2 & 5 \\ 3 & 7 \\ 4 & 9 \end{pmatrix}$$

is defined, a fact that demonstrates that matrix multiplication is not commutative.

In elementary vector algebra we often find it convenient to designate a vector by a single symbol instead of expressing the vector in terms of its components. We shall frequently use a single symbol to refer to a matrix. Particular caution must be observed when using such abbreviations to be sure that the dimensions of the objects involved are appropriate in terms of the definitions in our algebra. For example, the equation

$$\frac{dX}{dt} = X' = AX,$$

can be interpreted as a system of first-order linear equations with constant coefficients, if we interpret X as an n-dimensional column vector function of t

and A as an $n \times n$ matrix of real numbers. Thus

$$\frac{dx}{dt} = 2x + y,$$

$$\frac{dy}{dt} = x - y,$$

can be written

$$X' = AX,$$

where

$$A = \begin{pmatrix} 2 & 1 \\ 1 & -1 \end{pmatrix}, \qquad X = \begin{pmatrix} x \\ y \end{pmatrix}, \qquad X' = \begin{pmatrix} \dfrac{dx}{dt} \\ \dfrac{dy}{dt} \end{pmatrix}.$$

We shall use capital letters here for matrices and lower case letters for numbers. In particular, we shall use a large zero for a matrix, all of whose elements are zero wherever the dimension of that zero matrix is clear from the context.

EXAMPLE (g):   If

$$A = \begin{pmatrix} 2 & 1 & 1 \\ 3 & 2 & 1 \\ 1 & 4 & 1 \end{pmatrix}$$

and $X' - AX = O$, then X must be a three-dimensional column vector function and O must be the three-dimensional column vector

$$\begin{pmatrix} 0 \\ 0 \\ 0 \end{pmatrix}.$$

We shall single out one further kind of matrix for special attention. If we multiply any 2×2 matrix A by the matrix

$$I = \begin{pmatrix} 1 & 0 \\ 0 & 1 \end{pmatrix},$$

we clearly obtain the results

$$AI = A \qquad \text{and} \qquad IA = A.$$

A similar observation can be made for any $n \times n$ matrix providing I is interpreted as the $n \times n$ matrix modelled after the two-dimensional case, namely

$$I = (a_{ij}), \qquad \text{where} \quad a_{ij} = 0, \text{ if } i \neq j,$$
$$\text{and} \qquad a_{ij} = 1, \text{ if } i = j.$$

Again, we shall use the symbol I wherever the dimension is clear from the context.

The algebraic structure that one obtains as a consequence of the above definitions is called the algebra of matrices. Some of the basic theorems of this algebra are listed below. Most are easily proved in the cases where the dimensions are low and rather tedious for large matrices. We shall ask the student to prove some of the theorems in the exercises that follow. In each theorem it is presumed that the matrices involved have dimensions for which the indicated operations are defined.

$$(A + B) + C = A + (B + C). \tag{6}$$

$$A + B = B + A. \tag{7}$$

$$A + O = A. \tag{8}$$

$$A + (-1)A = O. \tag{9}$$

$$(AB)C = A(BC). \tag{10}$$

$$A(B + C) = AB + AC. \tag{11}$$

$$IA = AI = A. \tag{12}$$

$$k(A + B) = kA + kB. \tag{13}$$

$$kO = O. \tag{14}$$

$$k(AB) = (kA)B = A(kB). \tag{15}$$

As a final theorem for this section we state, without proof, a basic theorem from elementary algebra that we have often used in previous chapters.

THEOREM 24:   *Let A be an $n \times n$ matrix of constant real numbers and let X be an n-dimensional column vector. The system of equations*

$$AX = O$$

has nontrivial solutions, that is $X \neq O$, if and only if the determinant of A is zero.

Exercises

In exercises 1 through 8 find the requested matrix given the following matrices:

$$A = \begin{pmatrix} 1 & 2 \\ 3 & 1 \end{pmatrix}, \qquad B = \begin{pmatrix} 2 & 0 \\ 1 & -1 \end{pmatrix}, \qquad C = \begin{pmatrix} 1 & -1 \\ 1 & 2 \end{pmatrix}.$$

1. $A + 2B$.
   ANS. $\begin{pmatrix} 5 & 2 \\ 5 & -1 \end{pmatrix}$.

2. $2A + C$.

3. $AB + 2I$.
   ANS. $\begin{pmatrix} 6 & -2 \\ 7 & 1 \end{pmatrix}$.

4. $AC + BI$.

5. $C - 2I$.
   ANS. $\begin{pmatrix} -1 & -1 \\ 1 & 0 \end{pmatrix}$.

6. $AB + C$.

7. $AC - B$.
   ANS. $\begin{pmatrix} 1 & 3 \\ 3 & 0 \end{pmatrix}$.

8. AB and BA.
9. Prove $D + E = E + D$ for any matrices D and E for which the sum is defined.
10. Prove the distributive law of equation (11) above for 2×2 matrices.
11. Prove the theorem of equation (8).
12. Prove the theorem of equation (9).
13. Prove the theorem of equation (13).
14. Prove the theorem of equation (15) for 2×2 matrices.
15. Show that the system of equations

$$\begin{pmatrix} 1 & 2 \\ 1 & -1 \end{pmatrix} \begin{pmatrix} x \\ y \end{pmatrix} = O$$

has no nontrivial solutions.

Find all of the solutions of the systems in exercises 16 through 21.

16. $\begin{pmatrix} 1 & 2 \\ 2 & 4 \end{pmatrix} \begin{pmatrix} x \\ y \end{pmatrix} = O$.
    ANS. $\begin{pmatrix} x \\ y \end{pmatrix} = c \begin{pmatrix} -2 \\ 1 \end{pmatrix}$.

17. $\begin{pmatrix} 2 & 1 \\ -4 & -2 \end{pmatrix} \begin{pmatrix} x \\ y \end{pmatrix} = O$.
    ANS. $\begin{pmatrix} x \\ y \end{pmatrix} = c \begin{pmatrix} 1 \\ -2 \end{pmatrix}$.

18. $\begin{pmatrix} 1 & 4 \\ -1 & 1 \end{pmatrix} \begin{pmatrix} x \\ y \end{pmatrix} = O$.
    ANS. $\begin{pmatrix} x \\ y \end{pmatrix} = O$.

19. $\begin{pmatrix} 1 & 2 & 1 \\ 1 & -1 & 1 \\ 2 & 1 & 2 \end{pmatrix} \begin{pmatrix} x \\ y \\ z \end{pmatrix} = O$.
    ANS. $\begin{pmatrix} x \\ y \\ z \end{pmatrix} = c \begin{pmatrix} 1 \\ 0 \\ -1 \end{pmatrix}$.

20. $\begin{pmatrix} 1 & -1 & 2 \\ 2 & 1 & 3 \\ 0 & -1 & 1 \end{pmatrix} \begin{pmatrix} x \\ y \\ z \end{pmatrix} = O.$

ANS. $\begin{pmatrix} x \\ y \\ z \end{pmatrix} = O.$

21. $\begin{pmatrix} 2 & 1 & 3 \\ -1 & 1 & 2 \\ 5 & 1 & 4 \end{pmatrix} \begin{pmatrix} x \\ y \\ z \end{pmatrix} = O.$

ANS. $\begin{pmatrix} x \\ y \\ z \end{pmatrix} = c \begin{pmatrix} -1 \\ -7 \\ 3 \end{pmatrix}.$

In exercises 22 through 25, write the given system of differential equations as a matrix equation.

22. $\dfrac{dx}{dt} = 2x + 3y,$

$\dfrac{dy}{dt} = x - y.$

ANS. $X' = AX$, where

$$A = \begin{pmatrix} 2 & 3 \\ 1 & -1 \end{pmatrix}.$$

23. $\dfrac{dx}{dt} = x - y + z + t,$

$\dfrac{dy}{dt} = x + 2y - z + 1,$

$\dfrac{dz}{dt} = 2x - y + z + e^t.$

ANS. $X' = AX + B$, where

$$A = \begin{pmatrix} 1 & -1 & 1 \\ 1 & 2 & -1 \\ 2 & -1 & 1 \end{pmatrix} \quad \text{and} \quad B = \begin{pmatrix} t \\ 1 \\ e^t \end{pmatrix}.$$

24. $\dfrac{dx}{dt} = 2x - y + e^t,$

$\dfrac{dy}{dt} = x + y + t.$

25. $\dfrac{dx}{dt} = tx + y + \sin t,$

$\dfrac{dy}{dt} = t^2 x + ty + 1.$

79. First-order systems revisited

We return now to the consideration of first-order linear systems of equations with constant coefficients. Let X be an n-dimensional column vector function of t and let A be an $n \times n$ matrix of real numbers. Further suppose $B(t)$ is a column vector whose components are known functions of t. Then the vector equation

$$X' = AX + B \tag{1}$$

represents a system of n equations in the n unknown component functions of X. If $B = O$, we say that the system is homogeneous. For the present we will restrict our attention to homogeneous systems and return to the non-homogeneous case in a later section.

From our past experience we have reason to believe that the homogeneous system

$$X' = AX \tag{2}$$

may have solutions of the form

$$X = C e^{mt}, \tag{3}$$

where C is a constant column vector and m is some number that we wish to determine. Substitution of X into system (2) yields

$$Cm \, e^{mt} = AC \, e^{mt},$$

which can be written

$$(AC - mC) \, e^{mt} = O.$$

We can rewrite this equation again, remembering that $C = IC$, in the form

$$(A - mI)C \, e^{mt} = O. \tag{4}$$

Equation (4) is to be satisfied for all real values of t, a condition that can be satisfied only if

$$(A - mI)C = O. \tag{5}$$

The theorem at the end of the previous section states that the algebraic system (5) has nontrivial solutions only if the determinant of $A - mI$ is zero; that is,

$$|A - mI| = 0. \tag{6}$$

Equation (6) is a polynomial equation of degree n in the unknown number m. We observe that the polynomial $|A - mI|$ depends only on the matrix A.

The polynomial $|A - mI|$ is called the *characteristic polynomial* of the matrix A and equation (6) is called the *characteristic equation* of the matrix A.

The roots of the characteristic equation of A are called *eigenvalues* of the matrix A.

A nonzero vector C_1, which is a solution of equation (5) for a particular eigenvalue m_1, is called an *eigenvector* of the matrix A corresponding to the eigenvalue m_1.

Thus we see that the first step in solving the homogeneous system (2) is to find the eigenvalues and the corresponding eigenvectors of the matrix A.

We may also suspect from our experience with the roots of the auxiliary equation in Chapter 6, that the nature of the solutions of equation (3) will depend on whether the eigenvalues are real and distinct, complex, or repeated. We shall deal with these cases separately.

EXAMPLE (a): Let us now reconsider the system of equations of Section 77. In matrix notation we have

$$X' = AX, \qquad \text{where } A = \begin{pmatrix} 0 & 1 \\ -2 & 3 \end{pmatrix}. \tag{7}$$

The characteristic equation of A is

$$\left| \begin{pmatrix} 0 & 1 \\ -2 & 3 \end{pmatrix} - m \begin{pmatrix} 1 & 0 \\ 0 & 1 \end{pmatrix} \right| = \begin{vmatrix} -m & 1 \\ -2 & 3-m \end{vmatrix} = m^2 - 3m + 2 = 0.$$

The eigenvalues are the distinct real numbers $m_1 = 1$ and $m_2 = 2$.

For $m_1 = 1$, equation (5) becomes

$$\begin{pmatrix} -1 & 1 \\ -2 & 2 \end{pmatrix} \begin{pmatrix} c_1 \\ c_2 \end{pmatrix} = O,$$

so that $-c_1 + c_2 = 0$ and we have

$$C = \begin{pmatrix} c_1 \\ c_2 \end{pmatrix} = \begin{pmatrix} c_1 \\ c_1 \end{pmatrix} = c_1 \begin{pmatrix} 1 \\ 1 \end{pmatrix}.$$

Thus corresponding to the eigenvalue $m_1 = 1$, there is a set of eigenvectors whose elements are scalar multiples of the column vector $\begin{pmatrix} 1 \\ 1 \end{pmatrix}$.

Similarly for $m_2 = 2$ equation (5) is

$$\begin{pmatrix} -2 & 1 \\ -2 & 1 \end{pmatrix} \begin{pmatrix} c_1 \\ c_2 \end{pmatrix} = O,$$

so that $-2c_1 + c_2 = 0$ and

$$C = \begin{pmatrix} c_1 \\ c_2 \end{pmatrix} = \begin{pmatrix} c_1 \\ 2c_1 \end{pmatrix} = c_1 \begin{pmatrix} 1 \\ 2 \end{pmatrix}.$$

That is, the eigenvectors are multiples of the vector $\begin{pmatrix} 1 \\ 2 \end{pmatrix}$.

Thus we have obtained two distinct sets of solutions for system (7),

$$X_1 = b_1 \begin{pmatrix} 1 \\ 1 \end{pmatrix} e^t \qquad \text{and} \qquad X_2 = b_2 \begin{pmatrix} 1 \\ 2 \end{pmatrix} e^{2t}, \tag{8}$$

where b_1 and b_2 are arbitrary constants. It is a simple matter to verify that these vector functions are solutions of system (7). It is quite another matter to make the claim that every solution of system (7) is some combination of the two solutions we have found. The fact that this is true requires the support of several important definitions and theorems. The student should observe that these definitions and theorems closely parallel the theoretical development in Chapter 5. We shall state the appropriate theorems without proof.

A set of m constant vectors of dimension n

$$\{X_1, X_2, \ldots, X_m\}$$

is *linearly independent* if

$$c_1 X_1 + c_2 X_2 + \cdots + c_m X_m = O$$

implies that $c_1 = c_2 = \cdots = c_m = 0$.

A set of vector functions of t

$$\{X_1(t), X_2(t), \ldots, X_m(t)\}$$

is linearly independent on an interval $a < t < b$ if

$$c_1 X_1(t) + c_2 X_2(t) + \cdots + c_m X_m(t) = O$$

for all t on the interval implies that

$$c_1 = c_2 = \cdots = c_m = 0.$$

THEOREM 25: *If $X_1(t), \ldots, X_m(t)$ are each solutions of a homogeneous linear system $X' = AX$ then $c_1 X_1(t) + c_2 X_2(t) + \cdots + c_m X_m(t)$ is a solution of the same system for arbitrary constants $c_1, \ldots, c_m$.*

THEOREM 26: *If A is an $n \times n$ matrix of real numbers and $\{X_1, \ldots, X_n\}$ is a linearly independent set of solutions of the system $X' = AX$ on the interval $a < t < b$, then any solution of the system is a unique linear combination of the set $\{X_1, \ldots, X_n\}$.*

If the set of n-dimensional column vectors $\{X_1(t), \ldots, X_n(t)\}$ is considered as an $n \times n$ matrix

$$(X_1(t) \qquad X_2(t) \ldots X_n(t))$$

then the determinant

$$|X_1(t) \qquad X_2(t) \ldots X_n(t)|$$

is called the *Wronskian* of the set of vectors.

THEOREM 27: *The set of n-dimensional column vectors $\{X_1(t), \ldots, X_n(t)\}$ is linearly independent at $t = t_0$ if, and only if, the Wronskian of the set is not zero at $t = t_0$; that is,*

$$W\{X_1(t_0), X_2(t_0), \ldots, X_n(t_0)\} = |X_1(t_0) \qquad X_2(t_0) \ldots X_n(t_0)| \neq 0.$$

THEOREM 28: *If the vector functions $X_1(t), \ldots, X_n(t)$ are solutions of the system $X' = AX$ for all t on the interval $a < t < b$ where A is an $n \times n$ matrix,* *then the set $\{X_1(t), \ldots, X_n(t)\}$ is linearly independent on $a < t < b$ if, and only if, $W\{(X_1(t_0), \ldots, X_n(t_0)\} \neq 0$ for some t_0 on the interval $a < t < b$.*

THEOREM 29: *If $X_1(t), \ldots, X_n(t)$ are linearly independent solutions of the n-dimensional homogeneous system $X' = AX$ on the interval $a < t < b$ and if $X_p(t)$ is any solution of the nonhomogeneous system $X' = AX + B(t)$ on the interval $a < t < b$, then any solution of the nonhomogeneous system can be written*

$$X(t) = c_1 X_1(t) + \cdots + c_n X_n(t) + X_p(t)$$

for a unique choice of the constants $c_1, \ldots, c_n$.

We return to Example (a). In equation (8) we presented two sets of solutions of the system (7). If we pick $b_1 = b_2 = 1$ and consider the Wronskian of the resulting set, we have

$$W\{X_1, X_2\} = \begin{vmatrix} e^t & e^{2t} \\ e^t & 2e^{2t} \end{vmatrix} = \begin{vmatrix} 1 & 1 \\ 1 & 2 \end{vmatrix} e^{3t} = e^{3t}.$$

Since the Wronskian does not vanish for any value of t, we conclude that the solutions X_1 and X_2 are linearly independent on any interval and it follows that the general solution of the system (7) is

$$X(t) = c_1 \begin{pmatrix} 1 \\ 1 \end{pmatrix} e^t + c_2 \begin{pmatrix} 1 \\ 2 \end{pmatrix} e^{2t}.$$

EXAMPLE (b):   Consider the system

$$X' = AX \qquad \text{for } A = \begin{pmatrix} 4 & -1 \\ -4 & 4 \end{pmatrix}. \tag{9}$$

The characteristic equation of A is

$$\begin{vmatrix} 4 - m & -1 \\ -4 & 4 - m \end{vmatrix} = m^2 - 8m + 12 = 0.$$

Therefore the eigenvalues of A are $m_1 = 2$ and $m_2 = 6$.
   For $m_1 = 2$, we compute nontrivial solutions of

$$\begin{pmatrix} 2 & -1 \\ -4 & 2 \end{pmatrix} \begin{pmatrix} c_1 \\ c_2 \end{pmatrix} = 0.$$

Thus $2c_1 - c_2 = 0$. One such solution is obtained by choosing $c_1 = 1$ to give the eigenvector $\begin{pmatrix} 1 \\ 2 \end{pmatrix}$. It follows that $X_1 = \begin{pmatrix} 1 \\ 2 \end{pmatrix} e^{2t}$ is a solution of the system (9).

For $m_2 = 6$, the system

$$\begin{pmatrix} -2 & -1 \\ -4 & -2 \end{pmatrix}\begin{pmatrix} c_1 \\ c_2 \end{pmatrix} = O,$$

upon choosing $c_1 = 1$, leads to an eigenvector $\begin{pmatrix} 1 \\ -2 \end{pmatrix}$ for the matrix A and a second solution

$$X_2 = \begin{pmatrix} 1 \\ -2 \end{pmatrix} e^{6t}.$$

The Wronskian of X_1 and X_2 is

$$W\{X_1, X_2\} = \begin{vmatrix} e^{2t} & e^{6t} \\ 2\,e^{2t} & -2\,e^{6t} \end{vmatrix} = -4\,e^{8t}.$$

Because $W\{X_1, X_2\}$ is never zero, it follows from Theorem 28 that X_1 and X_2 are linearly independent. By Theorem 26 we see that the general solution of the system $X' = AX$ is

$$X = c_1 \begin{pmatrix} 1 \\ 2 \end{pmatrix} e^{2t} + c_2 \begin{pmatrix} 1 \\ -2 \end{pmatrix} e^{6t}.$$

EXAMPLE (c):　Solve the system

$$X' = AX \qquad \text{for } A = \begin{pmatrix} 1 & -1 & -1 \\ 0 & 1 & 3 \\ 0 & 3 & 1 \end{pmatrix}. \tag{10}$$

The characteristic equation of A is

$$\begin{vmatrix} 1-m & -1 & -1 \\ 0 & 1-m & 3 \\ 0 & 3 & 1-m \end{vmatrix} = (1-m)(m-4)(m+2) = 0. \tag{11}$$

Choosing the eigenvalue $m_1 = 1$ leads to

$$\begin{pmatrix} 0 & -1 & -1 \\ 0 & 0 & 3 \\ 0 & 3 & 0 \end{pmatrix}\begin{pmatrix} c_1 \\ c_2 \\ c_3 \end{pmatrix} = O,$$

which requires that $c_2 = c_3 = 0$ but leaves c_1 arbitrary. Thus

$$X_1 = \begin{pmatrix} 1 \\ 0 \\ 0 \end{pmatrix} e^t.$$

is one solution of (10).

The eigenvalue $m_2 = 4$ requires

$$\begin{pmatrix} -3 & -1 & -1 \\ 0 & -3 & 3 \\ 0 & 3 & -3 \end{pmatrix} \begin{pmatrix} c_1 \\ c_2 \\ c_3 \end{pmatrix} = O,$$

or

$$3c_1 + c_2 + c_3 = 0,$$
$$-c_2 + c_3 = 0.$$

One solution of this system is the eigenvector $\begin{pmatrix} 2 \\ -3 \\ -3 \end{pmatrix}$ from which we obtain

$X_2 = \begin{pmatrix} 2 \\ -3 \\ -3 \end{pmatrix} e^{4t}$ as a second solution to the system (10).

Finally, choosing $m_3 = -2$ from equation (11) we have

$$\begin{pmatrix} 3 & -1 & -1 \\ 0 & 3 & 3 \\ 0 & 3 & 3 \end{pmatrix} \begin{pmatrix} c_1 \\ c_2 \\ c_3 \end{pmatrix} = O,$$

which yields the eigenvector $\begin{pmatrix} 0 \\ 1 \\ -1 \end{pmatrix}$ and the solution

$$X_3 = \begin{pmatrix} 0 \\ 1 \\ -1 \end{pmatrix} e^{-2t}.$$

To establish the linear independence of the three solutions X_1, X_2, X_3 we compute the Wronskian at $t = 0$; that is,

$$W\{X_1(0), X_2(0), X_3(0)\} = \begin{vmatrix} 1 & 2 & 0 \\ 0 & -3 & 1 \\ 0 & -3 & -1 \end{vmatrix} = 6.$$

Because $W \neq 0$, it follows from Theorem 28 that the solutions are linearly independent on any interval. Thus the general solution of system (10) is

$$X(t) = c_1 \begin{pmatrix} 1 \\ 0 \\ 0 \end{pmatrix} e^t + c_2 \begin{pmatrix} 2 \\ -3 \\ -3 \end{pmatrix} e^{4t} + c_3 \begin{pmatrix} 0 \\ 1 \\ -1 \end{pmatrix} e^{-2t}.$$

EXAMPLE (d):   We conclude this section with an example to illustrate why the use of the word *Wronskian*, in the context of a set of solutions of a system of first-order linear differential equations, is consistent with the usage of the same word, made in Section 27 in the context of a set of solutions of a single *n*th-order linear differential equation.

Consider the second-order linear equation

$$[D^2 - (a + b)D + ab]x = 0, \qquad a \neq b, \tag{12}$$

in which $D = d/dt$. The operator factors into $(D - a)(D - b)$ and the functions e^{at} and e^{bt} are therefore solutions of equation (12). The Wronskian of these solutions, as defined in Section 27, is the determinant

$$W\{e^{at}, e^{bt}\} = \begin{vmatrix} e^{at} & e^{bt} \\ ae^{at} & be^{bt} \end{vmatrix}, \tag{13}$$

where the functions in the second row are the derivatives of the functions in the first row.

Equation (12) may be converted to a system of first-order equations by setting $Dx = y$ so that (12) becomes

$$D^2x = Dy = (a + b)y - abx.$$

Thus (12) is equivalent to the system

$$x' = y$$

$$y' = -abx + (a + b)y,$$

or

$$\begin{pmatrix} x \\ y \end{pmatrix}' = \begin{pmatrix} 0 & 1 \\ -ab & a + b \end{pmatrix} \begin{pmatrix} x \\ y \end{pmatrix}. \tag{14}$$

If we apply the technique of the current section, we write the characteristic equation of the matrix of system (14)

$$\begin{vmatrix} -m & 1 \\ -ab & a + b - m \end{vmatrix} = 0,$$

which reduces to

$$m^2 - (a + b)m + ab = 0. \tag{15}$$

It is important to observe that the characteristic polynomial of (15) and the operator polynomial of (12) have the same form, hence the same roots.

From equation (15) we obtain the eigenvalues $m_1 = a$ and $m_2 = b$. Choosing the eigenvalue $m_1 = a$ leads to

$$\begin{pmatrix} -a & 1 \\ -ab & b \end{pmatrix} \begin{pmatrix} c_1 \\ c_2 \end{pmatrix} = 0,$$

so that $c_2 = ac_1$ and

$$\binom{c_1}{c_2} = \binom{c_1}{ac_1} = c_1\binom{1}{a}.$$

Thus $\binom{1}{a} e^{at}$ is one solution of (14).

Choosing the eigenvalue $m_2 = b$ leads to

$$\begin{pmatrix} -b & 1 \\ -ab & a \end{pmatrix}\begin{pmatrix} c_1 \\ c_2 \end{pmatrix} = O,$$

so that $c_2 = bc_1$ and

$$\binom{c_1}{c_2} = \binom{c_1}{bc_1} = c_1\binom{1}{b}.$$

Thus $\binom{1}{b} e^{bt}$ is a second solution of (14).

In the context of the current section, the Wronskian of these two solutions is

$$W\{X_1(t), X_2(t)\} = \begin{vmatrix} e^{at} & e^{bt} \\ ae^{at} & be^{bt} \end{vmatrix}. \tag{16}$$

Thus we see that the expressions given in (13) and (16), while coming from entirely different contexts are the same, and the word *Wronskian* is used in both contexts for that expression.

The examples that we have considered have all involved matrices whose eigenvalues are distinct real numbers. In each case the eigenvectors corresponding to distinct eigenvalues turned out to be linearly independent. That was no accident. It is possible to prove a theorem to that effect.

THEOREM 30:   *If $m_1, m_2, \ldots, m_s$ are distinct eigenvalues of an $n \times n$ matrix A and if $X_1, X_2, \ldots, X_s$ are corresponding eigenvectors, then the set $\{X_1, \ldots, X_s\}$ is linearly independent.*

The definitions and theorems of this section have been stated without providing the proof required to understand them clearly. It is hoped that this will serve as a motivation for a study of linear algebra, where the definitions and theorems become more easily understood.

Exercises

In exercises 1 through 7, find the general solution of the system $X' = AX$ for the given matrix A. In each case check on the linear independence of solutions by examining the Wronskian.

1. $A = \begin{pmatrix} 8 & -3 \\ 16 & -8 \end{pmatrix}.$ ANS. $X = c_1 \begin{pmatrix} 3 \\ 4 \end{pmatrix} e^{4t} + c_2 \begin{pmatrix} 1 \\ 4 \end{pmatrix} e^{-4t}.$

2. $A = \begin{pmatrix} 1 & 0 \\ -2 & 2 \end{pmatrix}.$ ANS. $X = c_1 \begin{pmatrix} 1 \\ 2 \end{pmatrix} e^{t} + c_2 \begin{pmatrix} 0 \\ 1 \end{pmatrix} e^{2t}.$

3. $A = \begin{pmatrix} 4 & 3 \\ -4 & -4 \end{pmatrix}.$ ANS. $X = c_1 \begin{pmatrix} 3 \\ -2 \end{pmatrix} e^{2t} + c_2 \begin{pmatrix} 1 \\ -2 \end{pmatrix} e^{-2t}.$

4. $A = \begin{pmatrix} 3 & 3 \\ -1 & -1 \end{pmatrix}.$ ANS. $X = c_1 \begin{pmatrix} 1 \\ -1 \end{pmatrix} + c_2 \begin{pmatrix} 3 \\ -1 \end{pmatrix} e^{2t}.$

5. $A = \begin{pmatrix} 2 & 3 \\ 1 & -2 \end{pmatrix}.$ ANS. $X = c_1 \begin{pmatrix} 3 \\ -2 + \sqrt{7} \end{pmatrix} e^{\sqrt{7}t} + c_2 \begin{pmatrix} -3 \\ 2 + \sqrt{7} \end{pmatrix} e^{-\sqrt{7}t}.$

6. $A = \begin{pmatrix} 12 & -15 \\ 4 & -4 \end{pmatrix}.$ ANS. $X = c_1 \begin{pmatrix} 3 \\ 2 \end{pmatrix} e^{2t} + c_2 \begin{pmatrix} 5 \\ 2 \end{pmatrix} e^{6t}.$

7. $A = \begin{pmatrix} 1 & 2 & -1 \\ 0 & -1 & 3 \\ 0 & 0 & 2 \end{pmatrix}.$ ANS. $X = c_1 \begin{pmatrix} 1 \\ 0 \\ 0 \end{pmatrix} e^{t} + c_2 \begin{pmatrix} 1 \\ -1 \\ 0 \end{pmatrix} e^{-t} + c_3 \begin{pmatrix} 1 \\ 1 \\ 1 \end{pmatrix} e^{2t}.$

8. $A = \begin{pmatrix} 1 & 2 & -1 \\ 2 & 1 & 1 \\ -1 & 1 & 0 \end{pmatrix}.$ ANS. $X = c_1 \begin{pmatrix} 1 \\ -1 \\ -2 \end{pmatrix} e^{t} + c_2 \begin{pmatrix} 1 \\ 1 \\ 0 \end{pmatrix} e^{3t} + c_3 \begin{pmatrix} 1 \\ -1 \\ 1 \end{pmatrix} e^{-2t}.$

80. Complex eigenvalues

In Section 79 we carefully avoided systems for which the eigenvalues were complex numbers. We now consider some examples in which complex numbers occur.

EXAMPLE (a): Solve the system

$$\begin{pmatrix} x \\ y \end{pmatrix}' = \begin{pmatrix} 2 & -5 \\ 2 & -4 \end{pmatrix} \begin{pmatrix} x \\ y \end{pmatrix}. \tag{1}$$

The characteristic equation of the matrix in system (1) is

$$\begin{vmatrix} 2 - m & -5 \\ 2 & -4 - m \end{vmatrix} = m^2 + 2m + 2 = 0, \tag{2}$$

with eigenvalues $m_1 = -1 + i$ and $m_2 = -1 - i$.
  For $m_1 = -1 + i$ we must satisfy the system

$$\begin{pmatrix} 3 - i & -5 \\ 2 & -3 - i \end{pmatrix} \begin{pmatrix} c_1 \\ c_2 \end{pmatrix} = O,$$

which requires that

$$c_2 = \frac{3 - i}{5} c_1 .$$

One solution is obtained by choosing $c_1 = 5$. Thus an eigenvector corres-ponding to the eigenvalue m_1 is $\begin{pmatrix} 5 \\ 3 - i \end{pmatrix}$ with the complex vector function

$$X_1 = \begin{pmatrix} 5 \\ 3 - i \end{pmatrix} e^{(-1 + i)t}, \tag{3}$$

at least formally a solution of system (1).

The second eigenvalue $m_2 = -1 - i$ leads in a similar way to a second solution,

$$X_2 = \begin{pmatrix} 5 \\ 3 + i \end{pmatrix} e^{(-1 - i)t} . \tag{4}$$

The two solutions can be combined to give

$$X = c_1 \begin{pmatrix} 5 \\ 3 - i \end{pmatrix} e^{(-1 + i)t} + c_2 \begin{pmatrix} 5 \\ 3 + i \end{pmatrix} e^{(-1 - i)t} . \tag{5}$$

The presentation of a solution in this form should be reminiscent of the situation in Chapter 6, where we were solving single linear equations with constant coefficients. We will proceed in much the same way as we did there by making use of Euler's formula

$$e^{(a + bi)t} = e^{at}(\cos bt + i \sin bt). \tag{6}$$

Formally changing the form of equation (5) gives us

$$X = c_1 \begin{pmatrix} 5 \\ 3 - i \end{pmatrix} e^{-t}(\cos t + i \sin t) + c_2 \begin{pmatrix} 5 \\ 3 + i \end{pmatrix} e^{-t}(\cos t - i \sin t),$$

and after combining real and imaginary parts

$$X = e^{-t} \left[(c_1 + c_2) \begin{pmatrix} 5 \cos t \\ 3 \cos t + \sin t \end{pmatrix} + i(c_1 - c_2) \begin{pmatrix} 5 \sin t \\ -\cos t + 3 \sin t \end{pmatrix} \right]. \tag{7}$$

If we let $b_1 = c_1 + c_2$ and $b_2 = i(c_1 - c_2)$ equation (7) can be written finally as

$$X = e^{-t} \left[b_1 \left\{ \begin{pmatrix} 5 \\ 3 \end{pmatrix} \cos t - \begin{pmatrix} 0 \\ -1 \end{pmatrix} \sin t \right\} + b_2 \left\{ \begin{pmatrix} 0 \\ -1 \end{pmatrix} \cos t + \begin{pmatrix} 5 \\ 3 \end{pmatrix} \sin t \right\} \right]. \tag{8}$$

The linear independence of the two solutions in (8) can be established by computing the Wronskian at $t = 0$. The student should show that $W(0) = -5$.

We make the following observations from the above example:

(a) Since the matrix of system (1) is real, the eigenvalues occur in conjugate pairs.

(b) The eigenvectors corresponding to conjugate eigenvalues are also conjugates of one another.

(c) The first eigenvector,

$$B = \begin{pmatrix} 5 + 0i \\ 3 - 1i \end{pmatrix} = \begin{pmatrix} 5 \\ 3 \end{pmatrix} + \begin{pmatrix} 0 \\ -1 \end{pmatrix} i = \operatorname{Re} B + i \operatorname{Im} B,$$

appears in the solution (8) in the form

$$X = e^{-t}[b_1\{\operatorname{Re} B \cos t - \operatorname{Im} B \sin t\} + b_2\{\operatorname{Im} B \cos t + \operatorname{Re} B \sin t\}]. \tag{9}$$

(d) The Wronskian of the two solutions in (9) at $t = 0$ is given by the determinant $W = |\operatorname{Re} B \ \operatorname{Im} B|$.

In exercises 12 to 14 below the student is asked to show the applicability of observations (a) to (d) to the general system of two equations in two unknowns.

EXAMPLE (b):   Solve the system

$$\begin{pmatrix} x \\ y \end{pmatrix}' = \begin{pmatrix} 2 & 1 \\ -4 & 2 \end{pmatrix}\begin{pmatrix} x \\ y \end{pmatrix}, \tag{10}$$

making use of the observations made in Example (a).

The characteristic equation

$$\begin{vmatrix} 2 - m & 1 \\ -4 & 2 - m \end{vmatrix} = m^2 - 4m + 8 = 0,$$

has conjugate roots $m_1 = 2 + 2i$ and $m_2 = 2 - 2i$. An eigenvector corresponding to m_1 is

$$\begin{pmatrix} 1 \\ 2i \end{pmatrix} = \begin{pmatrix} 1 \\ 0 \end{pmatrix} + \begin{pmatrix} 0 \\ 2 \end{pmatrix} i.$$

If we accept the results of the observations in Example (a), observing that

$$W(0) = \begin{vmatrix} 1 & 0 \\ 0 & 2 \end{vmatrix} = 2 \neq 0,$$

we conclude that the general solution of system (10) is

$$X = e^{2t}\left[b_1\left\{\begin{pmatrix}1\\0\end{pmatrix}\cos 2t - \begin{pmatrix}0\\2\end{pmatrix}\sin 2t\right\} + b_2\left\{\begin{pmatrix}0\\2\end{pmatrix}\cos 2t + \begin{pmatrix}1\\0\end{pmatrix}\sin 2t\right\}\right].$$

EXAMPLE (c): We now consider a system of three equations in three unknowns given by

$$X' = AX, \quad \text{where } A = \begin{pmatrix}1 & 2 & -1\\0 & 1 & 1\\0 & -1 & 1\end{pmatrix}. \tag{11}$$

The characteristic equation

$$(1 - m)(m^2 - 2m + 2) = 0$$

has roots $m_1 = 1$, $m_2 = 1 + i$, and $m_3 = 1 - i$.

The eigenvalue $m_1 = 1$ has the vector $\begin{pmatrix}1\\0\\0\end{pmatrix}$ as an eigenvector, giving one solution of (11) as

$$X_1 = \begin{pmatrix}1\\0\\0\end{pmatrix}e^t.$$

The eigenvalue $m_2 = 1 + i$ yields an eigenvector

$$\begin{pmatrix}2 - i\\0 + i\\-1 + 0i\end{pmatrix} = \begin{pmatrix}2\\0\\-1\end{pmatrix} + i\begin{pmatrix}-1\\1\\0\end{pmatrix}.$$

The general solution can be written

$$X = c_1\begin{pmatrix}1\\0\\0\end{pmatrix}e^t + e^t\left[c_2\left\{\begin{pmatrix}2\\0\\-1\end{pmatrix}\cos t - \begin{pmatrix}-1\\1\\0\end{pmatrix}\sin t\right\}\right.$$

$$\left. + c_3\left\{\begin{pmatrix}-1\\1\\0\end{pmatrix}\cos t + \begin{pmatrix}2\\0\\-1\end{pmatrix}\sin t\right\}\right].$$

The linear independence of the solutions at $t = 0$ is guaranteed by evaluating the Wronskian at $t = 0$. Its value is

$$W(0) = \begin{vmatrix} 1 & 2 & -1 \\ 0 & 0 & 1 \\ 0 & -1 & 0 \end{vmatrix} = 1 \neq 0.$$

Exercises

In exercises 1 through 7 find the general solution of the system $X' = AX$ for the given matrix A.

1. $A = \begin{pmatrix} 4 & 5 \\ -4 & -4 \end{pmatrix}$.

ANS. $X = c_1 \left[\begin{pmatrix} 5 \\ -4 \end{pmatrix} \cos 2t - \begin{pmatrix} 0 \\ 2 \end{pmatrix} \sin 2t \right] + c_2 \left[\begin{pmatrix} 0 \\ 2 \end{pmatrix} \cos 2t + \begin{pmatrix} 5 \\ -4 \end{pmatrix} \sin 2t \right]$.

2. $A = \begin{pmatrix} 4 & 1 \\ -8 & 8 \end{pmatrix}$.

ANS. $X = c_1 e^{6t} \left[\begin{pmatrix} 1 \\ 2 \end{pmatrix} \cos 2t + \begin{pmatrix} 0 \\ -2 \end{pmatrix} \sin 2t \right] + c_2 e^{6t} \left[\begin{pmatrix} 0 \\ 2 \end{pmatrix} \cos 2t + \begin{pmatrix} 1 \\ 2 \end{pmatrix} \sin 2t \right]$.

3. $A = \begin{pmatrix} 4 & -13 \\ 2 & -6 \end{pmatrix}$.

ANS. $X = c_1 e^{-t} \left[\begin{pmatrix} 13 \\ 5 \end{pmatrix} \cos t - \begin{pmatrix} 0 \\ -1 \end{pmatrix} \sin t \right] + c_2 e^{-t} \left[\begin{pmatrix} 0 \\ -1 \end{pmatrix} \cos t + \begin{pmatrix} 13 \\ 5 \end{pmatrix} \sin t \right]$.

4. $A = \begin{pmatrix} 3 & 5 \\ -1 & -1 \end{pmatrix}$.

ANS. $X = c_1 e^t \left[\begin{pmatrix} 5 \\ -2 \end{pmatrix} \cos t - \begin{pmatrix} 0 \\ 1 \end{pmatrix} \sin t \right] + c_2 e^t \left[\begin{pmatrix} 0 \\ 1 \end{pmatrix} \cos t + \begin{pmatrix} 5 \\ -2 \end{pmatrix} \sin t \right]$.

5. $A = \begin{pmatrix} 12 & -17 \\ 4 & -4 \end{pmatrix}$.

ANS. $X = c_1 e^{4t} \left[\begin{pmatrix} 17 \\ 8 \end{pmatrix} \cos 2t - \begin{pmatrix} 0 \\ -2 \end{pmatrix} \sin 2t \right] + c_2 e^{4t} \left[\begin{pmatrix} 0 \\ -2 \end{pmatrix} \cos 2t + \begin{pmatrix} 17 \\ 8 \end{pmatrix} \sin 2t \right]$.

6. $A = \begin{pmatrix} 8 & -5 \\ 16 & -8 \end{pmatrix}$.

ANS. $X = c_1 \left[\begin{pmatrix} 5 \\ 8 \end{pmatrix} \cos 4t - \begin{pmatrix} 0 \\ -4 \end{pmatrix} \sin 4t \right] + c_2 \left[\begin{pmatrix} 0 \\ -4 \end{pmatrix} \cos 4t + \begin{pmatrix} 5 \\ 8 \end{pmatrix} \sin 4t \right]$.

7. $A = \begin{pmatrix} 1 & 0 & 0 \\ 2 & 1 & -2 \\ 3 & 2 & 1 \end{pmatrix}$.

$$\text{ANS.} \quad X = c_1 \begin{pmatrix} 2 \\ -3 \\ 2 \end{pmatrix} e^t + c_2\, e^t \left[\begin{pmatrix} 0 \\ 1 \\ 0 \end{pmatrix} \cos 2t - \begin{pmatrix} 0 \\ 0 \\ -1 \end{pmatrix} \sin 2t \right]$$

$$+ c_3\, e^t \left[\begin{pmatrix} 0 \\ 0 \\ -1 \end{pmatrix} \cos 2t + \begin{pmatrix} 0 \\ 1 \\ 0 \end{pmatrix} \sin 2t \right].$$

Following the example of Section 76, replace each of the following equations by a system of first-order equations. Solve that system using matrix techniques and check your answers by solving the original equation directly.

8. $y^{(4)} - y = 0$.
9. $y'' + 2y' + 2y = 0$.
10. $y''' - 3y'' + 4y' - 2y = 0$.
11. $y'' + 4y = 0$.

In exercises 12 through 15 we consider the general homogeneous system with real coefficients

$$\begin{pmatrix} x \\ y \end{pmatrix}' = \begin{pmatrix} a & b \\ c & d \end{pmatrix} \begin{pmatrix} x \\ y \end{pmatrix}. \tag{A}$$

12. Find the eigenvalues for the matrix of (A) and show that complex eigenvalues occur only if $(a - d)^2 + 4bc < 0$. In particular, note that complex eigenvalues occur as conjugate pairs and that they occur only if b and c are not zero.
13. For the system (A) suppose the eigenvalues are complex numbers $p + qi$ and $p - qi$, where $q \neq 0$. Show that the corresponding eigenvectors are conjugate pairs.
14. Show that the observation made in equation (9) above is true in the complex case.
15. Find the value of the Wronskian in the complex case for $t = 0$ and show that it is not zero.

81. Repeated eigenvalues

We now consider an example in which the characteristic equation has repeated roots.

EXAMPLE (a): Solve the system

$$X' = AX, \quad \text{for } A = \begin{pmatrix} 0 & 1 \\ -4 & 4 \end{pmatrix}. \tag{1}$$

The characteristic equation of A is

$$\begin{vmatrix} -m & 1 \\ -4 & 4 - m \end{vmatrix} = (m - 2)^2 = 0.$$

For the eigenvalue $m_1 = 2$ we obtain the solution

$$X_1 = \begin{pmatrix} 1 \\ 2 \end{pmatrix} e^{2t}. \tag{2}$$

A second solution X_2, independent of X_1, is not immediately available because the eigenvalue m_1 is a double root of the characteristic equation. From our experience with repeated roots in Chapter 6, we may be tempted to guess that a second solution has the form

$$X_2 = \begin{pmatrix} c_1 \\ c_2 \end{pmatrix} t\, e^{2t}. \tag{3}$$

However, a substitution back into equation (1) quickly shows that the only solution of this form is the trivial solution with $c_1 = c_2 = 0$.

Another suggestion might be made from our previous experience; that is, to try to find a second solution of the form

$$X_2 = \begin{pmatrix} c_1(t) \\ c_2(t) \end{pmatrix} e^{2t}, \tag{4}$$

where we are essentially using a variation of parameters technique. Direct substitution of (4) into (1) gives

$$\begin{pmatrix} c_1(t) \\ c_2(t) \end{pmatrix} 2\, e^{2t} + \begin{pmatrix} c'_1(t) \\ c'_2(t) \end{pmatrix} e^{2t} = \begin{pmatrix} 0 & 1 \\ -4 & 4 \end{pmatrix} \begin{pmatrix} c_1(t) \\ c_2(t) \end{pmatrix} e^{2t}.$$

We may rewrite this system of equations in the form

$$\begin{pmatrix} c'_1(t) \\ c'_2(t) \end{pmatrix} = \begin{pmatrix} -2 & 1 \\ -4 & 2 \end{pmatrix} \begin{pmatrix} c_1(t) \\ c_2(t) \end{pmatrix}. \tag{5}$$

System (5) can be rewritten

$$\begin{aligned} c'_1(t) &= -2c_1(t) + c_2(t), \\ c'_2(t) &= -4c_1(t) + 2c_2(t), \end{aligned} \tag{6}$$

from which we conclude that $c'_2(t) = 2c'_1(t)$. Integrating, we obtain $c_2(t) = 2c_1(t) + a$, for arbitrary constant a. Substituting back into the first equation of (6) gives

$$c'_1(t) = a, \quad \text{or} \quad c_1(t) = at + b.$$

Thus we obtain a set of solutions of equations (5)

$$c_1(t) = at + b, \quad \text{and} \quad c_2(t) = 2at + 2b + a,$$

with arbitrary constant values for a and b. Equation (4) now becomes

$$X_2 = \begin{pmatrix} 1 \\ 2 \end{pmatrix} a t\, e^{2t} + \begin{pmatrix} 1 \\ 2 \end{pmatrix} b\, e^{2t} + \begin{pmatrix} 0 \\ 1 \end{pmatrix} a\, e^{2t}.$$

If we were to choose $a = 0$ and $b = 1$, this solution would be the same as X_1. Instead, we will choose $a = 1$ and $b = 0$ to give

$$X_2 = \begin{pmatrix} 1 \\ 2 \end{pmatrix} t\, e^{2t} + \begin{pmatrix} 0 \\ 1 \end{pmatrix} e^{2t}.$$

At $t = 0$ the Wronskian

$$W\{X_1(0),\, X_2(0)\} = \begin{vmatrix} 1 & 0 \\ 2 & 1 \end{vmatrix} = 1 \neq 0,$$

so the two solutions are linearly independent.

The general solution of system (1) is therefore

$$X = c_1 \begin{pmatrix} 1 \\ 2 \end{pmatrix} e^{2t} + c_2 \left[\begin{pmatrix} 1 \\ 2 \end{pmatrix} t\, e^{2t} + \begin{pmatrix} 0 \\ 1 \end{pmatrix} e^{2t} \right].$$

In retrospect we note that the guess we made earlier in equation (3), although incorrect, was nevertheless not very far from the truth. We are now in a position to make a more reasonable assumption about the nature of a second solution in the case of repeated roots.

EXAMPLE (b):   Solve the system

$$\begin{pmatrix} x \\ y \end{pmatrix}' = \begin{pmatrix} 8 & -1 \\ 4 & 12 \end{pmatrix} \begin{pmatrix} x \\ y \end{pmatrix}. \tag{7}$$

The eigenvalues are the roots of the equation

$$\begin{vmatrix} 8 - m & -1 \\ 4 & 12 - m \end{vmatrix} = m^2 - 20m + 100 = (m - 10)^2 = 0.$$

Therefore, one solution is given by the eigenvalue $m_1 = 10$. This solution is

$$X_1 = \begin{pmatrix} 1 \\ -2 \end{pmatrix} e^{10t}. \tag{8}$$

Guided by our experience in Example (a) we now seek a second solution of the form

$$X_2 = \begin{pmatrix} 1 \\ -2 \end{pmatrix} t\, e^{10t} + \begin{pmatrix} c_3 \\ c_4 \end{pmatrix} e^{10t}. \tag{9}$$

Substitution of (9) into system (7) yields

$$\begin{pmatrix} 1 \\ -2 \end{pmatrix} 10t\, e^{10t} + \begin{pmatrix} 1 \\ -2 \end{pmatrix} e^{10t} + \begin{pmatrix} c_3 \\ c_4 \end{pmatrix} 10\, e^{10t}$$

$$= \begin{pmatrix} 8 & -1 \\ 4 & 12 \end{pmatrix} \begin{pmatrix} 1 \\ -2 \end{pmatrix} t\, e^{10t} + \begin{pmatrix} 8 & -1 \\ 4 & 12 \end{pmatrix} \begin{pmatrix} c_3 \\ c_4 \end{pmatrix} e^{10t}.$$

We note that the terms involving $t\, e^{10t}$ cancel each other leaving us with

$$\begin{pmatrix} -2 & -1 \\ 4 & 2 \end{pmatrix} \begin{pmatrix} c_3 \\ c_4 \end{pmatrix} e^{10t} = \begin{pmatrix} 1 \\ -2 \end{pmatrix} e^{10t},$$

or

$$\begin{pmatrix} -2 & -1 \\ 4 & 2 \end{pmatrix} \begin{pmatrix} c_3 \\ c_4 \end{pmatrix} = \begin{pmatrix} 1 \\ -2 \end{pmatrix}. \tag{10}$$

One solution of system (10) is $c_3 = 0$ and $c_4 = -1$. Therefore,

$$X_2 = \begin{pmatrix} 1 \\ -2 \end{pmatrix} t\, e^{10t} + \begin{pmatrix} 0 \\ -1 \end{pmatrix} e^{10t}$$

is a second solution of system (7). The general solution of system (7) is

$$X = c_1 \begin{pmatrix} 1 \\ -2 \end{pmatrix} e^{10t} + c_2\, e^{10t} \left[\begin{pmatrix} 1 \\ -2 \end{pmatrix} t + \begin{pmatrix} 0 \\ -1 \end{pmatrix} \right].$$

EXAMPLE (c): Solve the system

$$D^2 y + (D - 1)v = 0,$$

$$(2D - 1)y + (D - 1)w = 0, \tag{11}$$

$$(D + 3)y + (D - 4)v + 3w = 0.$$

In order to reduce (11) to a system of first-order equations, we let $Dy = u$. Then system (11) can be written

$$
\begin{aligned}
Du &= u - 3v + 3w + 3y, \\
Dv &= -u + 4v - 3w - 3y, \\
Dw &= -2u + w + y, \\
Dy &= u.
\end{aligned}
\tag{12}
$$

The matrix of system (12) has characteristic equation $m(m - 4)(m - 1)^2 = 0$. The eigenvalues $m_1 = 0$, $m_2 = 4$, and $m_3 = 1$ give rise to solutions

$$X_1 = \begin{pmatrix} 0 \\ 0 \\ -1 \\ 1 \end{pmatrix}, \qquad X_2 = \begin{pmatrix} 12 \\ -16 \\ -7 \\ 3 \end{pmatrix} e^{4t}, \qquad X_3 = \begin{pmatrix} 0 \\ 1 \\ 1 \\ 0 \end{pmatrix} e^{t}.$$

If we assume that the repeated root $m_3 = 1$ will yield a solution of the form

$$X_4 = \begin{pmatrix} 0 \\ 1 \\ 1 \\ 0 \end{pmatrix} t \, e^t + \begin{pmatrix} c_5 \\ c_6 \\ c_7 \\ c_8 \end{pmatrix} e^t,$$

a direct substitution into (12) will yield one set of values for the constants c_5 to c_8,

$$\begin{pmatrix} c_5 \\ c_6 \\ c_7 \\ c_8 \end{pmatrix} = \begin{pmatrix} -1 \\ 0 \\ 1 \\ -1 \end{pmatrix}.$$

Thus the desired solution is

$$X_4 = \begin{pmatrix} 0 \\ 1 \\ 1 \\ 0 \end{pmatrix} t \, e^t + \begin{pmatrix} -1 \\ 0 \\ 1 \\ -1 \end{pmatrix} e^t.$$

Finally, the solution of system (11) is

$$\begin{pmatrix} v \\ w \\ y \end{pmatrix} = c_1 \begin{pmatrix} 1 \\ 1 \\ 0 \end{pmatrix} e^t + c_2 \left[\begin{pmatrix} 1 \\ 1 \\ 0 \end{pmatrix} t \, e^t + \begin{pmatrix} 0 \\ 1 \\ -1 \end{pmatrix} e^t \right] + c_3 \begin{pmatrix} 0 \\ -1 \\ 1 \end{pmatrix} + c_4 \begin{pmatrix} -16 \\ -7 \\ 3 \end{pmatrix} e^{4t}.$$

The student is asked to fill in the details of this example in exercise 10 below.

EXAMPLE (d): We now consider an example in which a repeated root of the characteristic equation of a matrix gives rise to two linearly independent eigenvectors of that matrix and thus avoids the complications encountered in Examples (a), (b), and (c) above.

The eigenvalues of the matrix of the system

$$\begin{pmatrix} x \\ y \\ z \end{pmatrix}' = \begin{pmatrix} 0 & 1 & 1 \\ 1 & 0 & 1 \\ 1 & 1 & 0 \end{pmatrix} \begin{pmatrix} x \\ y \\ z \end{pmatrix} \tag{13}$$

are the roots of the equation

$$\begin{vmatrix} -m & 1 & 1 \\ 1 & -m & 1 \\ 1 & 1 & -m \end{vmatrix} = -(m + 1)^2 (m - 2) = 0.$$

For the repeated root $m = -1$ we seek nontrivial solutions of the system

$$\begin{pmatrix} 1 & 1 & 1 \\ 1 & 1 & 1 \\ 1 & 1 & 1 \end{pmatrix} \begin{pmatrix} c_1 \\ c_2 \\ c_3 \end{pmatrix} = O,$$

that is, solutions of $c_1 + c_2 + c_3 = 0$. Thus

$$C = \begin{pmatrix} c_1 \\ c_2 \\ c_3 \end{pmatrix} = \begin{pmatrix} c_1 \\ c_2 \\ -c_1 - c_2 \end{pmatrix} = c_1 \begin{pmatrix} 1 \\ 0 \\ -1 \end{pmatrix} + c_2 \begin{pmatrix} 0 \\ 1 \\ -1 \end{pmatrix}.$$

It follows that both

$$\begin{pmatrix} 1 \\ 0 \\ -1 \end{pmatrix} e^{-t} \quad \text{and} \quad \begin{pmatrix} 0 \\ 1 \\ -1 \end{pmatrix} e^{-t}$$

are solutions of system (13).

The second eigenvalue, $m = 2$, requires us to find nontrivial solutions of the system of equations

$$\begin{pmatrix} -2 & 1 & 1 \\ 1 & -2 & 1 \\ 1 & 1 & -2 \end{pmatrix} \begin{pmatrix} c_1 \\ c_2 \\ c_3 \end{pmatrix} = O.$$

That is,

$$\begin{aligned} -2c_1 + c_2 + c_3 &= 0 \\ c_1 - 2c_2 + c_3 &= 0 \\ c_1 + c_2 - 2c_3 &= 0. \end{aligned}$$

Elementary elimination of c_2 from the first and last equations and c_1 from the second and third equations leave us with

$$c_1 = c_3 \quad \text{and} \quad c_2 = c_3$$

so that

$$C = \begin{pmatrix} c_1 \\ c_2 \\ c_3 \end{pmatrix} = c_1 \begin{pmatrix} 1 \\ 1 \\ 1 \end{pmatrix}.$$

Hence a third solution of system (13) is

$$\begin{pmatrix} 1 \\ 1 \\ 1 \end{pmatrix} e^{2t}.$$

The linear independence of the three solutions is established by examining their Wronskian at $t = 0$,

$$W\{X_1(0), X_2(0), X_3(0)\} = \begin{vmatrix} 1 & 0 & 1 \\ 0 & 1 & 1 \\ -1 & -1 & 1 \end{vmatrix} = 3 \neq 0.$$

The general solution is therefore

$$\begin{pmatrix} x \\ y \\ z \end{pmatrix} = \left[c_1 \begin{pmatrix} 1 \\ 0 \\ -1 \end{pmatrix} + c_2 \begin{pmatrix} 0 \\ 1 \\ -1 \end{pmatrix} \right] e^{-t} + c_3 \begin{pmatrix} 1 \\ 1 \\ 1 \end{pmatrix} e^{2t}.$$

EXAMPLE (e): Solve the system

$$\begin{pmatrix} x \\ y \\ z \end{pmatrix}' = \begin{pmatrix} 2 & 1 & 2 \\ 1 & 2 & 2 \\ 1 & 1 & 3 \end{pmatrix} \begin{pmatrix} x \\ y \\ z \end{pmatrix}. \tag{14}$$

The characteristic equation

$$\begin{vmatrix} 2-m & 1 & 2 \\ 1 & 2-m & 2 \\ 1 & 1 & 3-m \end{vmatrix} = -(m-1)^2(m-5) = 0$$

has roots 1 and 5. The repeated eigenvalue 1 gives rise to the equation $c_1 = -c_2 - 2c_3$. Consequently

$$C = \begin{pmatrix} c_1 \\ c_2 \\ c_3 \end{pmatrix} = \begin{pmatrix} -c_2 - 2c_3 \\ c_2 \\ c_3 \end{pmatrix} = c_2 \begin{pmatrix} -1 \\ 1 \\ 0 \end{pmatrix} + c_3 \begin{pmatrix} -2 \\ 0 \\ 1 \end{pmatrix},$$

and two solutions of system (14) are

$$\begin{pmatrix} -1 \\ 1 \\ 0 \end{pmatrix} e^t \quad \text{and} \quad \begin{pmatrix} -2 \\ -0 \\ 1 \end{pmatrix} e^t.$$

The eigenvalue 5 forces us to solve the system

$$\begin{aligned} -3c_1 + c_2 + 2c_3 &= 0 \\ c_1 - 3c_2 + 2c_3 &= 0 \\ c_1 + c_2 - 2c_3 &= 0, \end{aligned}$$

which by elementary elimination yields $c_1 = c_2 = c_3$ and the eigenvector

$\begin{pmatrix} 1 \\ 1 \\ 1 \end{pmatrix}$. The Wronskian of the three solutions

$$\begin{pmatrix} -1 \\ 1 \\ 0 \end{pmatrix} e^t, \qquad \begin{pmatrix} -2 \\ 0 \\ 1 \end{pmatrix} e^t, \qquad \begin{pmatrix} 1 \\ 1 \\ 1 \end{pmatrix} e^{5t}$$

at $t = 0$ has value 4. Thus the general solution of system (14) is

$$\begin{pmatrix} x \\ y \\ z \end{pmatrix} = c_1 \begin{pmatrix} -1 \\ 1 \\ 0 \end{pmatrix} e^t + c_2 \begin{pmatrix} -2 \\ 0 \\ 1 \end{pmatrix} e^t + c_3 \begin{pmatrix} 1 \\ 1 \\ 1 \end{pmatrix} e^{5t}.$$

Exercises

In exercises 1 through 9, solve the system $X' = AX$.

1. $A = \begin{pmatrix} 4 & 1 \\ -4 & 8 \end{pmatrix}$.
    ANS. $X = c_1 \begin{pmatrix} 1 \\ 2 \end{pmatrix} e^{6t} + c_2 \left[\begin{pmatrix} 1 \\ 2 \end{pmatrix} t + \begin{pmatrix} 0 \\ 1 \end{pmatrix} \right] e^{6t}$.

2. $A = \begin{pmatrix} 4 & -9 \\ 4 & -8 \end{pmatrix}$.
    ANS. $X = c_1 \begin{pmatrix} 3 \\ 2 \end{pmatrix} e^{-2t} + c_2 \left[\begin{pmatrix} 6 \\ 4 \end{pmatrix} t + \begin{pmatrix} 1 \\ 0 \end{pmatrix} \right] e^{-2t}$.

3. $A = \begin{pmatrix} 2 & -1 \\ 4 & 6 \end{pmatrix}$.
    ANS. $X = c_1 \begin{pmatrix} 1 \\ -2 \end{pmatrix} e^{4t} + c_2 \left[\begin{pmatrix} 1 \\ -2 \end{pmatrix} t + \begin{pmatrix} 0 \\ -1 \end{pmatrix} \right] e^{4t}$.

4. $A = \begin{pmatrix} 1 & -2 \\ 2 & -3 \end{pmatrix}$.
    ANS. $X = c_1 \begin{pmatrix} 1 \\ 1 \end{pmatrix} e^{-t} + c_2 \left[\begin{pmatrix} 2 \\ 2 \end{pmatrix} t + \begin{pmatrix} 2 \\ 1 \end{pmatrix} \right] e^{-t}$.

5. $A = \begin{pmatrix} 1 & 2 & -1 \\ 0 & 1 & 1 \\ 0 & 0 & 2 \end{pmatrix}$.
  ANS. $X = c_1 \begin{pmatrix} 1 \\ 1 \\ 1 \end{pmatrix} e^{2t} + c_2 \begin{pmatrix} 1 \\ 0 \\ 0 \end{pmatrix} e^t + c_3 \left[\begin{pmatrix} 2 \\ 0 \\ 0 \end{pmatrix} t + \begin{pmatrix} 0 \\ 1 \\ 0 \end{pmatrix} \right] e^t$.

6. $A = \begin{pmatrix} 2 & 1 & -1 \\ 0 & -1 & 2 \\ 0 & 0 & -1 \end{pmatrix}$.

    ANS. $X = c_1 \begin{pmatrix} 1 \\ 0 \\ 0 \end{pmatrix} e^{2t} + c_2 \begin{pmatrix} 1 \\ -3 \\ 0 \end{pmatrix} e^{-t} + c_3 \left[\begin{pmatrix} 6 \\ -18 \\ 0 \end{pmatrix} t + \begin{pmatrix} -1 \\ 0 \\ -9 \end{pmatrix} \right] e^{-t}$.

7. $A = \begin{pmatrix} 0 & 3 & 1 \\ 1 & 2 & 1 \\ 1 & 3 & 0 \end{pmatrix}$. ANS. $X = \left[c_1 \begin{pmatrix} 1 \\ 0 \\ -1 \end{pmatrix} + c_2 \begin{pmatrix} 0 \\ 1 \\ -3 \end{pmatrix} \right] e^{-t} + \left[c_3 \begin{pmatrix} 1 \\ 1 \\ 1 \end{pmatrix} e^{4t} \right]$.

8. $A = \begin{pmatrix} 12 & 2 & -2 \\ 5 & 3 & -1 \\ 5 & 1 & 1 \end{pmatrix}$. ANS. $X = c_1 \begin{pmatrix} 2 \\ 1 \\ 1 \end{pmatrix} e^{12t} + \left[c_2 \begin{pmatrix} 1 \\ 0 \\ 5 \end{pmatrix} + c_3 \begin{pmatrix} 1 \\ -5 \\ 0 \end{pmatrix} \right] e^{2t}$.

9. $A = \begin{pmatrix} 0 & -1 & 3 \\ 2 & -3 & 3 \\ 2 & -1 & 1 \end{pmatrix}$. ANS. $X = c_1 \begin{pmatrix} 1 \\ 1 \\ 1 \end{pmatrix} e^{2t} + \left[c_2 \begin{pmatrix} 3 \\ 0 \\ -2 \end{pmatrix} + c_3 \begin{pmatrix} 1 \\ 2 \\ 0 \end{pmatrix} \right] e^{-2t}$.

10. Complete the details in Example (c) of this section.

11. Discuss in complete detail the possible solutions of the system $X' = AX$ if A is the diagonal matrix

$$ A = \begin{pmatrix} a & 0 & 0 \\ 0 & b & 0 \\ 0 & 0 & c \end{pmatrix}. $$

12. Consider the system $X' = AX$ for

$$ A = \begin{pmatrix} a & b \\ c & d \end{pmatrix}. $$

(a) Show that the characteristic equation of A has a repeated root only if $(a - d)^2 + 4bc = 0$.

(b) Show that if $a \neq d$ and if $(a - d)^2 + 4bc = 0$ then the complete solution of the system is

$$ X = c_1 \begin{pmatrix} 2b \\ d - a \end{pmatrix} e^{\frac{1}{2}(a+d)t} + c_2 \left[\begin{pmatrix} 2b \\ d - a \end{pmatrix} t + \begin{pmatrix} 0 \\ 2 \end{pmatrix} \right] e^{\frac{1}{2}(a+d)t}. $$

(c) Completely discuss the solution in case

$$ (a - d)^2 + 4bc = 0 \text{ and } a = d. $$

82. Nonhomogeneous systems

Now that we have some understanding of homogeneous systems with constant coefficients we turn our attention to systems that are nonhomogeneous. Consider the system

$$ X' = AX + B, \tag{1} $$

where A is a constant $n \times n$ matrix and B is a vector function of t. Theorem 29 on page 251 indicates that we need to find a particular solution X_p of system (1) and add it to the general solution of the associated homogeneous system. We will use a variation of parameters technique to find the particular solution X_p.

EXAMPLE (a):   Consider the system

$$\begin{pmatrix} x \\ y \end{pmatrix}' = \begin{pmatrix} 0 & 1 \\ -2 & 3 \end{pmatrix}\begin{pmatrix} x \\ y \end{pmatrix} + \begin{pmatrix} f(t) \\ g(t) \end{pmatrix}. \tag{2}$$

In Example (a) of Section 79 we found the general solution of the homogeneous system

$$\begin{pmatrix} x \\ y \end{pmatrix}' = \begin{pmatrix} 0 & 1 \\ -2 & 3 \end{pmatrix}\begin{pmatrix} x \\ y \end{pmatrix} \tag{3}$$

to be

$$\begin{pmatrix} x \\ y \end{pmatrix}_c = a_1 \begin{pmatrix} 1 \\ 1 \end{pmatrix} e^t + a_2 \begin{pmatrix} 1 \\ 2 \end{pmatrix} e^{2t}, \tag{4}$$

where a_1 and a_2 are arbitrary constants.

   We now seek a solution of system (2) of the form

$$\begin{pmatrix} x \\ y \end{pmatrix}_p = a_1(t) \begin{pmatrix} 1 \\ 1 \end{pmatrix} e^t + a_2(t) \begin{pmatrix} 1 \\ 2 \end{pmatrix} e^{2t}. \tag{5}$$

Direct substitution into (2) gives

$$a_1(t)\begin{pmatrix} 1 \\ 1 \end{pmatrix} e^t + 2a_2(t)\begin{pmatrix} 1 \\ 2 \end{pmatrix} e^{2t} + a_1'(t)\begin{pmatrix} 1 \\ 1 \end{pmatrix} e^t + a_2'(t)\begin{pmatrix} 1 \\ 2 \end{pmatrix} e^{2t}$$
$$= \begin{pmatrix} 0 & 1 \\ -2 & 3 \end{pmatrix}\begin{pmatrix} 1 \\ 1 \end{pmatrix}a_1(t) e^t + \begin{pmatrix} 0 & 1 \\ -2 & 3 \end{pmatrix}\begin{pmatrix} 1 \\ 2 \end{pmatrix}a_2(t) e^{2t} + \begin{pmatrix} f(t) \\ g(t) \end{pmatrix}, \tag{6}$$

or more simply

$$a_1'(t)\begin{pmatrix} 1 \\ 1 \end{pmatrix} e^t + a_2'(t)\begin{pmatrix} 1 \\ 2 \end{pmatrix} e^{2t} = \begin{pmatrix} f(t) \\ g(t) \end{pmatrix}. \tag{7}$$

The other terms in (6) cancel each other precisely because $\begin{pmatrix} x \\ y \end{pmatrix}_c$ is a solution of the homogeneous system (3). Equation (7) can now be written

$$\begin{pmatrix} 1 & 1 \\ 1 & 2 \end{pmatrix}\begin{pmatrix} a_1'(t) e^t \\ a_2'(t) e^{2t} \end{pmatrix} = \begin{pmatrix} f(t) \\ g(t) \end{pmatrix}.$$

   Using Cramer's rule, we find

$$a_1'(t) e^t = \frac{\begin{vmatrix} f(t) & 1 \\ g(t) & 2 \end{vmatrix}}{\begin{vmatrix} 1 & 1 \\ 1 & 2 \end{vmatrix}} = 2f(t) - g(t),$$

$$a_2'(t) e^{2t} = \frac{\begin{vmatrix} 1 & f(t) \\ 1 & g(t) \end{vmatrix}}{\begin{vmatrix} 1 & 1 \\ 1 & 2 \end{vmatrix}} = g(t) - f(t).$$

Thus

$$a_1'(t) = [2f(t) - g(t)] e^{-t},$$

$$a_2'(t) = [g(t) - f(t)] e^{-2t}.$$

If, for example, $f(t) = e^t$ and $g(t) = 1$, we have

$$a_1'(t) = (2e^t - 1) e^{-t} = 2 - e^{-t},$$

$$a_2'(t) = (1 - e^t) e^{-2t} = e^{-2t} - e^{-t},$$

so that

$$a_1(t) = 2t + e^{-t},$$

$$a_2(t) = -\tfrac{1}{2} e^{-2t} + e^{-t}.$$

The particular solution (5) is

$$\begin{pmatrix} x \\ y \end{pmatrix}_p = (2t + e^{-t}) \begin{pmatrix} 1 \\ 1 \end{pmatrix} e^t + (-\tfrac{1}{2} e^{-2t} + e^{-t}) \begin{pmatrix} 1 \\ 2 \end{pmatrix} e^{2t},$$

or

$$X_p = \begin{pmatrix} x \\ y \end{pmatrix}_p = (2t e^t + 1) \begin{pmatrix} 1 \\ 1 \end{pmatrix} + (e^t - \tfrac{1}{2}) \begin{pmatrix} 1 \\ 2 \end{pmatrix}.$$

The general solution of system (2) for $f(t) = e^t$ and $g(t) = 1$ is therefore

$$X = a_1 \begin{pmatrix} 1 \\ 1 \end{pmatrix} e^t + a_2 \begin{pmatrix} 1 \\ 2 \end{pmatrix} e^{2t} + (2t e^t + 1) \begin{pmatrix} 1 \\ 1 \end{pmatrix} + (e^t - \tfrac{1}{2}) \begin{pmatrix} 1 \\ 2 \end{pmatrix}.$$

EXAMPLE (b):   Solve the system

$$X' = AX + B \qquad \text{for } A = \begin{pmatrix} 2 & 1 \\ -4 & 2 \end{pmatrix} \text{ and } B = \begin{pmatrix} 3 e^{2t} \\ t e^{2t} \end{pmatrix}. \tag{8}$$

The associated homogeneous problem is the same as Example (b) of Section 80. The general solution of the homogeneous system is

$$X_c = e^{2t}\left[b_1\left\{\begin{pmatrix}1\\0\end{pmatrix}\cos 2t - \begin{pmatrix}0\\2\end{pmatrix}\sin 2t\right\} + b_2\left\{\begin{pmatrix}0\\2\end{pmatrix}\cos 2t + \begin{pmatrix}1\\0\end{pmatrix}\sin 2t\right\}\right]. \quad (9)$$

We seek a particular solution of the nonhomogeneous system of the form

$$X_p = e^{2t}\left[b_1(t)\left\{\begin{pmatrix}1\\0\end{pmatrix}\cos 2t - \begin{pmatrix}0\\2\end{pmatrix}\sin 2t\right\} + b_2(t)\left\{\begin{pmatrix}0\\2\end{pmatrix}\cos 2t + \begin{pmatrix}1\\0\end{pmatrix}\sin 2t\right\}\right].$$

Omitting the terms that cancel one another when X_p is substituted into (8), we have

$$e^{2t}\left[b_1'(t)\left\{\begin{pmatrix}1\\0\end{pmatrix}\cos 2t - \begin{pmatrix}0\\2\end{pmatrix}\sin 2t\right\} + b_2'(t)\left\{\begin{pmatrix}0\\2\end{pmatrix}\cos 2t + \begin{pmatrix}1\\0\end{pmatrix}\sin 2t\right\}\right]$$
$$= \begin{pmatrix}3\,e^{2t}\\t\,e^{2t}\end{pmatrix}.$$

We may rewrite this system in the form

$$\begin{pmatrix}\cos 2t & \sin 2t \\ -2\sin 2t & 2\cos 2t\end{pmatrix}\begin{pmatrix}b_1'(t)\\b_2'(t)\end{pmatrix} = \begin{pmatrix}3\\t\end{pmatrix}.$$

Solving for $b_1'(t)$ and $b_2'(t)$ yields

$$b_1'(t) = \frac{1}{2}\begin{vmatrix}3 & \sin 2t \\ t & 2\cos 2t\end{vmatrix} = \frac{1}{2}(6\cos 2t - t\sin 2t),$$

$$b_2'(t) = \frac{1}{2}\begin{vmatrix}\cos 2t & 3 \\ -2\sin 2t & t\end{vmatrix} = \frac{1}{2}(t\cos 2t + 6\sin 2t).$$

Integration of these functions gives

$$b_1(t) = \tfrac{1}{8}(2t\cos 2t + 11\sin 2t),$$

$$b_2(t) = \tfrac{1}{8}(2t\sin 2t - 11\cos 2t).$$

One particular solution of system (8) is therefore

$$X_p = e^{2t}\begin{pmatrix}b_1(t)\cos 2t + b_2(t)\sin 2t \\ -2b_1(t)\sin 2t + 2b_2(t)\cos 2t\end{pmatrix} = e^{2t}\begin{pmatrix}\tfrac{1}{4}t \\ -\tfrac{11}{4}\end{pmatrix},$$

or more simply

$$X_p = \tfrac{1}{4}e^{2t}\begin{pmatrix}t \\ -11\end{pmatrix}.$$

The general solution of system (8) is

$$X = X_c + \tfrac{1}{4}e^{2t}\begin{pmatrix}t \\ -11\end{pmatrix},$$

where X_c is given in equation (9).

Exercises

Find the general solution of each of the following systems.

1. $\begin{pmatrix} x \\ y \end{pmatrix}' = \begin{pmatrix} 0 & 1 \\ -2 & 3 \end{pmatrix} \begin{pmatrix} x \\ y \end{pmatrix} + \begin{pmatrix} e^t \\ 2 \end{pmatrix}.$　See Example (a), Section 82.

ANS. $X_p = t e^t \begin{pmatrix} 2 \\ 2 \end{pmatrix} + e^t \begin{pmatrix} 1 \\ 2 \end{pmatrix} + \begin{pmatrix} 1 \\ 0 \end{pmatrix}.$

2. $\begin{pmatrix} x \\ y \end{pmatrix}' = \begin{pmatrix} 2 & 1 \\ -4 & 2 \end{pmatrix} \begin{pmatrix} x \\ y \end{pmatrix} + \begin{pmatrix} t\, e^{2t} \\ -e^{2t} \end{pmatrix}.$　See Example (b), Section 82.

ANS. $X_p = t e^{2t} \begin{pmatrix} 0 \\ -1 \end{pmatrix}.$

3. $\begin{pmatrix} x \\ y \end{pmatrix}' = \begin{pmatrix} 3 & 3 \\ -1 & -1 \end{pmatrix} \begin{pmatrix} x \\ y \end{pmatrix} + \begin{pmatrix} t \\ 1 \end{pmatrix}.$　ANS. $X_p = \frac{1}{8} \begin{pmatrix} -2t^2 - 18t - 6 \\ 2t^2 + 14t \end{pmatrix}.$

4. $\begin{pmatrix} x \\ y \end{pmatrix}' = \begin{pmatrix} 3 & 3 \\ -1 & -1 \end{pmatrix} \begin{pmatrix} x \\ y \end{pmatrix} + \begin{pmatrix} e^{-t} \\ e^{2t} \end{pmatrix}.$　See exercise 3.

ANS. $X_p = -\frac{1}{3} e^{-t} \begin{pmatrix} 0 \\ 1 \end{pmatrix} - \frac{1}{4} e^{2t} \begin{pmatrix} 3 - 6t \\ -3 + 2t \end{pmatrix}.$

5. $\begin{pmatrix} x \\ y \end{pmatrix}' = \begin{pmatrix} 0 & 1 \\ -2 & 3 \end{pmatrix} \begin{pmatrix} x \\ y \end{pmatrix} + \begin{pmatrix} 0 \\ 3 \end{pmatrix} e^t.$　ANS. $X_p = -3t e^t \begin{pmatrix} 1 \\ 1 \end{pmatrix} - 3 e^t \begin{pmatrix} 1 \\ 2 \end{pmatrix}.$

6. $\begin{pmatrix} x \\ y \end{pmatrix}' = \begin{pmatrix} 4 & 1 \\ -4 & 8 \end{pmatrix} \begin{pmatrix} x \\ y \end{pmatrix} + \begin{pmatrix} 1 \\ 6t \end{pmatrix} e^{6t}.$　ANS. $X_p = \begin{pmatrix} t^3 - t^2 + t \\ 2t^3 + t^2 \end{pmatrix} e^{6t}.$

83. Arms races

An interesting application that leads to a system of linear differential equations is the study of arms races. The presentation made here is often called the Richardson model since it was first proposed by the English meteorologist L. F. Richardson (1881–1953).*

Consider the problem of two countries with expenditures for armaments x and y measured in billions of dollars. We presume that x and y are functions of time measured in years. The Richardson model then makes the following assumptions:

* See, for instance, T. L. Saaty, *Mathematical Models of Arms Control and Disarmament* (New York: John Wiley & Sons, Inc., 1968).

(a) The expenditure for armaments of each country will increase at a rate that is proportional to the other country's expenditure.
(b) The expenditure for armaments of each country will decrease at a rate that is proportional to its own expenditure.
(c) The rate of change of arms expenditure for a country has a constant component that measures the level of antagonism of that country toward the other.
(d) The effects of the three previous assumptions are additive.

These assumptions lead to the system

$$\frac{dx}{dt} = ay - px + r,$$

$$\frac{dy}{dt} = bx - qy + s.$$

(1)

The constants a, b, p, and q are positive, but the numbers r and s may have any value, positive values arising if the countries have internal attitudes of distrust for each other.

In matrix notation the problem may be written

$$X' = AX + B, \qquad X(0) = \begin{pmatrix} x_0 \\ y_0 \end{pmatrix},$$

(2)

where

$$X(t) = \begin{pmatrix} x(t) \\ y(t) \end{pmatrix}, \qquad A = \begin{pmatrix} -p & a \\ b & -q \end{pmatrix}, \qquad B = \begin{pmatrix} r \\ s \end{pmatrix}.$$

As we have seen, the nature of the solutions of the system will depend on the eigenvalues of the matrix A, that is, on the roots of the characteristic equation

$$\begin{vmatrix} -p - m & a \\ b & -q - m \end{vmatrix} = m^2 + (p + q)m + (pq - ab) = 0.$$

These roots are

$$\frac{-(p + q) \pm \sqrt{(p + q)^2 - 4(pq - ab)}}{2} = \frac{-(p + q) \pm \sqrt{(p - q)^2 + 4ab}}{2},$$

and, since a and b are positive, the eigenvalues are real and distinct. Because $p > 0$ and $q > 0$, it follows that if $pq - ab > 0$, then the two eigenvalues are both negative, but if $pq - ab < 0$, then the eigenvalues will have opposite signs. The presence of a positive eigenvalue is disturbing since it will lead to an exponential function that becomes unbounded as time increases, a situation that may result in a runaway arms race.

We now examine several different examples to illustrate the possible consequences of Richardson's model.

EXAMPLE (a): Consider a situation in which the parameters in equations (2) are $a = 4$, $b = 2$, $p = 3$, $q = 1$, $r = 2$, $s = 2$, $x_0 = 4$, and $y_0 = 1$, that is

$$X' = \begin{pmatrix} -3 & 4 \\ 2 & -1 \end{pmatrix} X + \begin{pmatrix} 2 \\ 2 \end{pmatrix} \quad \text{and} \quad X(0) = \begin{pmatrix} 4 \\ 1 \end{pmatrix}.$$

The characteristic equation of the matrix is

$$\begin{vmatrix} -3 - m & 4 \\ 2 & -1 - m \end{vmatrix} = m^2 + 4m - 5 = 0,$$

so that the eigenvalues are $m_1 = 1$ and $m_2 = -5$.

For $m_1 = 1$, we compute the nontrivial solutions of the system

$$\begin{pmatrix} -4 & 4 \\ 2 & -2 \end{pmatrix}\begin{pmatrix} c_1 \\ c_2 \end{pmatrix} = O.$$

Thus $c_1 = c_2$. One such solution is obtained by taking $c_1 = 1$ to give the eigenvector $\begin{pmatrix} 1 \\ 1 \end{pmatrix}$.

For $m_2 = -5$, the system

$$\begin{pmatrix} 2 & 4 \\ 2 & 4 \end{pmatrix}\begin{pmatrix} c_1 \\ c_2 \end{pmatrix} = O$$

requires $c_1 + 2c_2 = 0$. Taking $c_2 = -1$ yields the eigenvector $\begin{pmatrix} 2 \\ -1 \end{pmatrix}$.

The general solution of the homogeneous system $X' = AX$ is therefore

$$X(t) = c_1 \begin{pmatrix} 1 \\ 1 \end{pmatrix} e^t + c_2 \begin{pmatrix} 2 \\ -1 \end{pmatrix} e^{-5t}.$$

The nonhomogeneous system $X' = AX + B$ has a constant solution of the form $\begin{pmatrix} e \\ f \end{pmatrix}$. Substitution into the system gives

$$\begin{pmatrix} -3 & 4 \\ 2 & -1 \end{pmatrix}\begin{pmatrix} e \\ f \end{pmatrix} + \begin{pmatrix} 2 \\ 2 \end{pmatrix} = O,$$

a system with solution $\begin{pmatrix} -2 \\ -2 \end{pmatrix}$. Thus the general solution of the nonhomogeneous system is

$$X(t) = c_1 \begin{pmatrix} 1 \\ 1 \end{pmatrix} e^t + c_2 \begin{pmatrix} 2 \\ -1 \end{pmatrix} e^{-5t} + \begin{pmatrix} -2 \\ -2 \end{pmatrix}.$$

The initial condition $X(0) = \begin{pmatrix} 4 \\ 1 \end{pmatrix}$ now requires that

$$\begin{pmatrix} 4 \\ 1 \end{pmatrix} = c_1 \begin{pmatrix} 1 \\ 1 \end{pmatrix} + c_2 \begin{pmatrix} 2 \\ -1 \end{pmatrix} + \begin{pmatrix} -2 \\ -2 \end{pmatrix},$$

so that $c_1 = 4$ and $c_2 = 1$. The final solution is therefore

$$X(t) = 4 \begin{pmatrix} 1 \\ 1 \end{pmatrix} e^t + \begin{pmatrix} 2 \\ -1 \end{pmatrix} e^{-5t} + \begin{pmatrix} -2 \\ -2 \end{pmatrix},$$

or

$$x(t) = 4e^t + 2e^{-5t} - 2,$$

$$y(t) = 4e^t - e^{-5t} - 2.$$

We have a runaway arms race.

EXAMPLE (b): As a second example of an arms race, we take the following values for the parameters in equation (2): $a = 4, b = 2, p = 3, q = 1$, $r = -2, s = -2, x_0 = 2, y_0 = \frac{1}{2}$. The system of differential equations has the same solution as in Example (a) except for the sign of the particular solution. Thus the general solution is

$$X(t) = c_1 \begin{pmatrix} 1 \\ 1 \end{pmatrix} e^t + c_2 \begin{pmatrix} 2 \\ -1 \end{pmatrix} e^{-5t} + \begin{pmatrix} 2 \\ 2 \end{pmatrix}.$$

The initial conditions now require that

$$\begin{pmatrix} 2 \\ \frac{1}{2} \end{pmatrix} = c_1 \begin{pmatrix} 1 \\ 1 \end{pmatrix} + c_2 \begin{pmatrix} 2 \\ -1 \end{pmatrix} + \begin{pmatrix} 2 \\ 2 \end{pmatrix},$$

from which we obtain $c_1 = -1$ and $c_2 = \frac{1}{2}$. The solution is

$$x(t) = -e^t + e^{-5t} + 2,$$

$$y(t) = -e^t - \tfrac{1}{2}e^{-5t} + 2,$$

and each party will eventually decrease its arms expenditure to zero, a condition of disarmament.

EXAMPLE (c): Let us now change the values of the parameters in system (2) to $a = 3, b = 1, p = 4, q = 2, r = 6, s = 1$ with initial conditions $x_0 = 0$ and $y_0 = 0$. The system to be solved becomes

$$\begin{pmatrix} x \\ y \end{pmatrix}' = \begin{pmatrix} -4 & 3 \\ 1 & -2 \end{pmatrix} \begin{pmatrix} x \\ y \end{pmatrix} + \begin{pmatrix} 6 \\ 1 \end{pmatrix}.$$

A particular solution of this system is the vector $\begin{pmatrix} 3 \\ 2 \end{pmatrix}$ and the eigenvalues of the matrix are -1 and -5 with corresponding eigenvectors $\begin{pmatrix} 1 \\ 1 \end{pmatrix}$ and $\begin{pmatrix} 3 \\ -1 \end{pmatrix}$.

The general solution of the system is therefore

$$\begin{pmatrix} x \\ y \end{pmatrix} = c_1 \begin{pmatrix} 1 \\ 1 \end{pmatrix} e^{-t} + c_2 \begin{pmatrix} 3 \\ -1 \end{pmatrix} e^{-5t} + \begin{pmatrix} 3 \\ 2 \end{pmatrix}.$$

The initial conditions $x_0 = y_0 = 0$ require

$$\begin{pmatrix} 0 \\ 0 \end{pmatrix} = c_1 \begin{pmatrix} 1 \\ 1 \end{pmatrix} + c_2 \begin{pmatrix} 3 \\ -1 \end{pmatrix} + \begin{pmatrix} 3 \\ 2 \end{pmatrix},$$

an equation that is satisfied only if $c_1 = -\frac{9}{4}$ and $c_2 = -\frac{1}{4}$. Thus the solution of the initial value problem is

$$\begin{pmatrix} x \\ y \end{pmatrix} = \frac{-9}{4} \begin{pmatrix} 1 \\ 1 \end{pmatrix} e^{-t} - \frac{1}{4} \begin{pmatrix} 3 \\ -1 \end{pmatrix} e^{-5t} + \begin{pmatrix} 3 \\ 2 \end{pmatrix}. \tag{3}$$

It is also true that

$$\begin{pmatrix} x \\ y \end{pmatrix}' = \frac{9}{4} \begin{pmatrix} 1 \\ 1 \end{pmatrix} e^{-t} + \frac{5}{4} \begin{pmatrix} 3 \\ -1 \end{pmatrix} e^{-5t}. \tag{4}$$

We may now interpret equations (3) and (4) as an arms race with each party starting with zero expenditure but with $dx/dt = 6$ and $dy/dt = 1$, both positive quantities. Because of the negative exponents, the rates at which the expenditures are changing will tend toward zero and the arms expenditures will approach $x = 3$ and $y = 2$. There will be a stabilized arms race.

Exercises

For Richardson's model as described by equations (2), solve the following special cases, noting in each exercise whether there will be a stable arms race, a runaway arms race, or disarmament.

1. $a = 2, b = 4, p = 5, q = 3, r = 1, s = 2, x_0 = 8, y_0 = 7$.
   ANS. $x(t) = 4e^{-t} + 3e^{-7t} + 1$, $y(t) = 8e^{-t} - 3e^{-7t} + 2$. Stable arms race.
2. What effect does the changing of the initial values x_0 and y_0 have on the stability of the solution in exercise 1?
3. $a = 4, b = 4, p = 2, q = 2, r = 8, s = 2, x_0 = 5, y_0 = 2$.
   ANS. $x(t) = e^{-6t} + 6e^{2t} - 2$, $y(t) = -e^{-6t} + 6e^{2t} - 3$. Runaway arms race.
4. Show that the solution in exercise 3 will remain unstable if the initial values are changed to any other nonnegative values.
5. For $a = 4, b = 4, p = 2, q = 2, r = -2, s = -2$, show that there will be disarmament if $x_0 + y_0 < 2$ and a runaway arms race if $x_0 + y_0 > 2$.

6. For $a = 4$, $b = 4$, $p = 2$, $q = 2$, $r > 0$, $s > 0$, show that there will be a runaway arms race for any nonnegative x_0 and y_0.

7. For $a = 4$, $b = 4$, $p = 2$, $q = 2$, $r < 0$, $s < 0$, show that there will be a runaway arms race if $x_0 + y_0 > \dfrac{-r - s}{2}$.

8. Show that if $pq - ab > 0$, $r > 0$, and $s > 0$, there will be a stable solution to the arms race.

9. Show that if $pq - ab < 0$, $r > 0$, and $s > 0$, there will be a runaway arms race.

10. Show that if $pq - ab > 0$, $r < 0$, and $s < 0$, there will be disarmament.

84. The Laplace transform

The Laplace operator can be used to transform a system of linear differential equations with constant coefficients into a system of algebraic equations.

EXAMPLE (a): Solve the system of equations

$$x''(t) - x(t) + 5y'(t) = t, \tag{1}$$

$$y''(t) - 4y(t) - 2x'(t) = -2, \tag{2}$$

with the initial conditions

$$x(0) = 0, \qquad x'(0) = 0, \qquad y(0) = 0, \qquad y'(0) = 0. \tag{3}$$

Let $L\{x(t)\} = u(s)$ and $L\{y(t)\} = v(s)$. Then, application of the Laplace operator transforms the problem into that of solving a pair of simultaneous algebraic equations:

$$(s^2 - 1)u(s) + 5sv(s) = \frac{1}{s^2}, \tag{4}$$

$$-2su(s) + (s^2 - 4)v(s) = -\frac{2}{s}. \tag{5}$$

We solve equations (4) and (5) to obtain

$$u(s) = \frac{11s^2 - 4}{s^2(s^2 + 1)(s^2 + 4)}, \tag{6}$$

$$v(s) = \frac{-2s^2 + 4}{s(s^2 + 1)(s^2 + 4)}. \tag{7}$$

Seeking the inverse transforms of u and v, we first expand the right members of (6) and (7) into partial fractions:

$$u(s) = -\frac{1}{s^2} + \frac{5}{s^2 + 1} - \frac{4}{s^2 + 4}, \tag{8}$$

$$v(s) = \frac{1}{s} - \frac{2s}{s^2 + 1} + \frac{s}{s^2 + 4}. \tag{9}$$

Since $x(t) = L^{-1}\{u(s)\}$ and $y(t) = L^{-1}\{v(s)\}$, we get the desired result

$$x(t) = -t + 5 \sin t - 2 \sin 2t, \tag{10}$$

$$y(t) = 1 - 2 \cos t + \cos 2t, \tag{11}$$

which is easily verified by direct substitution into (1), (2), and (3).

The foregoing procedure is simple in concept, but of course its practical use will depend on our ability to find the inverse transforms of $u(s)$ and $v(s)$. On the other hand, the use of the transform theory can help us gain insight into the general theory of systems of linear differential equations. We illustrate this idea in the following example.

EXAMPLE (b):   Solve the system

$$\frac{dx}{dt} = y + F(t), \tag{12}$$

$$\frac{dy}{dt} = x + G(t). \tag{13}$$

Here we presume that the transforms of the functions $x(t)$, $y(t)$, $F(t)$, and $G(t)$ all exist and are given by $u(s)$, $v(s)$, $f(s)$, and $g(s)$, respectively. We then have

$$su(s) - c_1 = v(s) + f(s), \tag{14}$$

$$sv(s) - c_2 = u(s) + g(s), \tag{15}$$

where c_1 and c_2 represent the initial values of $x(t)$ and $y(t)$. We can rewrite equations (14) and (15) in the form

$$su(s) - v(s) = c_1 + f(s), \tag{16}$$

$$-u(s) + sv(s) = c_2 + g(s). \tag{17}$$

Using Cramer's rule, the solution of the algebraic system (16) and (17) may be written

$$u(s) = \frac{\begin{vmatrix} c_1 + f(s) & -1 \\ c_2 + g(s) & s \end{vmatrix}}{s^2 - 1} = \frac{\begin{vmatrix} c_1 & -1 \\ c_2 & s \end{vmatrix}}{s^2 - 1} + \frac{\begin{vmatrix} f(s) & -1 \\ g(s) & s \end{vmatrix}}{s^2 - 1}, \tag{18}$$

$$v(s) = \frac{\begin{vmatrix} s & c_1 + f(s) \\ -1 & c_2 + g(s) \end{vmatrix}}{s^2 - 1} = \frac{\begin{vmatrix} s & c_1 \\ -1 & c_2 \end{vmatrix}}{s^2 - 1} + \frac{\begin{vmatrix} s & f(s) \\ -1 & g(s) \end{vmatrix}}{s^2 - 1}. \tag{19}$$

A careful examination of equations (18) and (19) reveals several important properties of the pair of functions, $x(t)$ and $y(t)$, we are seeking. First, each of these functions can be considered as the sum of two functions; that is,

$$x(t) = x_c(t) + x_p(t),$$

$$y(t) = y_c(t) + y_p(t).$$

The notation used is intended to remind us of a similar situation we met when we were dealing with a single linear differential equation with one dependent variable. We note that the pair of functions, $x_c(t)$ and $y_c(t)$, is a solution of the system (12) and (13) in the case where $F(t) = G(t) = 0$ (and of course $f(s) = g(s) = 0$). Moreover, each of the functions $x_c(t)$ and $y_c(t)$ involves the constants c_1 and c_2 that are the initial values of $x(t)$ and $y(t)$.

On the other hand, the functions $x_p(t)$ and $y_p(t)$, although independent of the initial conditions, are intimately related to the functions $F(t)$ and $G(t)$.

Although it is possible to obtain the functions $x(t)$ and $y(t)$ from (18) and (19) by using convolution integrals (see exercise 9 below), let us simplify the problem by considering a particular case.

EXAMPLE (c):   Solve the system of equations (12) and (13) if $F(t) = 1$ and $G(t) = t$.

Equations (18) and (19) now become

$$u(s) = \frac{\begin{vmatrix} c_1 & -1 \\ c_2 & s \end{vmatrix}}{s^2 - 1} + \frac{\begin{vmatrix} \dfrac{1}{s} & -1 \\ \dfrac{1}{s^2} & s \end{vmatrix}}{s^2 - 1}, \tag{20}$$

$$v(s) = \frac{\begin{vmatrix} s & c_1 \\ -1 & c_2 \end{vmatrix}}{s^2 - 1} + \frac{\begin{vmatrix} s & \dfrac{1}{s} \\ -1 & \dfrac{1}{s^2} \end{vmatrix}}{s^2 - 1}. \tag{21}$$

These equations may be written

$$u(s) = \frac{c_1 + c_2}{2(s - 1)} + \frac{c_1 - c_2}{2(s + 1)} - \frac{1}{s^2} + \frac{1}{s - 1} - \frac{1}{s + 1}, \tag{22}$$

$$v(s) = \frac{c_1 + c_2}{2(s - 1)} - \frac{c_1 - c_2}{2(s + 1)} - \frac{2}{s} + \frac{1}{s - 1} + \frac{1}{s + 1}, \tag{23}$$

where we have taken care to keep the two parts of each of these functions separate. The inversion of equations (22) and (23) yields

$$x(t) = \frac{c_1 + c_2}{2} e^t + \frac{c_1 - c_2}{2} e^{-t} - t + e^t - e^{-t}, \tag{24}$$

$$y(t) = \frac{c_1 + c_2}{2} e^t - \frac{c_1 - c_2}{2} e^{-t} - 2 + e^t + e^{-t}. \tag{25}$$

The reader should now verify directly that the pair of functions, $x_c(t)$ and $y_c(t)$, which involve the initial values c_1 and c_2, is a solution of the system

$$\frac{dx}{dt} = y \quad \text{and} \quad \frac{dy}{dt} = x.$$

This system results from placing $F(t) = G(t) = 0$ in equations (12) and (13). The reader should also verify that the remaining part of the solution, $x_p(t)$ and $y_p(t)$, is a particular solution of the system (12) and (13) with $F(t) = 1$ and $G(t) = t$. Finally, it is easy to show that $x(0) = c_1$ and $y(0) = c_2$.

Exercises

In exercises 1 through 8, use the Laplace transform method to solve the given system.

1. $x''(t) - 3x'(t) - y'(t) + 2y(t) = 14t + 3,$
$x'(t) - 3x(t) + y'(t) = 1; x(0) = 0, x'(0) = 0, y(0) = 6.5.$

                     ANS. $x(t) = 2 - \frac{1}{2} e^t - \frac{1}{2} e^{3t} - e^{-2t},$
                                 $y(t) = 7t + 5 - e^t + \frac{5}{2} e^{-2t}.$

2. $2x'(t) + 2x(t) + y'(t) - y(t) = 3t,$
$x'(t) + x(t) + y'(t) + y(t) = 1; x(0) = 1, y(0) = 3.$

                     ANS. $x(t) = t + 3 e^{-t} - 2 e^{-3t},$
                                 $y(t) = 1 - t + 2 e^{-3t}.$

3. $x'(t) - 2x(t) - y'(t) - y(t) = 6 e^{3t},$
$2x'(t) - 3x(t) + y'(t) - 3y(t) = 6 e^{3t}; x(0) = 3, y(0) = 0.$

                     ANS. $x(t) = (1 + 2t) e^t + 2 e^{3t},$
                                 $y(t) = (1 - t) e^t - e^{3t}.$

4. $x''(t) + 2x(t) - y'(t) = 2t + 5,$
$x'(t) - x(t) + y'(t) + y(t) = -2t - 1; x(0) = 3, x'(0) = 0, y(0) = -3.$

                     ANS. $x(t) = t + 2 + e^{-2t} + \sin t,$
                                 $y(t) = 1 - t - 3 e^{-2t} - \cos t.$

5. The equations of Example (a) of Section 84 with initial conditions $x(0) = 0,$ $x'(0) = 0, y(0) = 1, y'(0) = 0.$

                     ANS. $x(t) = -t - \frac{5}{3} \sin t + \frac{4}{3} \sin 2t,$
                                 $y(t) = 1 + \frac{2}{3} \cos t - \frac{2}{3} \cos 2t.$

6. The equations of Example (a) of Section 84 with initial conditions $x(0) = 9$, $x'(0) = 2$, $y(0) = 1$, $y'(0) = 0$.

> ANS. $x(t) = -t + 15 \cos t - 5 \sin t - 6 \cos 2t + 4 \sin 2t$,
> $y(t) = 1 + 2 \cos t + 6 \sin t - 2 \cos 2t - 3 \sin 2t$.

7. $x''(t) + y'(t) - y(t) = 0$,
   $2x'(t) - x(t) + z'(t) - z(t) = 0$,
   $x'(t) + 3x(t) + y'(t) - 4y(t) + 3z(t) = 0$;
   $x(0) = 0$, $x'(0) = 1$, $y(0) = 0$, $z(0) = 0$.

8. $x''(t) - x(t) + 5y'(t) = \beta(t)$,
   $y''(t) - 4y(t) - 2x'(t) = 0$,
   in which $\beta(t) = 6t$, $0 \le t \le 2$,
   $\qquad\qquad = 12$, $t > 2$;
   $x(0) = 0$, $x'(0) = 0$, $y(0) = 0$, $y'(0) = 0$.

> ANS. $x(t) = -2(3t - 5 \sin t + \sin 2t)$
> $\qquad + 2[3(t - 2) - 5 \sin (t - 2) + \sin 2(t - 2)]\alpha(t - 2)$,
> $y(t) = 3 - 4 \cos t + \cos 2t$
> $\qquad - [3 - 4 \cos (t - 2) + \cos 2(t - 2)]\alpha(t - 2)$.

9. Write the solution of the system of Example (b) in Section 84 in terms of convolution integrals.

10. Use the results of exercise 9 to obtain the solution of Example (c) in Section 84.

In exercises 11 and 12, write the solution in terms of convolution integrals.

11. $x'(t) - 2y(t) = F(t)$,
    $y'(t) + x(t) = G(t)$; $x(0) = 1$, $y(0) = 0$.

12. $2x'(t) + 3y(t) = F(t)$,
    $y'(t) + 2x(t) = G(t)$; $x(0) = 2$, $y(0) = 1$.

13. Consider the initial value problem
    $x'(t) = ax + by + f(t)$,
    $y'(t) = cx + dy + g(t)$; $x(0) = c_1$, $y(0) = c_2$,
    where a, b, c, d, and c_1, c_2 are constants. Use an argument similar to that of Example (b) of Section 84 to show that the solution, if it exists, should have the form

$$x(t) = x_c(t) + x_p(t),$$
$$y(t) = y_c(t) + y_p(t),$$

where $x_c(t)$ and $y_c(t)$ depend on c_1 and c_2 whereas $x_p(t)$ and $y_p(t)$ depend on $f(t)$ and $g(t)$.

14. For the initial value problem of exercise 13, state a reasonable definition for a first-order homogeneous linear system with constant coefficients. Now state a theorem that expresses the results of exercise 13 for the solution of a nonhomogeneous system.

15. Consider the initial value problem

$$x'(t) + 2y(t) = 0,$$
$$x''(t) + 2y'(t) + 2y(t) = 2 e^t;$$
$$x(0) = 1, x'(0) = 0, y(0) = 0.$$

(a) Show that the Laplace transform method produces

$$x = 3 - 2 e^t, \qquad y = e^t.$$

(b) Verify that these functions satisfy the differential equations but do not satisfy the initial conditions.

(c) By elementary elimination, show that a solution of the system of differential equations has the form

$$x = c_1 - 2 e^t, \qquad y = e^t,$$

and thus the initial conditions given are not compatible with the system of equations.

Electric Circuits and Networks

85. Circuits

The basic laws governing the flow of electric current in a circuit or a network will be given here without derivation. The notation used is common to most texts in electrical engineering and is:

t (sec) = time
Q (coulombs) = quantity of electricity; for example, charge on a capacitor
I (amperes) = current, time rate of flow of electricity
E (volts) = electromotive force or voltage
R (ohms) = resistance
L (henrys) = inductance
C (farads) = capacitance.

By the definition of Q and I it follows that

$$I(t) = Q'(t).$$

The current at each point in a network may be determined by solving the equations that result from applying Kirchhoff's laws:

(a) *The sum of the currents into (or away from) any point is zero.*
(b) *Around any closed path the sum of the instantaneous voltage drops in a specified direction is zero.*

A circuit is treated as a network containing only one closed path. Figure 29 exhibits an "*RLC* circuit" with some of the customary conventions for indicating various elements.

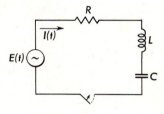

FIGURE 29

For a circuit, Kirchhoff's current law (a) indicates merely that the current is the same throughout. That law plays a larger role in networks, as we shall see later.

To apply Kirchhoff's voltage law (b), it is necessary to know the contributions of each of the idealized elements in Figure 29. The voltage drop across the resistance is RI, that across the inductance is $LI'(t)$, and that across the capacitor is $C^{-1}Q(t)$. The impressed electromotive force $E(t)$ is contributing a voltage rise.

Assume that at time $t = 0$ the switch shown in Figure 29 is to be closed. At $t = 0$ there is no current flowing, $I(0) = 0$ and, if the capacitor is initially without charge, $Q(0) = 0$. From Kirchhoff's law (b), we get the differential equation

$$LI'(t) + RI(t) + C^{-1}Q(t) = E(t), \tag{1}$$

in which

$$I(t) = Q'(t). \tag{2}$$

Equations (1) and (2), with the initial conditions

$$I(0) = 0, \qquad Q(0) = 0, \tag{3}$$

constitute the problem to be solved.

The function $I(t)$ may be eliminated from (1), (2), (3) to obtain the initial value problem

$$LQ''(t) + RQ'(t) + C^{-1}Q(t) = E(t); \qquad Q(0) = 0, Q'(0) = 0. \tag{4}$$

It follows that the circuit problem is equivalent to a problem in damped vibrations of a spring (Section 54). The resistance term $RQ'(t)$ corresponds to the damping term in vibration problems. The analogies between electrical and mechanical systems are useful in practice.

Initial value problems of the type given in equation (4) may be solved either by the general theory of linear equations with constant coefficients or by the use of the Laplace transform. We present here an example using each technique.

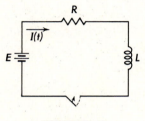

FIGURE 30

EXAMPLE (a):   In the RL circuit shown in Figure 30, find the current $I(t)$ if the current at $t = 0$ is zero and E is a constant.

From equations (1) and (3) we have

$$LI'(t) + RI(t) = E; \qquad I(0) = 0. \tag{5}$$

This first-order linear equation may be written

$$\left(D + \frac{R}{L}\right)I = \frac{E}{L}$$

for which the general solution is

$$I(t) = \frac{E}{R} + c_1 \exp\left(-\frac{R}{L}t\right).$$

The initial condition $I(0) = 0$ requires that

$$0 = \frac{E}{R} + c_1,$$

so that finally

$$I(t) = \frac{E}{R}\left[1 - \exp\left(\frac{-R}{L}t\right)\right].$$

EXAMPLE (b):   In the RL circuit with the schematic diagram shown in Figure 30, let the switch be closed at $t = 0$. At some later time, $t = t_0$, the direct current element, the constant E, is to be removed from the circuit, which remains closed. Find the current for all $t > 0$.

The initial value problem to be solved is

$$LI'(t) + RI(t) = E(t); \qquad I(0) = 0, \tag{6}$$

$$E(t) = E[1 - \alpha(t - t_0)]. \tag{7}$$

Let the Laplace transform of $I(t)$ be $i(s)$. We know the transform of $E(t)$. Therefore we obtain the transformed problem

$$sLi(s) + Ri(s) = \frac{E}{s}[1 - \exp(-t_0 s)], \tag{8}$$

from which

$$i(s) = \frac{E[1 - \exp(-t_0 s)]}{s(sL + R)}. \tag{9}$$

Now

$$\frac{1}{s(sL + R)} = \frac{1}{R}\left(\frac{1}{s} - \frac{1}{s + RL^{-1}}\right),$$

so

$$i(s) = \frac{E}{R}\left(\frac{1}{s} - \frac{1}{s + RL^{-1}}\right)[1 - \exp(-t_0 s)].$$

Therefore

$$I(t) = \frac{E}{R}\left[1 - \alpha(t - t_0) - \exp\left(-\frac{R}{L}t\right) + \exp\left\{-\frac{R}{L}(t - t_0)\right\}\alpha(t - t_0)\right]. \tag{10}$$

The student should verify (10) and show that it can be written

$$\text{For } 0 \leqq t \leqq t_0, \qquad I(t) = \frac{E}{R}\left[1 - \exp\left(-\frac{Rt}{L}\right)\right]; \tag{11}$$

$$\text{For } t > t_0, \qquad I(t) = I(t_0)\exp\left[-\frac{R}{L}(t - t_0)\right]. \tag{12}$$

86. Simple networks

Systems of equations occur naturally in the application of Kirchhoff's laws to electric networks. We consider in this section two extremely simple networks to indicate how the techniques of Chapter 13 can be applied. In each example we will first use matrix techniques to determine the solution and then solve the problem again using the Laplace transformation.

EXAMPLE (a):   Determine the character of the currents $I_1(t)$, $I_2(t)$, and $I_3(t)$ in the network having the schematic diagram shown in Figure 31, under the assumption that when the switch is closed the currents are each zero.

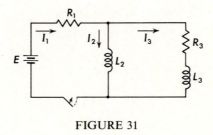

FIGURE 31

In a network, we apply Kirchhoff's laws, page 285, to obtain a system of equations to determine the currents. Since there are three dependent variables, I_1, I_2, I_3, we need three equations.

From the current law it follows that

$$I_1 = I_2 + I_3. \tag{1}$$

Application of the voltage law to the circuit on the left in Figure 31 yields

$$R_1 I_1 + L_2 I_2' = E. \tag{2}$$

Using the voltage law on the outside circuit, we get

$$R_1 I_1 + R_3 I_3 + L_3 I_3' = E. \tag{3}$$

Still another equation can be obtained from the circuit on the right in Figure 31:

$$R_3 I_3 + L_3 I_3' - L_2 I_2' = 0. \tag{4}$$

Equation (4) also follows at once from equations (2) and (3); it may be used instead of either (2) or (3).

We wish to obtain the currents from the initial value problem consisting of equations (1), (2), and (3) and the conditions $I_1(0) = 0$, $I_2(0) = 0$, and $I_3(0) = 0$. One of the three initial conditions is redundant because of equation (1).

If we eliminate I_1 from equations (1), (2), (3) we can write

$$I_2' = -\frac{R_1}{L_2} I_2 - \frac{R_1}{L_2} I_3 + \frac{E}{L_2},$$

$$I'_3 = -\frac{R_1}{L_3}I_2 - \frac{R_1 + R_3}{L_3}I_3 + \frac{E}{L_3},$$

or in matrix notation,

$$\begin{pmatrix} I_2 \\ I_3 \end{pmatrix}' = \begin{pmatrix} -\dfrac{R_1}{L_2} & -\dfrac{R_1}{L_2} \\[2ex] -\dfrac{R_1}{L_3} & -\dfrac{R_1 + R_3}{L_3} \end{pmatrix} \begin{pmatrix} I_2 \\ I_3 \end{pmatrix} + \begin{pmatrix} \dfrac{E}{L_2} \\[2ex] \dfrac{E}{L_3} \end{pmatrix} \qquad (5)$$

The characteristic equation of the matrix of system (5) is therefore

$$\begin{vmatrix} -\dfrac{R_1}{L_2} - m & -\dfrac{R_1}{L_2} \\[3ex] -\dfrac{R_1}{L_3} & -\dfrac{R_1 + R_3}{L_3} - m \end{vmatrix} = 0,$$

or

$$\Delta = L_2 L_3 m^2 + (R_1 L_2 + R_3 L_2 + R_1 L_3)m + R_1 R_3 = 0. \qquad (6)$$

We are interested in the factors of the characteristic polynomial Δ. Equation (6) has no positive roots. The discriminant of Δ is

$$(R_1 L_2 + R_3 L_2 + R_1 L_3)^2 - 4 L_2 L_3 R_1 R_3$$

and may be written

$$(R_1 L_2)^2 + 2R_1 L_2(R_3 L_2 + R_1 L_3) + (R_3 L_2 + R_1 L_3)^2 - 4 L_2 L_3 R_1 R_3,$$

which equals

$$(R_1 L_2)^2 + 2R_1 L_2(R_3 L_2 + R_1 L_3) + (R_3 L_2 - R_1 L_3)^2$$

and is therefore positive. Thus we see that equation (6) has two distinct negative roots. Call them $(-a_1)$ and $(-a_2)$. It follows that

$$\Delta = L_2 L_3(m + a_1)(m + a_2)$$

and that the eigenvalues of the matrix of system (5) are $(-a_1)$ and $(-a_2)$. Corresponding to these eigenvalues, we obtain eigenvectors

$$\begin{pmatrix} R_1 \\ a_1 L_2 - R_1 \end{pmatrix} \quad \text{and} \quad \begin{pmatrix} R_1 \\ a_2 L_2 - R_1 \end{pmatrix}. \qquad (7)$$

It follows that the general solution of the homogeneous system associated with (5) is

$$\begin{pmatrix} I_2 \\ I_3 \end{pmatrix}_c = c_1 \begin{pmatrix} R_1 \\ a_1 L_2 - R_1 \end{pmatrix} e^{-a_1 t} + c_2 \begin{pmatrix} R_1 \\ a_2 L_2 - R_1 \end{pmatrix} e^{-a_2 t}.$$

It should be clear in system (5) that there exists a particular solution of the form

$$\begin{pmatrix} I_2 \\ I_3 \end{pmatrix}_p = \begin{pmatrix} B_1 \\ B_2 \end{pmatrix},$$

where B_1 and B_2 are constants. Direct substitution into (5) yields

$$\begin{pmatrix} I_2 \\ I_3 \end{pmatrix}_p = \frac{E}{R_1} \begin{pmatrix} 1 \\ 0 \end{pmatrix}.$$

We have therefore found the general solution of system (5) to be

$$\begin{pmatrix} I_2 \\ I_3 \end{pmatrix} = c_1 \begin{pmatrix} R_1 \\ a_1 L_2 - R_1 \end{pmatrix} e^{-a_1 t} + c_2 \begin{pmatrix} R_1 \\ a_2 L_2 - R_1 \end{pmatrix} e^{-a_2 t} + \frac{E}{R_1} \begin{pmatrix} 1 \\ 0 \end{pmatrix}. \tag{8}$$

The initial conditions $I_1(0) = I_2(0) = 0$ now require

$$c_1 \begin{pmatrix} R_1 \\ a_1 L_2 - R_1 \end{pmatrix} + c_2 \begin{pmatrix} R_1 \\ a_2 L_2 - R_1 \end{pmatrix} = -\frac{E}{R_1} \begin{pmatrix} 1 \\ 0 \end{pmatrix} \tag{9}$$

as the system which must be satisfied by c_1 and c_2. The solution of equation (9) is

$$c_1 = -\frac{E(a_2 L_2 - R_1)}{R_1^2 L_2 (a_2 - a_1)} \quad \text{and} \quad c_2 = \frac{E(a_1 L_2 - R_1)}{R_1^2 L_2 (a_2 - a_1)}. \tag{10}$$

The solution of the initial value problem is given by the insertion of these constants into equation (8). Finally, the current I_1 is easily obtained as the sum of I_2 and I_3

$$I_1 = -\frac{E a_1 (a_2 L_2 - R_1)}{R_1^2 (a_2 - a_1)} e^{-a_1 t} + \frac{E a_2 (a_1 L_2 - R_1)}{R_1^2 (a_2 - a_1)} e^{-a_2 t} + \frac{E}{R_1}. \tag{11}$$

EXAMPLE (b):   Determine the current $I_1(t)$ of Example (a) by using Laplace transform techniques.

To retain the conventional symbol L for the number of henrys inductance of the circuit, we shall in this section denote by L_t the Laplace operator for which L is used in all other sections of the book.

Let $L_t\{I_k(t)\} = i_k(s)$ for each $k = 1, 2, 3$. Then the operator L_t transforms the problem of solving equations (1), (2), and (3) into the algebraic problem of solving the equations

$$i_1 - i_2 - i_3 = 0, \tag{12}$$

$$R_1 i_1 + s L_2 i_2 = \frac{E}{s}, \tag{13}$$

$$R_1 i_1 + (R_3 + s L_3) i_3 = \frac{E}{s}. \tag{14}$$

Since we desire only $i_1(s)$, let us use Cramer's rule to write the solution

$$i_1(s) = \frac{\begin{vmatrix} 0 & -1 & -1 \\ \dfrac{E}{s} & sL_2 & 0 \\ \dfrac{E}{s} & 0 & R_3 + sL_3 \end{vmatrix}}{\Delta} = \frac{E}{s} \cdot \frac{R_3 + s(L_2 + L_3)}{\Delta}, \tag{15}$$

in which

$$\Delta = \begin{vmatrix} 1 & -1 & -1 \\ R_1 & sL_2 & 0 \\ R_1 & 0 & R_3 + sL_3 \end{vmatrix},$$

or

$$\Delta = L_2 L_3 s^2 + (R_1 L_2 + R_3 L_2 + R_1 L_3)s + R_1 R_3. \tag{16}$$

It is important to recognize that this polynomial in s is the same as the characteristic polynomial obtained in Example (a) in equation (6). Therefore the remarks made there concerning the roots of Δ hold here also. That is,

$$\Delta = L_2 L_3 (s + a_1)(s + a_2),$$

where a_1 and a_2 are distinct positive real numbers. We therefore have, from (15),

$$i_1(s) = \frac{E}{s} \cdot \frac{R_3 + s(L_2 + L_3)}{L_2 L_3 (s + a_1)(s + a_2)}. \tag{17}$$

The right member of equation (17) has a partial fractions expansion

$$i_1(s) = \frac{A_0}{s} + \frac{A_1}{s + a_1} + \frac{A_2}{s + a_2}, $$

so that

$$I_1(t) = A_0 + A_1 e^{-a_1 t} + A_2 e^{-a_2 t}. \tag{18}$$

Some rather tedious algebra will determine the constants A_0, A_1, A_2 and show that equation (18) is identical to equation (11).

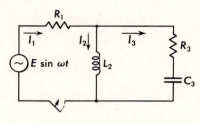

FIGURE 32

EXAMPLE (c): For the network shown in Figure 32, set up the equations for the determination of the currents I_1, I_2, I_3, and the charge Q_3. Assume that when the switch is closed all currents and charges are zero. Find the characteristic polynomial for the matrix of the resultant system.

Using Kirchhoff's laws we write the equations

$$I_1 = I_2 + I_3, \tag{19}$$

$$R_1 I_1 + L_2 \frac{dI_2}{dt} = E \sin \omega t, \tag{20}$$

$$R_1 I_1 + R_3 I_3 + \frac{1}{C_3} Q_3 = E \sin \omega t; \tag{21}$$

and the definition of current as time rate of change of charge yields

$$I_3 = \frac{dQ_3}{dt}. \tag{22}$$

Our problem consists of the four equations (19) through (22) with the initial conditions that

$$I_2(0) = 0, \qquad I_3(0) = 0, \qquad Q_3(0) = 0. \tag{23}$$

If we use equations (19) and (22) to eliminate I_3 and Q_3 from the system, we obtain

$$\frac{dI_1}{dt} = -\frac{C_3 R_1 R_3 + L_2}{C_3 L_2 (R_1 + R_3)} I_1 + \frac{1}{C_3 (R_1 + R_3)} I_2$$

$$+ \frac{E\omega}{R_1 + R_3} \cos \omega t + \frac{ER_3}{L_2(R_1 + R_3)} \sin \omega t,$$

$$\frac{dI_2}{dt} = -\frac{R_1}{L_2} I_1 + \frac{E}{L_2} \sin \omega t.$$

The matrix of the associated homogeneous system is

$$\begin{pmatrix} -\dfrac{C_3R_1R_3 + L_2}{C_3L_2(R_1 + R_3)} & \dfrac{1}{C_3(R_1 + R_3)} \\ -\dfrac{R_1}{L_2} & 0 \end{pmatrix}.$$

Thus the characteristic polynomial is

$$\begin{vmatrix} -\dfrac{C_3R_1R_3 + L_2}{C_3L_2(R_1 + R_3)} - m & \dfrac{1}{C_3(R_1 + R_3)} \\ -\dfrac{R_1}{L_2} & -m \end{vmatrix}$$

$$= m^2 + \frac{C_3R_1R_3 + L_2}{C_3L_2(R_1 + R_3)}m + \frac{R_1}{C_3L_2(R_1 + R_3)}. \tag{24}$$

EXAMPLE (d): In Example (c) obtain the Laplace transform of the initial value problem.

Let $L_t\{I_k(t)\} = i_k(s), k = 1, 2, 3,$ and $L_t\{Q_3(t)\} = q_3(s)$. Then the transformed problem becomes

$$i_1 - i_2 - i_3 = 0, \tag{25}$$

$$R_1i_1 + sL_2i_2 = \frac{E\omega}{s^2 + \omega^2}, \tag{26}$$

$$R_1i_1 + R_3i_3 + \frac{1}{C_3}q_3 = \frac{E\omega}{s^2 + \omega^2}, \tag{27}$$

$$i_3 = sq_3. \tag{28}$$

If we eliminate q_3 by substitution from equation (28) into equation (27), we obtain a system of three equations in the three unknowns i_1, i_2, i_3. The nature of the solutions of this system is governed by the determinant

$$\Delta = \begin{vmatrix} 1 & -1 & -1 \\ R_1 & sL_2 & 0 \\ R_1 & 0 & R_3 + \dfrac{1}{C_3s} \end{vmatrix}.$$

Expanding this determinant gives

$$\Delta = \frac{L_2(R_1 + R_3)}{s}\left[s^2 + \frac{C_3R_1R_3 + L_2}{C_3L_2(R_1 + R_3)}s + \frac{R_1}{C_3L_2(R_1 + R_3)}\right]. \tag{29}$$

The student should compare the quadratic polynomial in equation (29) with the characteristic polynomial of equation (24) in Example (c) and realize that the nature of the solutions of the initial value problem is dictated in each example by the same polynomial.

Exercises

1. For the *RL* circuit of Figure 30, page 286, find the current *I* if the direct current element *E* is not removed from the circuit.

   ANS.　$I = ER^{-1}[1 - \exp(-RtL^{-1})]$.

2. Solve exercise 1 if the direct-current element is replaced by an alternating-current element *E* cos *ωt*. For convenience, use the notation

$$Z^2 = R^2 + \omega^2 L^2,$$

in which *Z* is called the steady-state impedance of this circuit.

   ANS.　$I = EZ^{-2}[\omega L \sin \omega t + R \cos \omega t - R \exp(-RtL^{-1})]$.

3. Solve exercise 2, replacing *E* cos *ωt* with *E* sin *ωt*.

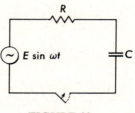

FIGURE 33

4. Figure 33 shows an *RC* circuit with an alternating-current element inserted. Assume that the switch is closed at *t* = 0, at which time *Q* = 0 and *I* = 0. Use the notation

$$Z^2 = R^2 + (\omega C)^{-2},$$

where *Z* is the steady-state impedance of this circuit. Find *I* for *t* > 0.

   ANS.　$I = EZ^{-2}[R \sin \omega t + (\omega C)^{-1} \cos \omega t - (\omega C)^{-1} \exp(-tR^{-1}C^{-1})]$.

5. In Figure 33, replace the alternating-current element with a direct-current element *E* = 50 volts and use *R* = 10 ohms, *C* = 4(10)$^{-4}$ farad. Assume that when the switch is closed (at *t* = 0) the charge on the capacitor is 0.015 coulomb. Find the initial current in the circuit and the current for *t* > 0.

   ANS.　$I(0) = 1.25$(amp), $I(t) = 1.25 \exp(-250t)$(amp).

6. In Figure 29, page 285, find *I(t)* if *E(t)* = 60 volts, *R* = 40 ohms, *C* = 5(10)$^{-5}$ farad, *L* = 0.02 henry. Assume that *I(0)* = 0, *Q(0)* = 0.

   ANS.　$I = 3000t \exp(-1000t)$(amp).

7. In exercise 6, find the maximum current.　　　ANS.　$I_{max} = 3 e^{-1}$(amp).

In exercises 8 through 11, use Figure 29, page 285, with *E(t)* = *E* sin *ωt* and with the following notations used to simplify the appearance of the formulas:

$$a = \frac{R}{2L}, \qquad b^2 = a^2 - \frac{1}{LC}, \qquad \beta^2 = \frac{1}{LC} - a^2,$$

$$\gamma = \omega L - \frac{1}{\omega C}, \qquad Z^2 = R^2 + \gamma^2.$$

The quantity Z is the steady-state impedance for an RLC circuit. In each of exercises 8 to 11, find $I(t)$ assuming that $I(0) = 0$ and $Q(0) = 0$.

8. Assume that $4L < R^2C$.
   ANS. $I = EZ^{-2}(R \sin \omega t - \gamma \cos \omega t) + \frac{1}{2}Eb^{-1}Z^{-2}[\{\gamma(a + b)$
   $- \omega R\} \exp\{-(a - b)t\} + \{\omega R - \gamma(a - b)\} \exp\{-(a + b)t\}].$

9. Assume that $R^2C < 4L$.
   ANS. $I = EZ^{-2}(R \sin \omega t - \gamma \cos \omega t)$
   $+ E\beta^{-1}Z^{-2}e^{-at}[\beta\gamma \cos \beta t - a(\gamma + 2\omega^{-1}C^{-1}) \sin \beta t].$

10. Assume that $R^2C = 4L$.
    ANS. $I = EZ^{-2}(R \sin \omega t - \gamma \cos \omega t) + E\omega^{-1}Z^{-2}e^{-at}[\gamma\omega - a(\gamma\omega + aR)t].$

11. Show that the answer to exercise 10 can be put in the form

$$I = EZ^{-2}(R \sin \omega t - \gamma \cos \omega t) + EZ^{-2}e^{-at}[\gamma + (a\gamma - R\omega)t].$$

12. In exercise 4, replace the alternating current element $E \sin \omega t$ with

$$E[\alpha(t - t_0) - \alpha(t - t_1)], \qquad t_1 > t_0 > 0.$$

Graph the new electromotive force (emf). Determine the current in the circuit.

$$\text{ANS.} \quad I(t) = \frac{E}{R}\left[\exp\left(-\frac{t - t_0}{RC}\right)\alpha(t - t_0) - \exp\left(-\frac{t - t_1}{RC}\right)\alpha(t - t_1)\right].$$

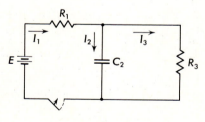

FIGURE 34

13. In Figure 34, let $E = 60$ volts, $R_1 = 10$ ohms, $R_3 = 20$ ohms, and $C_2 = 5(10)^{-4}$ farad. Determine the currents if, when the switch is closed, the capacitor carries a charge of 0.03 coulomb.
    ANS. $I_1 = 2(1 - e^{-300t})$, $I_2 = -3e^{-300t}$, $I_3 = 2 + e^{-300t}$.

14. In exercise 13, let the initial charge on the capacitor be 0.01 coulomb, but leave the rest of the problem unchanged.
    ANS. $I_1 = 2(1 + e^{-300t})$, $I_2 = 3e^{-300t}$, $I_3 = 2 - e^{-300t}$.

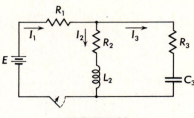

FIGURE 35

15. For the network in Figure 35, set up the equations for the determination of the
charge Q_3 and the currents I_1, I_2, I_3. Assume all four of those quantities to be
zero at time zero. Use either the Laplace transform or matrix algebra to show that
the nature of the solutions depends on the nature of the roots of the polynomial

$$C_3 L_2 (R_1 + R_3) m^2 + [C_3 (R_1 R_2 + R_2 R_3 + R_3 R_1) + L_2] m + R_1 + R_3.$$

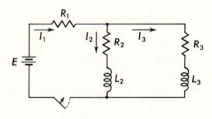

FIGURE 36

16. For the network in Figure 36 set up the equations for the determination of the
currents. Assume all currents to be zero at time zero. Use either the Laplace transform or matrix algebra to discuss the character of $I_1(t)$ without explicitly finding the
function.

The Existence and Uniqueness of Solutions

87. Preliminary remarks

The methods of Chapter 2 are strictly dependent on certain special properties (variables separable, exactness, and so on), that may or may not be possessed by an individual equation. It is intuitively plausible that no collection of methods can be found that would permit the explicit solution, in the sense of Chapter 2, of all first-order differential equations. We may seek solutions in other forms, employing infinite series or other limiting processes; we may resort to numerical approximations.

Confronted with this situation, a mathematician reacts by searching for what is known as an existence theorem. He seeks to determine conditions sufficient to insure the existence of a solution that has certain properties. In Chapter 2 we stated such a theorem, and we now wish to examine it more closely.

88. An existence and uniqueness theorem

Consider the equation of order one

$$\frac{dy}{dx} = f(x, y). \tag{1}$$

Let T denote the rectangular region defined by

$$|x - x_0| \leqq a \quad \text{and} \quad |y - y_0| \leqq b,$$

a region with the point (x_0, y_0) at its center. Let the function f in equation (1) and the function $\partial f / \partial y$ be continuous at each point in T. Then there exists an interval, $|x - x_0| \leqq h$, and a function $\phi(x)$ that have the following properties:

(a) $y = \phi(x)$ is a solution of equation (1) on the interval $|x - x_0| \leqq h$.
(b) On the interval $|x - x_0| \leqq h$, $\phi(x)$ satisfies the inequality

$$|\phi(x) - y_0| \leqq b.$$

(c) $\phi(x_0) = y_0$.
(d) $\phi(x)$ is unique on the interval $|x - x_0| \leqq h$ in the sense that it is the only function that has all of the properties (a), (b), and (c).

The interval $|x - x_0| \leqq h$ may or may not need to be smaller than the interval $|x - x_0| \leqq a$ over which conditions were imposed on f and $\partial f / \partial y$.

In rough language the theorem states that, if $f(x, y)$ is sufficiently well behaved near the point (x_0, y_0), the differential equation (1) has a solution that passes through the point (x_0, y_0) and that solution is unique near (x_0, y_0).

A proof of this fundamental theorem is presented in the next three sections. In essence the proof involves showing that a certain sequence of functions has a limit and that the limiting function is the desired solution. The sequence considered will be defined as follows:

$$y_0(x) = y_0,$$

$$y_1(x) = y_0 + \int_{x_0}^{x} f(t, y_0(t)) \, dt,$$

$$y_2(x) = y_0 + \int_{x_0}^{x} f(t, y_1(t)) \, dt, \tag{2}$$

$$\vdots$$

$$y_n(x) = y_0 + \int_{x_0}^{x} f(t, y_{n-1}(t)) \, dt.$$

So that the proof may appear more reasonable, we first consider some examples of the proof for special differential equations.

EXAMPLE (a):   Show that the sequence of functions defined in equations (2) converges to a solution for the initial value problem

$$\frac{dy}{dx} = y; \qquad x_0 = 0, y_0 = 1. \tag{3}$$

We find that

$$y_0(x) = 1,$$

$$y_1(x) = 1 + \int_0^x dt = 1 + x,$$

$$y_2(x) = 1 + \int_0^x (1 + t)\, dt = 1 + x + \frac{x^2}{2},$$

$$y_3(x) = 1 + \int_0^x \left(1 + t + \frac{t^2}{2}\right) dt = 1 + x + \frac{x^2}{2} + \frac{x^3}{3!}.$$

From the pattern that is developing, it is easy to conjecture that

$$y_n(x) = \sum_{k=0}^n \frac{x^k}{k!}.$$

Indeed, this is easy to prove by induction. Moreover, the limit of this sequence exists for every real number x because the limit is nothing more than the Maclaurin series expansion for e^x, which converges for every x. That is,

$$\phi(x) = \lim_{n \to \infty} y_n(x) = \sum_{k=0}^\infty \frac{x^k}{k!} = e^x.$$

It is a simple matter to verify that e^x is a solution to the initial value problem (3).

EXAMPLE (b):   Find a solution of the initial value problem

$$\frac{dy}{dx} = x^2; \qquad x_0 = 2, y_0 = 1. \tag{4}$$

The sequence defined in (2) above now becomes

$$y_0(x) = 1$$

$$y_1(x) = 1 + \int_2^x t^2\, dt = \frac{x^3}{3} - \frac{5}{3},$$

$$y_2(x) = 1 + \int_2^x t^2 \, dt = \frac{x^3}{3} - \frac{5}{3},$$

$$\vdots$$

$$y_n(x) = 1 + \int_2^x t^2 \, dt = \frac{x^3}{3} - \frac{5}{3}.$$

Clearly the limit of this sequence is $x^3/3 - \frac{5}{3}$, and this function is a solution of (4).

Exercises

In each of the following exercises, determine the limit of the sequence defined in (2) above. Verify that the function you obtain is a solution of the initial value problem.

1. $y' = x$; $x_0 = 2, y_0 = 1$.
2. $y' = y$; $x_0 = 0, y_0 = 2$.
3. $y' = 2y$; $x_0 = 0, y_0 = 1$.
4. $y' = x + y$; $x_0 = 0, y_0 = 1$.

89. A Lipschitz condition

We have assumed in the hypothesis of the foregoing existence theorem that the function f and its derivative $\partial f/\partial y$ are continuous in the rectangle T. Thus if (x, y_1) and (x, y_2) are points in T, the mean value theorem applies to f as a function of y. Hence, there exists a number y^* between y_1 and y_2 such that

$$f(x, y_1) - f(x, y_2) = \frac{\partial f}{\partial y}(x, y^*)(y_1 - y_2).$$

The assumption that $\partial f/\partial y$ is continuous in T allows us to assert that $\partial f/\partial y$ is bounded there. That is, there exists a number $K > 0$ such that

$$\left| \frac{\partial f}{\partial y} \right| \leq K,$$

for every point in T. Since (x, y^*) is in T, it follows that

$$|f(x, y_1) - f(x, y_2)| = \left| \frac{\partial f}{\partial y}(x, y^*) \right| \cdot |y_1 - y_2|,$$

$$|f(x, y_1) - f(x, y_2)| \leq K|y_1 - y_2|, \tag{1}$$

for every pair of points (x, y_1) and (x, y_2) in T.

The inequality (1) is called a "Lipschitz condition" for the function f. We have shown that under the hypotheses of our existence theorem, the Lipschitz condition (1) holds for every pair of points (x, y_1) and (x, y_2) in T.

In the proof in Section 90 we shall actually use the Lipschitz condition rather than the hypothesized continuity of $\partial f / \partial y$. Thus, we could restate the existence theorem in terms of condition (1) instead of assuming $\partial f / \partial y$ is continuous in T.

90. A proof of the existence theorem

One hypothesis of the existence theorem of Section 88 is that f is continuous in the rectangle T. It follows that f must be bounded in T. Let $M > 0$ be a number such that $|f(x, y)| \leq M$ for every point in T. We now take h to be the smaller of the two numbers a and b/M, and define the rectangle R to be the set of points (x, y) for which

$$|x - x_0| \leq h \quad \text{and} \quad |y - y_0| \leq b.$$

Clearly R is a subset of T.

As indicated in Section 88, we now consider the sequence of functions

$$y_n(x) = y_0 + \int_{x_0}^{x} f(t, y_{n-1}(t))\, dt \tag{1}$$

and prove the following lemma:

LEMMA 1:   *If* $|x - x_0| \leq h$ *then*

$$|y_n(x) - y_0| \leq b,$$

for $n = 1, 2, 3, \ldots$.

The proof of this lemma will be accomplished by induction. First of all, if $|x - x_0| \leq h$ we have

$$|y_1(x) - y_0| = \left| \int_{x_0}^{x} f(t, y_0)\, dt \right|$$

$$\leq M \left| \int_{x_0}^{x} dt \right|$$

$$\leq M|x - x_0|$$

$$\leq Mh$$

$$\leq b.$$

If we now assume that for $|x - x_0| \leq h$, $|y_k(x) - y_0| \leq b$, it follows that the point $[x, y_k(x)]$ is in R so that $|f(x, y_k(x))| \leq M$. Thus

$$|y_{k+1}(x) - y_0| \leq \left| \int_{x_0}^{x} f(t, y_k(t)) \, dt \right|$$

$$\leq M \left| \int_{x_0}^{x} dt \right|$$

$$\leq Mh$$

$$\leq b.$$

By induction we can now assert the validity of the lemma.

Lemma 1 may be stated in a slightly different way: if $|x - x_0| \leq h$, then the points $[x, y_n(x)]$, $n = 0, 1, 2, \ldots$, are in R. The Lipschitz condition of Section 89 may now be used to deduce the following lemma.

LEMMA 2:   *If $|x - x_0| \leq h$, then*

$$|f(x, y_n(x)) - f(x, y_{n-1}(x))| \leq K|y_n(x) - y_{n-1}(x)|,$$

for $n = 1, 2, 3, \ldots$.

We are now in a position to give an inductive proof of still another lemma.

LEMMA 3:   *If $|x - x_0| \leq h$, then*

$$|y_n(x) - y_{n-1}(x)| \leq \frac{MK^{n-1}|x - x_0|^n}{n!} \leq \frac{MK^{n-1}h^n}{n!},$$

for $n = 1, 2, 3, \ldots$.

For the case $n = 1$, we have from the proof of Lemma 1,

$$|y_1(x) - y_0| \leq M|x - x_0|.$$

Assuming that

$$|y_{n-1}(x) - y_{n-2}(x)| \leq \frac{MK^{n-2}|x - x_0|^{n-1}}{(n-1)!}, \tag{2}$$

we now must show that

$$|y_n(x) - y_{n-1}(x)| \leq \frac{MK^{n-1}|x - x_0|^n}{n!}.$$

We will prove this for the case $x_0 \leq x \leq x_0 + h$. From Lemma 2 we have

$$|y_n(x) - y_{n-1}(x)| = \left| \int_{x_0}^x [f(t, y_{n-1}(t)) - f(t, y_{n-2}(t))] \, dt \right|$$

$$\leq \int_{x_0}^x |f(t, y_{n-1}(t)) - f(t, y_{n-2}(t))| \, dt$$

$$\leq K \int_{x_0}^x |y_{n-1}(t) - y_{n-2}(t)| \, dt.$$

Using the hypothesis (2) we conclude that

$$|y_n(x) - y_{n-1}(x)| \leq \frac{MK^{n-1}}{(n-1)!} \int_{x_0}^x (t - x_0)^{n-1} \, dt,$$

or

$$|y_n(x) - y_{n-1}(x)| \leq \frac{MK^{n-1}}{n!} |x - x_0|^n. \tag{3}$$

For the case $x_0 - h \leq x \leq x_0$, the same type of argument will yield the same result. The proof of Lemma 3 is thus complete.

To utilize the results of Lemma 3 we now compare the two infinite series

$$\sum_{n=1}^{\infty} [y_n(x) - y_{n-1}(x)] \quad \text{and} \quad \sum_{n=1}^{\infty} \frac{MK^{n-1}h^n}{n!}.$$

The second of these series is an absolutely convergent series. Moreover, by Lemma 3, the second series dominates the first series. Hence, by the Weierstrass M test the series

$$\sum_{n=1}^{\infty} [y_n(x) - y_{n-1}(x)] \tag{4}$$

converges absolutely and uniformly on the interval $|x - x_0| \leq h$. If we consider the kth partial sum of the series (4)

$$\sum_{n=1}^{k} [y_n(x) - y_{n-1}(x)] = [y_1(x) - y_0(x)] + [y_2(x) - y_1(x)] + \cdots$$

$$+ [y_k(x) - y_{k-1}(x)],$$

we see that

$$\sum_{n=1}^{k} [y_n(x) - y_{n-1}(x)] = y_k(x).$$

That is, the statement that the series (4) converges absolutely and uniformly is equivalent to the statement that the sequence $y_n(x)$ converges uniformly on the interval

$$|x - x_0| \leq h.$$

If we now define

$$\phi(x) = \lim_{n \to \infty} y_n(x)$$

and recall from the definition of the sequence $y_n(x)$ that each $y_n(x)$ is continuous on $|x - x_0| \leq h$, it follows (since the convergence is uniform) that $\phi(x)$ is also continuous and

$$\phi(x) = \lim_{n \to \infty} y_n(x) = y_0 + \lim_{n \to \infty} \int_{x_0}^{x} f(t, y_{n-1}(t)) \, dt.$$

Because of the continuity of f and the uniform convergence of the sequence $y_n(x)$, we may interchange the order of the two limiting processes to show that $\phi(x)$ is a solution of the integral equation

$$\phi(x) = y_0 + \int_{x_0}^{x} f(t, \phi(t)) \, dt. \tag{5}$$

It follows immediately upon differentiation of equation (5) that $\phi(x)$ is a solution of the differential equation $dy/dx = f(x, y)$ on the interval $|x - x_0| \leq h$. Furthermore, it is clear from equation (5) that $\phi(x_0) = y_0$.

Finally, since we have shown in Lemma 1 that $|y_n(x) - y_0| \leq b$ for each n and for $|x - x_0| \leq h$, it follows that the same inequality must hold for $\phi(x) = \lim_{n \to \infty} y_n(x)$. That is, if $|x - x_0| \leq h$, then $|\phi(x) - y_0| \leq b$.

Thus we have completed the proof of parts (a), (b), and (c) of the existence theorem of Section 88.

91. A proof of the uniqueness theorem

We must now show that the function $\phi(x)$ obtained in Section 90 is unique. Suppose there is another function $Y(x)$, such that $dY/dx = f[x, Y(x)]$, $Y(x_0) = y_0$, and $|Y(x) - y_0| \leq b$ for $|x - x_0| \leq h$. Then we may write

$$Y(x) = y_0 + \int_{x_0}^{x} f(t, Y(t)) \, dt.$$

If we compare $Y(x)$ to the functions of the sequence $y_n(x)$ of Section 90 we see that

$$|Y(x) - y_n(x)| \leq \left| \int_{x_0}^{x} [f(t, Y(t)) - f(t, y_{n-1}(t))] \, dt \right|. \tag{1}$$

We shall now show that as $n \to \infty$ the integral on the right side of (1) approaches zero for $|x - x_0| \leq h$. It will then follow that $Y(x) = \lim\limits_{n \to \infty} y_n(x)$, so that finally $Y(x) \equiv \phi(x)$ on the interval $|x - x_0| \leq h$.

For any x on the interval $|x - x_0| \leq h$ it is true that $[x, Y(x)]$ and $[x, y_{n-1}(x)]$ are in the rectangle R, hence the Lipschitz condition of Section 89 will allow us to change (1) into

$$|Y(x) - y_n(x)| \leq K \int_{x_0}^{x} |Y(t) - y_{n-1}(t)| \, dt. \tag{2}$$

We now proceed by an inductive proof and limit our attention to values of x greater than x_0. (A similar argument obtains the same result for $x_0 - h \leq x \leq x_0$.) For $n = 1$, we have

$$|Y(x) - y_1(x)| \leq K \int_{x_0}^{x} |Y(t) - y_0| \, dt$$

$$\leq Kb(x - x_0).$$

We wish also to show that the assumption

$$|Y(x) - y_{n-1}(x)| \leq \frac{K^{n-1}b(x - x_0)^{n-1}}{(n-1)!}$$

leads to the conclusion

$$|Y(x) - y_n(x)| \leq \frac{K^n b(x - x_0)^n}{n!}. \tag{3}$$

This will complete an inductive argument for the relation (3). We have for $x_0 \leq x \leq x_0 + h$,

$$|Y(x) - y_n(x)| \leq \int_{x_0}^{x} |f(t, Y(t)) - f(t, y_{n-1}(t))| \, dt$$

$$\leq K \int_{x_0}^{x} |Y(t) - y_{n-1}(t)| \, dt$$

$$\leq \frac{K^n b}{(n-1)!} \int_{x_0}^{x} (t - x_0)^{n-1} \, dt$$

$$\leq \frac{K^n b}{n!} (x - x_0)^n,$$

thus completing the proof of relation (3).

For $|x - x_0| \leq h$ we have from the inequality (3),

$$|Y(x) - y_n(x)| \leq \frac{K^n b h^n}{n!}. \tag{4}$$

As $n \to \infty$ the expression on the right side of relation (4) approaches zero. Hence it follows that for $|x - x_0| \leq h$, $y_n(x) \to Y(x)$. Thus $Y(x)$ must be the same function $\phi(x)$ we obtained in Section 90. That is, the solution $\phi(x)$ is unique.

92. Other existence theorems

The existence theorem we have proved in the preceding sections for a first-order equation can be extended to equations of higher order. The simplest such extension is to equations of second order that can be written in the form

$$y'' = f(x, y, y').$$
(1)

It is natural to expect the theorem to involve continuity requirements on the function f and its partial derivatives. The theorem may be stated as follows:

THEOREM 31:   *If the function f of equation* (1) *and its partial derivatives with respect to y and y' are continuous functions in a region T defined by*

$$|x - x_0| \leq a, \qquad |y - y_0| \leq b, \qquad |y' - y_0'| \leq c,$$

then there exists an interval $|x - x_0| \leq h$ and a unique function $\phi(x)$ such that $\phi(x)$ is a solution of (1) *for all x in the interval $|x - x_0| \leq h$, $\phi(x_0) = y_0$, and $\phi'(x_0) = y_0'$.*

A proof of this theorem that is quite similar to the proof given in Sections 90 and 91 can be found in Ince.* The generalization of the theorem to equations of higher order is direct.

---

*E. L. Ince, *Ordinary Differential Equations* (London: Longmans, Green & Co., 1927), Chapter 3.

Nonlinear Equations

93. Preliminary remarks

The existence and uniqueness theorem of Chapter 15 made no distinction between linear and nonlinear differential equations. We know from our study in the earlier chapters of this book, however, that the methods we have found for actually determining solutions of a given equation often depend on the equation being a linear one. For example, in Chapter 2 we found that certain particular kinds of first-order nonlinear equations can be solved; that is, if the equation is exact, separable, homogeneous, and so on. On the other hand, if a first-order equation is linear, we have a method which can produce all possible solutions of the differential equation.

The fact is, there is no general method for solving first-order nonlinear differential equations, even if the existence of such solutions can be shown by the theorems of Chapter 15. Indeed, the determination of such solutions is often difficult if not impossible.

. In this chapter we shall briefly discuss a few of the special difficulties that arise with nonlinear equations and a few techniques that will find solutions for certain particular types of equations.

94. Factoring the left member

To give illustration to the kind of complexity that may arise in nonlinear situations, we consider first a relatively simple complication. For an equation of the form

$$f(x, y, y') = 0, \tag{1}$$

it may be possible to factor the left member. The problem of solving (1) is then replaced by two or more problems of simpler type. The latter may be capable of solution by the methods of Chapters 2 and 4.

Since y' will be raised to powers in the example and exercises, let us simplify the printing and writing by a common device, using p for y':

$$p = \frac{dy}{dx}.$$

EXAMPLE: Solve the differential equation

$$xyp^2 + (x + y)p + 1 = 0. \tag{2}$$

The left member of equation (2) is readily factored. Thus (2) leads to

$$(xp + 1)(yp + 1) = 0,$$

from which it follows that either

$$yp + 1 = 0 \tag{3}$$

or

$$xp + 1 = 0. \tag{4}$$

From equation (3) in the form

$$y\, dy + dx = 0$$

it follows that

$$y^2 = -2(x - c_1). \tag{5}$$

Equation (4) may be written

$$x\, dy + dx = 0$$

from which, for $x \neq 0$,

$$dy + \frac{dx}{x} = 0,$$

so

$$y = -\ln |c_2 x|. \tag{6}$$

We say, and it is very rough language, that the solutions of (2) are (5) and (6). Particular solutions may be made up from these solutions; they may be drawn from (5) alone, from (6) alone, or conceivably pieced together by using (5) in some intervals and (6) in others. At a point where a solution from (5) is to be joined with a solution from (6), the slope must remain continuous (see exercise 21 below), so the piecing together must take place along the line $y = x$. Note (see exercise 24) that the second derivative, which does not enter the differential equation, need not be continuous.

The existence of these three sets of particular solutions of (2), that is, solutions from (5), from (6), or from (5) and (6), leads to an interesting phenomenon in initial value problems. Consider the problem of finding a solution of (2) such that the solution passes through the point $(-\frac{1}{2}, 2)$. If the result is to be valid for the interval $-1 < x < -\frac{1}{4}$, there are two answers, which will be found in exercise 25, following. If the result is to be valid for $-1 < x < \frac{1}{2}$, there is only one answer (exercise 26), one of the two answers to exercise 25. If the result is to be valid in $-1 < x < 2$, there is only one answer (exercise 27).

Exercises

In exercises 1 through 18, find the solutions in the sense of (5) and (6) above.

1. $x^2p^2 - y^2 = 0.$ ANS. $y = c_1x, xy = c_2.$

2. $xp^2 - (2x + 3y)p + 6y = 0.$ ANS. $y = c_1x^3, y = 2x + c_2.$

3. $x^2p^2 - 5xyp + 6y^2 = 0.$ ANS. $y = c_1x^2, y = c_2x^3.$

4. $x^2p^2 + xp - y^2 - y = 0.$ ANS. $y = c_1x, x(y + 1) = c_2.$

5. $xp^2 + (1 - x^2y)p - xy = 0.$ ANS. $y = c_1 \exp(\frac{1}{2}x^2), y = -\ln|c_2x|.$

6. $p^2 - (x^2y + 3)p + 3x^2y = 0.$ ANS. $y = 3x + c_1, x^3 = 3\ln|c_2y|.$

7. $xp^2 - (1 + xy)p + y = 0.$ ANS. $y = \ln|c_1x|, x = \ln|c_2y|.$

8. $p^2 - x^2y^2 = 0.$ ANS. $x^2 = 2\ln|c_1y|, x^2 = -2\ln|c_2y|.$

9. $(x + y)^2p^2 = y^2.$ ANS. $x = y\ln|c_1y|, y(2x + y) = c_2.$

10. $yp^2 + (x - y^2)p - xy = 0.$ ANS. $x^2 + y^2 = c_1^2, y = c_2 e^x.$

11. $p^2 - xy(x + y)p + x^3y^3 = 0.$ ANS. $y(x^2 + c_1) = -2, x^3 = 3\ln|c_2y|.$

12. $(4x - y)p^2 + 6(x - y)p + 2x - 5y = 0.$

   ANS. $x + y = c_1, (2x + y)^2 = c_2(y - x).$

13. $(x - y)^2p^2 = y^2.$ ANS. $x = -y\ln|c_1y|, y(2x - y) = c_2.$

14. $xyp^2 + (xy^2 - 1)p - y = 0.$ ANS. $y^2 = 2\ln|c_1x|, x = -\ln|c_2y|.$

15. $(x^2 + y^2)^2p^2 = 4x^2y^2.$ ANS. $y^2 - x^2 = c_1y, y(3x^2 + y^2) = c_2.$

16. $(y + x)^2p^2 + (2y^2 + xy - x^2)p + y(y - x) = 0.$

   ANS. $y^2 + 2xy = c_1, y^2 + 2xy - x^2 = c_2.$

17. $xy(x^2 + y^2)(p^2 - 1) = p(x^4 + x^2y^2 + y^4).$

   ANS. $y^2(y^2 + 2x^2) = c_1, y^2 = 2x^2\ln|c_2x|.$

18. $xp^3 - (x^2 + x + y)p^2 + (x^2 + xy + y)p - xy = 0.$

   ANS. $y = c_1x, y = x + c_2, x^2 = 2(y - c_3).$

Exercises 19 through 27 refer to the example of this section. There the differential equation

$$xyp^2 + (x + y)p + 1 = 0 \tag{2}$$

was shown to have the solutions

$$y^2 = -2(x - c_1) \tag{5}$$

and

$$y = -\ln |c_2 x|. \tag{6}$$

19. Show that of the family (5) above, the only curve that passes through the point (1, 1) is $y = (3 - 2x)^{1/2}$ and that this solution is valid for $x < \frac{3}{2}$.

20. Show that of the family (6) above, the only curve that passes through the point (1, 1) is $y = 1 - \ln x$ and that this solution is valid for $0 < x$.

21. Show that the function defined by

$$y = (2 - 2x)^{1/2} \qquad \text{for } x \leqq 1,$$
$$y = -\ln x \qquad \text{for } x \geqq 1,$$

is a solution of equation (2) for all $x \neq 1$, but fails to have a derivative at $x = 1$ and is therefore not a solution there.

22. Show that if a solution of (2) is to be pieced together from (5) and (6), then the slopes of the curves must be equal where the pieces join. Show that the pieces must therefore be joined at a point on the line $y = x$.

23. Show that the function determined by

$$y = (3 - 2x)^{1/2} \qquad \text{for } x \leqq 1,$$
$$y = 1 - \ln x \qquad \text{for } 1 \leqq x$$

is a solution of equation (2) and is valid for all x. The interesting portion of this curve is shown in Figure 37.

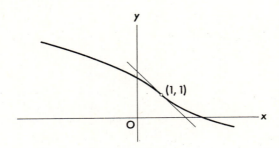

FIGURE 37

24. Show for the solution given in exercise 23 that y'' is not continuous at $x = 1$. Show that as $x \to 1^-$, $y'' \to -1$, and as $x \to 1^+$, $y'' \to +1$.

25. Find those solutions of (2) which are valid in $-1 < x < -\frac{1}{4}$ and each of which has its graph passing through the point $(-\frac{1}{2}, 2)$.

ANS. $y = (3 - 2x)^{1/2}$; and $y = 2 - \ln 2 - \ln(-x)$.

26. Find that solution of (2) which is valid for $-1 < x < \frac{1}{2}$ and has its graph passing through the point $(-\frac{1}{2}, 2)$.

ANS. $y = (3 - 2x)^{1/2}$.

27. Find that solution of (2) which is valid for $-1 < x < 2$ and has its graph passing through the point $(-\frac{1}{2}, 2)$.

ANS. $y = (3 - 2x)^{1/2}$ for $x \leq 1$.
$y = 1 - \ln x$ for $1 \leq x$.

95. Singular solutions

Let us solve the differential equation

$$y^2 p^2 - a^2 + y^2 = 0. \tag{1}$$

Here

$$yp = \pm\sqrt{a^2 - y^2}$$

so we may write

$$\frac{y\,dy}{\sqrt{a^2 - y^2}} = dx, \tag{2}$$

or

$$-\frac{y\,dy}{\sqrt{a^2 - y^2}} = dx, \tag{3}$$

or (if the division by $\sqrt{a^2 - y^2}$ cannot be effected)

$$a^2 - y^2 = 0. \tag{4}$$

From (2) it follows that

$$x = c_1 - \sqrt{a^2 - y^2}, \tag{5}$$

while from (3) that

$$x = c_2 + \sqrt{a^2 - y^2}, \tag{6}$$

and from (4) that

$$y = a \qquad \text{or} \qquad y = -a. \tag{7}$$

Graphically, the solutions (5) are left-hand semicircles with radius a and centered on the x-axis; the solutions (6) are right-hand semicircles of radius

a, centered on the x-axis. We may combine (5) and (6) into

$$(x - c)^2 + y^2 = a^2, \tag{8}$$

which we might be tempted to call the "general" solution of (1). However, from either of equations (7) we get $p = 0$, so that $y = a$ and $y = -a$ are both solutions of equation (1), but neither of these functions is a special case of (8).

We therefore see that the use of the term "general" solution for the functions defined implicitly by (8) is not consistent with the usage of the term as applied to linear differential equations. For linear equations any solution was a particular case of the general solution. It is perhaps unfortunate that the words "general solution" are used for the one-parameter family of solutions defined by (8). The particular solutions $y = a$ and $y = -a$ are called singular solutions. (It should be clear that linear equations cannot have singular solutions.)

A *singular solution* of a nonlinear first-order differential equation is any solution that

(a) is not a special case of the general solution, and
(b) is, at each of its points, tangent to some element of the one-parameter family that is the general solution.

Figure 38 shows several elements of the family of circles given by equation (8) and also shows the two lines representing $y = a$ and $y = -a$. At each point of either line, the line is tangent to an element of the family of circles. A curve which at each of its points is tangent to an element of a one-parameter family of curves is called an *envelope* of that family.

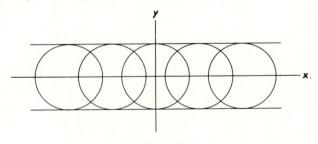

FIGURE 38

96. The c-discriminant equation

Consider the differential equation of first order,

$$f(x, y, p) = 0; \qquad p = \frac{dy}{dx}, \tag{1}$$

in which the left member is a polynomial in x, y, and p. It may not be possible to factor the left member into factors that are themselves polynomials in x, y, and p. Then the equation is said to be irreducible.

The general solution of (1) will be a one-parameter family,

$$\phi(x, y, c) = 0. \tag{2}$$

A singular solution, if it exists, for equation (1) must be an envelope of the family (2). Each point on the envelope is a point of tangency of the envelope with some element of the family (2) and is determined by the value of c that identifies that element of the family. Then the envelope has parametric equations, $x = x(c)$ and $y = y(c)$, with the c of equation (2) as the parameter. The functions $x(c)$ and $y(c)$ are as yet unknown to us. But the x and y of the point of contact must also satisfy equation (2), from which we get, by differentiation with respect to c, the equation

$$\frac{\partial \phi}{\partial x} \frac{dx}{dc} + \frac{\partial \phi}{\partial y} \frac{dy}{dc} + \frac{\partial \phi}{\partial c} = 0. \tag{3}$$

The slope of the envelope and the slope of the family element concerned must be equal at the point of contact. That slope can be determined by differentiating equation (2) with respect to x, keeping c constant. Thus it follows that

$$\frac{\partial \phi}{\partial x} + \frac{\partial \phi}{\partial y} \frac{dy}{dx} = 0. \tag{4}$$

Equations (3) and (4) both hold at the point of contact and from them it follows that

$$\frac{\partial \phi}{\partial c} = 0. \tag{5}$$

We now have two equations, $\phi = 0$ and $\partial \phi / \partial c = 0$, which must be satisfied by x, y, and c. These two equations may be taken as the desired parametric equations. They contain any envelope which may exist for the original family of curves, $\phi = 0$. Fortunately, there is no need for us to put these equations into the form $x = x(c)$, $y = y(c)$.

The equation that results from the elimination of c from the equations $\phi = 0$ and $\partial \phi / \partial c = 0$ is called the *c-discriminant equation** of the family $\phi = 0$. It is a necessary and sufficient condition that the equation

$$\phi(x, y, c) = 0, \tag{2}$$

considered as an equation in c, have at least two of its roots equal.

---

* The c-discriminant equation may contain a locus of cusps of the elements of the general solution and a locus of nodes of those elements, as well as the envelope which aroused our interest in it.

There is nothing in our work to guarantee that the c-discriminant equation, or any part of it, will yield a solution of the differential equation. To get the c-discriminant equation we need the general solution. During the process of obtaining the general solution, we find also the singular solution, if there is one.

97. The p-discriminant equation

Suppose that in the irreducible differential equation

$$f(x, y, p) = 0 \tag{1}$$

the polynomial f is of degree n in p. There will be n roots of equation (1), each yielding a result of the form

$$p = g(x, y). \tag{2}$$

If at a point (x_0, y_0) the equation (1) has, as an equation in p, all its roots distinct, then near (x_0, y_0) there will be n distinct equations of the type of equation (2). Near (x_0, y_0) the right members of these n equations will be single-valued and may satisfy the conditions of the existence theorem described in Chapter 15. But if at (x_0, y_0) equation (1) has at least two of its roots equal, then at least two of the n equations like (2) will have right members assuming the same value at (x_0, y_0). For such equations there is no region, no matter how small, surrounding (x_0, y_0) in which the right member is single-valued. Hence the existence theorem of Chapter 15 cannot be applied when equation (1) has two or more equal roots as an equation in p. Therefore we must give separate consideration to the locus of points (x, y) for which (1) has at least two of its roots equal.

The condition that equation (1) have at least two equal roots as an equation in p is that both $f = 0$ and $\partial f / \partial p = 0$. These two equations in the three variables x, y, and p are parametric equations of a curve in the xy-plane with p playing the role of parameter. The equation that results when p is eliminated from the parametric equations $f = 0$ and $\partial f / \partial p = 0$ is called the p-discriminant equation.

If an envelope of the general solution of $f = 0$ exists, it will be contained in the p-discriminant equation. No proof is included here.* For us the p-discriminant equation is useful in two ways. When a singular solution is obtained in the natural course of solving an equation, the p-discriminant equation furnishes us with a check. If none of our methods of attack leads to a general solution, then the p-discriminant equation offers functions that

---

* For more detail on singular solutions and the discriminants see E. L. Ince, *Ordinary Differential Equations* (London: Longmans, Green & Co., 1927), pp. 82–92.

may be particular (including singular) solutions of the differential equation. Then the *p*-discriminant equation should be tested for possible solutions of the differential equation. Such particular solutions make, of course, no contribution toward finding the general solution.

The *p*-discriminant equation may contain singular solutions, solutions that are not singular, and functions that are not solutions at all.

Exercises

1. For the quadratic equation

$$f = Ap^2 + Bp + C = 0,$$

with A, B, C functions of x and y, show that the *p*-discriminant obtained by eliminating p from $f = 0$ and $\partial f / \partial p = 0$ is the familiar equation

$$B^2 - 4AC = 0.$$

2. For the cubic $p^3 + Ap + B = 0$, show that the *p*-discriminant equation is $4A^3 + 27B^2 = 0$.

3. For the cubic $p^3 + Ap^2 + B = 0$, show that the *p*-discriminant equation is $B(4A^3 + 27B) = 0$.

4. Set up the condition that the equation $x^3 p^2 + x^2 yp + 4 = 0$ have equal roots as a quadratic in p. Compare with the singular solution $xy^2 = 16$.

5. Show that the condition that the equation $xyp^2 + (x + y)p + 1 = 0$ of the example of Section 94 have equal roots in p is $(x - y)^2 = 0$ and that the latter equation does not yield a solution of the differential equation. Was there a singular solution?

6. For the equation $y^2 p^2 - a^2 + y^2 = 0$ of Section 95, find the condition for equal roots in p and compare with the singular solution.

7. For the differential equation of exercise 6 show that the function defined by

$$y = [a^2 - (x + 2a)^2]^{1/2} \qquad \text{for } -3a < x \leqq -2a,$$

$$y = a \qquad\qquad\qquad \text{for } -2a \leqq x \leqq 2a,$$

$$y = [a^2 - (x - 2a)^2]^{1/2} \qquad \text{for } 2a \leqq x < 3a,$$

is a solution. Sketch the graph and show how it was pieced together from the general solution and the singular solution given in equations (5), (6), and (7) of Section 95.

In exercises 8 through 16, obtain (a) the *p*-discriminant equation and (b) those solutions of the differential equation that are contained in the *p*-discriminant.

8. $xp^2 - 2yp + 4x = 0.$ ANS. (a) $y^2 - 4x^2 = 0$; (b) $y = 2x, y = -2x.$

9. $3x^4 p^2 - xp - y = 0.$ ANS. (a) $x^2(1 + 12x^2 y) = 0$; (b) $x = 0, 12x^2 y = -1.$

10. $p^2 - xp - y = 0.$ ANS. (a) $x^2 + 4y = 0$; (b) None.

11. $p^2 - xp + y = 0.$ ANS. (a) $x^2 - 4y = 0$; (b) $x^2 - 4y = 0.$

12. $p^2 + 4x^5 p - 12x^4 y = 0.$ ANS. (a) $x^4(x^6 + 3y) = 0$; (b) $3y = -x^6.$

13. $4y^3 p^2 - 4xp + y = 0.$ ANS. (a) $(y^2 - x)(y^2 + x) = 0$; (b) Same as (a).

14. $4y^3p^2 + 4xp + y = 0$. See also exercise 13.   ANS.   (a) $(y^2 - x)(y^2 + x) = 0$;
                                                                           (b) None.

15. $p^3 + xp^2 - y = 0$.   ANS.   (a) $y(4x^3 - 27y) = 0$; (b) $y = 0$.
16. $y^4p^3 - 6xp + 2y = 0$.   ANS.   (a) $y^8(y^2 - 2x)(y^4 + 2xy^2 + 4x^2) = 0$;
                                                                (b) $y = 0$, $y^2 = 2x$.

17. For the differential equation of exercise 4 above, the general solution will be found to be $cxy + 4x + c^2 = 0$. Find the condition that this quadratic equation in c have equal roots. Compare that condition with the singular solution.

98. Eliminating the dependent variable

Suppose the equation

$$f(x, y, p) = 0; \qquad p = \frac{dy}{dx}, \tag{1}$$

is of a form such that we can readily solve it for the dependent variable y and write

$$y = g(x, p). \tag{2}$$

We can differentiate equation (2) with respect to x and, since $dy/dx = p$, get an equation

$$h\left(x, p, \frac{dp}{dx}\right) = 0 \tag{3}$$

involving only x and p. If we can solve equation (3), we will have two equations relating x, y, and p, namely, equation (2) and the solution of (3). These together form parametric equations of the solution of (1) with p now considered a parameter. Or, if p be eliminated between (2) and the solution of (3), then a solution in the nonparametric form is obtained.

EXAMPLE: Solve the differential equation

$$xp^2 - 3yp + 9x^2 = 0, \qquad \text{for } x > 0. \tag{4}$$

Rewrite (4) as

$$3y = xp + 9x^2/p. \tag{5}$$

Then differentiate both members of (5) with respect to x, using the fact that $dy/dx = p$, thus getting

$$3p = p + \frac{18x}{p} + \left(x - \frac{9x^2}{p^2}\right)\frac{dp}{dx},$$

or

$$2p\left(1 - \frac{9x}{p^2}\right) = x\left(1 - \frac{9x}{p^2}\right)\frac{dp}{dx}. \tag{6}$$

From (6) it follows that either

$$1 - \frac{9x}{p^2} = 0 \tag{7}$$

or

$$2p = x\frac{dp}{dx}. \tag{8}$$

First consider (8), which leads to

$$2\frac{dx}{x} = \frac{dp}{p}$$

so that

$$p = cx^2. \tag{9}$$

Therefore equations (4) and (9), with p as a parameter, constitute a solution of (4) looked upon as a differential equation with $p = dy/dx$.

In this example it is easy to eliminate p from equations (4) and (9), so we perform that elimination. The result is

$$x \cdot c^2x^4 - 3y \cdot cx^2 + 9x^2 = 0.$$

Since $x > 0$, we have

$$c^2x^3 - 3cy + 9 = 0,$$

or

$$3cy = c^2x^3 + 9.$$

Now put $c = 3k$ to get

$$ky = k^2x^3 + 1. \tag{10}$$

Equation (10) with k as an arbitrary constant is called the general solution of the differential equation (4).

We have yet to deal with equation (7). Note that (7) is an algebraic relation between x and p in contrast to the differential relation (8), which we have already used. We reason that the elimination of p from (7) and (4) may lead to a solution of the differential equation (4) and that the solution will not involve an arbitrary constant. From (7) it is seen that $p = 3x^{1/2}$ or $p = -3x^{1/2}$. Either of these expressions for p may be substituted into (4) and will lead to

$$y^2 = 4x^3. \tag{11}$$

It is not difficult to show that equation (11) defines two solutions of the differential equation. These solutions are not special cases of the general solution (10). They are singular solutions. Equation (11) has for its graph the envelope of the family of curves given by equation (10). The solutions defined by (11) are also easily obtained from the p-discriminant equation.

99. Clairaut's equation

Any differential equation of the form

$$y = px + f(p), \tag{1}$$

where $f(p)$ contains neither x nor y explicitly, can be solved at once by the method of Section 98. Equation (1) is called *Clairaut's equation.*

Let us differentiate both members of (1) with respect to x, thus getting

$$p = p + [x + f'(p)]p',$$

$$[x + f'(p)]\frac{dp}{dx} = 0. \tag{2}$$

Then either

$$\frac{dp}{dx} = 0 \tag{3}$$

or

$$x + f'(p) = 0. \tag{4}$$

The solution of the differential equation (3) is, of course, $p = c$, where c is an arbitrary constant. Returning to the differential equation (1), we can now write its general solution as

$$y = cx + f(c), \tag{5}$$

a result easily verified by direct substitution into the differential equation (1). Note that (5) is the equation of a family of straight lines.

Now consider equation (4). Since $f(p)$ and $f'(p)$ are known functions of p, equations (4) and (1) together constitute a set of parametric equations giving x and y in terms of the parameter p. Indeed, from equation (4) it follows that

$$x = -f'(p), \tag{6}$$

which, combined with equation (1), yields

$$y = f(p) - pf'(p). \tag{7}$$

If $f(p)$ is not a linear function of p and not a constant, it can be shown (exercises 1 and 2 below) that (6) and (7) are parametric equations of a non-linear solution of the differential equation (1). Since the general solution (5) represents a straight line for each value of c, the solution (6) and (7) cannot be a special case of (5); it is a singular solution.

EXAMPLE (a): Solve the differential equation

$$y = px + p^3. \tag{8}$$

Since (8) is a Clairaut equation, we can write its general solution

$$y = cx + c^3$$

at once.

Then using (6) and (7) we obtain the parametric equations

$$x = -3p^2, \qquad y = -2p^3, \tag{9}$$

of the singular solutions. The parameter p may be eliminated from equations (9), yielding the form

$$27y^2 = -4x^3 \tag{10}$$

for the singular solutions. See Figure 39.

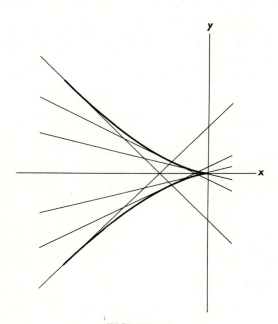

FIGURE 39

EXAMPLE (b):   Solve the differential equation

$$(x^2 - 1)p^2 - 2xyp + y^2 - 1 = 0. \tag{11}$$

Rewrite (11) as

$$x^2p^2 - 2xyp + y^2 - 1 - p^2 = 0.$$

Then it is clear that the equation is of the form

$$(y - xp)^2 - 1 - p^2 = 0 \tag{12}$$

and so could be broken up into two equations, each of Clairaut's form. Then the general solution of (11) is obtained by replacing p everywhere in it by an arbitrary constant c. That is,

$$(x^2 - 1)c^2 - 2xyc + y^2 - 1 = 0 \tag{13}$$

is the general solution of (11). The solution (13) is composed of two families of straight lines,

$$y = c_1 x + \sqrt{1 + c_1^2} \tag{14}$$

and

$$y = c_2 x - \sqrt{1 + c_2^2}. \tag{15}$$

From the p-discriminant equation for (11) we obtain at once the singular solutions defined by

$$x^2 + y^2 = 1. \tag{16}$$

Exercises

1. Let α be a parameter and prove that if $f''(\alpha)$ exists, then

$$x = -f'(\alpha), \qquad y = f(\alpha) - \alpha f'(\alpha) \tag{A}$$

is a solution of the differential equation $y = px + f(p)$. *Hint:* use dx and dy to get p in terms of α and then show that $y - px - f(p)$ vanishes identically.

2. Prove that if $f''(\alpha) \neq 0$, then (A) above is not a special case of the general solution $y = cx + f(c)$. *Hint:* show that the slope of the graph of one solution depends upon x whereas the slope of the graph of the other does not depend upon x.

In exercises 3 through 30, find the general solution and also the singular solution, if it exists.

3. $p^2 + x^3 p - 2x^2 y = 0.$ ANS. $c^2 + cx^2 = 2y$; sing. sol., $8y = -x^4$.
4. $p^2 + 4x^5 p - 12x^4 y = 0.$ ANS. $12y = c(c + 4x^3)$; sing. sol., $3y = -x^6$.
5. $2xp^3 - 6yp^2 + x^4 = 0.$ ANS. $2c^3 x^3 = 1 - 6c^2 y$; sing. sol., $2y = x^2$.
6. $p^2 - xp + y = 0.$ ANS. $y = cx - c^2$; sing. sol., $x^2 = 4y.$

7. $y = px + kp^2$.   ANS. $y = cx + kc^2$; sing. sol., $x^2 = -4ky$.

8. $x^8p^2 + 3xp + 9y = 0$.   ANS. $x^3(y + c^2) + c = 0$; sing. sol., $4x^6y = 1$.

9. $x^4p^2 + 2x^3yp - 4 = 0$.   ANS. $x^2(1 + cy) = c^2$.

10. $xp^2 - 2yp + 4x = 0$.   ANS. $x^2 = c(y - c)$; sing. sol., $y = 2x$ and $y = -2x$.

11. $3x^4p^2 - xp - y = 0$.   ANS. $xy = c(3cx - 1)$; sing. sol., $12x^2y = -1$.

12. $xp^2 + (x - y)p + 1 - y = 0$.

     ANS. $xc^2 + (x - y)c + 1 - y = 0$; sing. sol., $(x + y)^2 = 4x$.

13. $p(xp - y + k) + a = 0$.

     ANS. $c(xc - y + k) + a = 0$; sing. sol., $(y - k)^2 = 4ax$.

14. $x^6p^3 - 3xp - 3y = 0$.   ANS. $3xy = c(xc^2 - 3)$; sing. sol., $9x^3y^2 = 4$.

15. $y = x^6p^3 - xp$.   ANS. $xy = c(c^2x - 1)$; sing. sol., $27x^3y^2 = 4$.

16. $xp^4 - 2yp^3 + 12x^3 = 0$.   ANS. $2c^3y = c^4x^2 + 12$; sing. sol., $3y^2 = \pm 8x^3$.

17. $xp^3 - yp^2 + 1 = 0$.   ANS. $xc^3 - yc^2 + 1 = 0$; sing. sol., $4y^3 = 27x^2$.

18. $y = px + p^n$; for $n \neq 0$, $n \neq 1$.

     ANS. $y = cx + c^n$; sing. sol., $\left(\dfrac{x}{n}\right)^n = -\left(\dfrac{y}{n-1}\right)^{n-1}$.

19. $p^2 - xp - y = 0$.   ANS. $3x = 2p + cp^{-1/2}$ and $3y = p^2 - cp^{1/2}$.

20. $2p^3 + xp - 2y = 0$.   ANS. $x = 2p(3p + c)$ and $y = p^2(4p + c)$.

21. $2p^2 + xp - 2y = 0$.   ANS. $x = 4p \ln |pc|$ and $y = p^2[1 + 2 \ln |pc|]$.

22. $p^3 + 2xp - y = 0$.   ANS. $4x = -3p^2 + cp^{-2}$ and $2y = -p^3 + cp^{-1}$.

23. $4xp^2 - 3yp + 3 = 0$.   ANS. $2x = 3p^{-2} + cp^{-4}$ and $3y = 9p^{-1} + 2cp^{-3}$.

24. $p^3 - xp + 2y = 0$.   ANS. $x = p(c - 3p)$ and $2y = p^2(c - 4p)$.

25. $5p^2 + 6xp - 2y = 0$.   ANS. $p^3(x + p)^2 = c$ and $2y = 6xp + 5p^2$.

26. $2xp^2 + (2x - y)p + 1 - y = 0$.   ANS. $p^2x = (1 + p)^{-1} + \ln |c(1 + p)|$ and $py = 1 + (1 + p)^{-1} + 2 \ln |c(1 + p)|$.

27. $5p^2 + 3xp - y = 0$.   ANS. $p^3(x + 2p)^2 = c$ and $y = 3xp + 5p^2$.

28. $p^2 + 3xp - y = 0$.   ANS. $p^3(5x + 2p)^2 = c$ and $y = 3xp + p^2$.

29. $y = xp + x^3p^2$.   ANS. $x^2 = cp^{-4/3} - 2p^{-1}$ and $y = xp + x^3p^2$.

30. $8y = 3x^2 + p^2$.   ANS. $(p - 3x)^3 = c(p - x)$ and $8y = 3x^2 + p^2$.

100. Dependent variable missing

Consider a second-order equation,

$$f(x, y', y'') = 0, \tag{1}$$

which does not contain the dependent variable y explicitly. Let us put

$$y' = p.$$

Then

$$y'' = \frac{dp}{dx}$$

and equation (1) may be replaced by

$$f\left(x, p, \frac{dp}{dx}\right) = 0, \tag{2}$$

an equation of order one in p. If we can find p from equation (2), then y can be obtained from $y' = p$ by an integration.

EXAMPLE: Solve the equation

$$xy'' - (y')^3 - y' = 0 \tag{3}$$

of Example (b), page 12.

Because y does not appear explicitly in the differential equation (3), put $y' = p$. Then

$$y'' = \frac{dp}{dx},$$

so equation (3) becomes

$$x\frac{dp}{dx} - p^3 - p = 0.$$

Separation of variables leads to

$$\frac{dp}{p(p^2 + 1)} = \frac{dx}{x}$$

or

$$\frac{dp}{p} - \frac{p\,dp}{p^2 + 1} = \frac{dx}{x},$$

from which

$$\ln|p| - \tfrac{1}{2}\ln(p^2 + 1) + \ln|c_1| = \ln|x| \tag{4}$$

follows.

Equation (4) yields

$$c_1 p(p^2 + 1)^{-1/2} = x, \tag{5}$$

which we wish to solve for p. From (5) we conclude that

$$c_1^2 p^2 = x^2(1 + p^2),$$

$$p^2 = \frac{x^2}{c_1^2 - x^2}.$$

But $p = y'$, so we have

$$dy = \pm\frac{x\,dx}{\sqrt{c_1^2 - x^2}}. \tag{6}$$

The solutions of (6) are

$$y - c_2 = \mp (c_1^2 - x^2)^{1/2},$$

or

$$x^2 + (y - c_2)^2 = c_1^2. \tag{7}$$

Equation (7) is the desired general solution of the differential equation (3). Note that in dividing by p early in the work we might have discarded the solutions $y = k(p = 0)$, where k is constant. But (7) can be put in the form

$$c_3(x^2 + y^2) + c_4 y + 1 = 0, \tag{8}$$

with new arbitrary constants c_3 and c_4. Then the choice $c_3 = 0$, $c_4 = -1/k$ yields the solution $y = k$.

101. Independent variable missing

A second-order equation

$$f(y, y', y'') = 0 \tag{1}$$

in which the independent variable x does not appear explicitly can be reduced to a first-order equation in y and y'. Put

$$y' = p,$$

then

$$y'' = \frac{dp}{dx} = \frac{dy}{dx}\frac{dp}{dy} = p\frac{dp}{dy},$$

so equation (1) becomes

$$f\left(y, p, p\frac{dp}{dy}\right) = 0. \tag{2}$$

We try to determine p in terms of y from equation (2) and then substitute the result into $y' = p$.

EXAMPLE: Solve the equation

$$yy'' + (y')^2 + 1 = 0 \tag{3}$$

of exercise 8, page 14.

Since the independent variable does not appear explicitly in equation (3), we put $y' = p$ and obtain

$$y'' = p\frac{dp}{dy}$$

as before. Then equation (3) becomes

$$yp\frac{dp}{dy} + p^2 + 1 = 0, \qquad (4)$$

in which the variables p and y are easily separated.

From (4) it follows that

$$\frac{p\,dp}{p^2 + 1} + \frac{dy}{y} = 0,$$

from which

$$\tfrac{1}{2}\ln(p^2 + 1) + \ln|y| = \ln|c_1|,$$

so

$$p^2 + 1 = c_1^2 y^{-2}. \qquad (5)$$

We solve equation (5) for p and find that

$$p = \pm\frac{(c_1^2 - y^2)^{1/2}}{y}.$$

Therefore

$$\frac{dy}{dx} = \pm\frac{(c_1^2 - y^2)^{1/2}}{y}$$

or

$$\pm y(c_1^2 - y^2)^{-1/2}\,dy = dx.$$

Then

$$\mp(c_1^2 - y^2)^{1/2} = x - c_2$$

from which we obtain the final result

$$(x - c_2)^2 + y^2 = c_1^2.$$

Exercises

1. $y'' = x(y')^3$.   ANS. $x = c_1\sin(y + c_2)$.

2. $x^2 y'' + (y')^2 - 2xy' = 0$; when $x = 2$, $y = 5$, $y' = -4$.
       ANS. $y = \tfrac{1}{2}x^2 + 3x - 3 + 9\ln(3 - x)$.

3. $x^2 y'' + (y')^2 - 2xy' = 0$; when $x = 2$, $y = 5$, $y' = 2$.   ANS. $x^2 = 2(y - 3)$.

4. $yy'' + (y')^2 = 0$.   ANS.   See exercise 20, page 15.

5. $y^2 y'' + (y')^3 = 0$.   ANS. $x = c_1 y - \ln|c_2 y|$.

6. $(y + 1)y'' = (y')^2$.   ANS. $y + 1 = c_2\,e^{c_1 x}$.

7. $2ay'' + (y')^3 = 0$.   ANS. $(y - c_2)^2 = 4a(x - c_1)$.

8. Do exercise 7 by another method.

9. $xy'' = y' + x^5$; when $x = 1, y = \frac{1}{2}, y' = 1$. ANS. $24y = x^6 + 9x^2 + 2$.

10. $xy'' + y' + x = 0$; when $x = 2, y = -1, y' = -\frac{1}{2}$. ANS. $y = -\frac{1}{4}x^2 + \ln(\frac{1}{2}x)$.

11. $y'' = 2y(y')^3$. ANS. $y^3 = 3(c_2 - x - c_1 y)$.

12. $yy'' + (y')^3 - (y')^2 = 0$. ANS. $x = y - c_1 \ln|c_2 y|$.

13. $y'' + \beta^2 y = 0$. Check your result by solving the equation in two ways.

14. $yy'' + (y')^3 = 0$. ANS. $x = c_1 + y \ln|c_2 y|$.

15. $y'' \cos x = y'$. ANS. $y = c_2 + c_1 \ln(1 - \sin x)$.

16. $y'' = x(y')^2$; when $x = 2, y = \frac{1}{4}\pi, y' = -\frac{1}{4}$. ANS. $x = 2 \cot y$.

17. $y'' = x(y')^2$; when $x = 0, y = 1, y' = \frac{1}{2}$. ANS. $y = 1 + \frac{1}{2}\ln\dfrac{2+x}{2-x}$.

18. $y'' = -e^{-2y}$; when $x = 3, y = 0, y' = 1$. ANS. $y = \ln(x - 2)$.

19. $y'' = -e^{-2y}$; when $x = 3, y = 0, y' = -1$. ANS. $y = \ln(4 - x)$.

20. $2y'' = \sin 2y$; when $x = 0, y = \pi/2, y' = 1$. ANS. $x = -\ln(\csc y + \cot y)$.

21. $2y'' = \sin 2y$; when $x = 0, y = -\pi/2, y' = 1$. ANS. $x = \ln(-\csc y - \cot y)$.

22. Show that if you can perform the integrations encountered, then you can solve any equation of the form $y'' = f(y)$.

23. $x^3 y'' - x^2 y' = 3 - x^2$. ANS. $y = x^{-1} + x + c_1 x^2 + c_2$.

24. $y'' = (y')^2$. ANS. $y = -\ln|c_2(c_1 - x)|$; or $x = c_1 + c_3 e^{-y}$.

25. $y'' = e^x(y')^2$. ANS. $c_1 y + c_2 = -\ln|c_1 e^{-x} - 1|$.

26. $2y'' = (y')^3 \sin 2x$; when $x = 0, y = 1, y' = 1$. ANS. $y = 1 + \ln(\sec x + \tan x)$.

27. $x^2 y'' + (y')^2 = 0$. ANS. $c_1^2 y = c_1 x + \ln|c_2(c_1 x - 1)|$.

28. $y'' = 1 + (y')^2$. ANS. $e^y \cos(x + c_1) = c_2$.

29. Do exercise 28 by another method.

30. $y'' = [1 + (y')^2]^{3/2}$. Solve in three ways, by considering the geometric significance of the equation, and by the methods of this chapter.

31. $yy'' = (y')^2[1 - y' \sin y - yy' \cos y]$. ANS. $x = c_1 \ln|c_2 y| - \cos y$.

32. $(1 + y^2)y'' + (y')^3 + y' = 0$. ANS. $x = c_2 + c_1 y - (1 + c_1^2) \ln|y + c_1|$.

33. $[yy'' + 1 + (y')^2]^2 = [1 + (y')^2]^3$. ANS. $(y - c_1)^2 + (x - c_2)^2 = c_1^2$.

34. $x^2 y'' = y'(2x - y')$; when $x = -1, y = 5, y' = 1$.
ANS. $2y - 1 = (x - 2)^2 + 8 \ln(x + 2)$.

35. $x^2 y'' = y'(3x - 2y')$. ANS. $2y = x^2 + c_2 - c_1 \ln|x^2 + c_1|$.

36. $xy'' = y'(2 - 3xy')$. ANS. $3y = c_2 + \ln|x^3 + c_1|$.

37. $x^4 y'' = y'(y' + x^3)$; when $x = 1, y = 2, y' = 1$.
ANS. $y = 1 + x^2 - \ln\left(\dfrac{1 + x^2}{2}\right)$.

38. $y'' = 2x + (x^2 - y')^2$. ANS. $3y = x^3 + c_2 - 3\ln|x + c_1|$.

39. $(y'')^2 - 2y'' + (y')^2 - 2xy' + x^2 = 0$; when $x = 0, y = \frac{1}{2}$ and $y' = 1$.
ANS. $2y = 1 + x^2 + 2\sin x$.

40. $(y'')^2 - xy'' + y' = 0$. ANS. General solution: $2y = c_1 x^2 - 2c_1^2 x + c_2$; family of singular solutions: $12y = x^3 + k$.

41. $(y'')^3 = 12y'(xy'' - 2y')$. ANS. General solution: $y = c_1(x - c_1)^3 + c_2$; family of singular solutions: $9y = x^4 + k$.

42. $3yy'y'' = (y')^3 - 1$. ANS. $27c_1(y + c_1)^2 = 8(x + c_2)^3$.

43. $4y(y')^2 y'' = (y')^4 + 3$. ANS. $256c_1(y - c_1)^3 = 243(x - c_2)^4$.

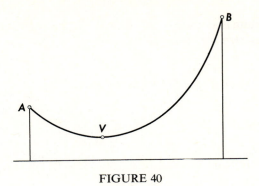

FIGURE 40

102. The catenary

Let a cable of uniformly distributed weight w (lb/ft) be suspended between two supports at points A and B as indicated in Figure 40. The cable will sag and there will be a lowest point V as indicated in the figure. We wish to determine the curve formed by the suspended cable. That curve is called the *catenary*.

Choose coordinate axes as shown in Figure 41, the y-axis vertical through the point V and the x-axis horizontal and passing at a distance y_0 (to be chosen later) below V. Let s represent length (ft) of the cable measured from V to the variable point P with coordinates (x, y). Then the portion of the cable from V to P is subject to the three forces shown in Figure 41. Those forces are: (a) the gravitational force ws (lb) acting downward through the center of gravity of the portion of the cable from V to P, (b) the tension T_1 (lb) acting tangentially at P, and (c) the tension T_2 (lb) acting horizontally (again tangentially) at V. The tension T_1 is a variable; the tension T_2 is constant.

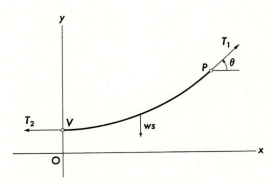

FIGURE 41

Since equilibrium is assumed, the algebraic sum of the vertical components of these forces is zero and the algebraic sum of the horizontal components of these forces is also zero. Therefore, if θ is the angle of inclination, from the horizontal, of the tangent to the curve at the point (x, y), we have

$$T_1 \sin \theta - ws = 0 \tag{1}$$

and

$$T_1 \cos \theta - T_2 = 0. \tag{2}$$

But $\tan \theta$ is the slope of the curve of the cable, so

$$\tan \theta = \frac{dy}{dx}. \tag{3}$$

We may eliminate the variable tension T_1 from equations (1) and (2) and obtain

$$\tan \theta = \frac{ws}{T_2}. \tag{4}$$

The constant T_2/w has the dimension of a length. Put $T_2/w = a$ (ft). Then equation (4) becomes

$$\tan \theta = \frac{s}{a}. \tag{5}$$

From equations (3) and (5) we see that

$$\frac{s}{a} = \frac{dy}{dx}. \tag{6}$$

Now we know from calculus that since s is the length of arc of the curve, then

$$\frac{ds}{dx} = \sqrt{1 + \left(\frac{dy}{dx}\right)^2}. \tag{7}$$

From (6) we get

$$\frac{1}{a}\frac{ds}{dx} = \frac{d^2y}{dx^2},$$

so the elimination of s yields the differential equation

$$\frac{d^2y}{dx^2} = \frac{1}{a}\sqrt{1 + \left(\frac{dy}{dx}\right)^2}. \tag{8}$$

The desired equation of the curve assumed by the suspended cable is that solution of the differential equation (8) which also satisfies the initial conditions

$$\text{when } x = 0, \qquad y = y_0 \text{ and } \frac{dy}{dx} = 0. \tag{9}$$

Equation (8) fits into either of the types studied in this chapter. It is left as an exercise for the student to solve the differential equation (8) with the conditions (9) and arrive at the result

$$y = a \cosh \frac{x}{a} + y_0 - a. \tag{10}$$

Then, of course, the sensible choice $y_0 = a$ is made, so the equation of the desired curve (the catenary) is

$$y = a \cosh \frac{x}{a}.$$

Miscellaneous Exercises

1. $x^3 p^2 + x^2 yp + 4 = 0.$ ANS. $cxy + 4x + c^2 = 0$; sing. sol., $xy^2 = 16.$

2. $6xp^2 - (3x + 2y)p + y = 0.$ ANS. $y^3 = c_1 x, 2y = x + c_2.$

3. $9p^2 + 3xy^4 p + y^5 = 0.$ ANS. $cy^3(x - c) = 1$; sing. sol., $x^2 y^3 = 4.$

4. $4y^3 p^2 - 4xp + y = 0.$ ANS. $y^4 = 4c(x - c)$; sing. sol., $y^2 = x.$

5. $x^6 p^2 - 2xp - 4y = 0.$ ANS. $x^2(y - c^2) = c$; sing. sol., $4x^4 y = -1.$

6. $5p^2 + 6xp - 2y = 0.$ ANS. $x = cp^{-3/2} - p$ and $2y = 6cp^{-1/2} - p^2.$

7. Do exercise 6 by another method.

8. $y^2 p^2 - y(x + 1)p + x = 0.$ ANS. $x^2 - y^2 = c_1, y^2 = 2(x - c_2).$

9. $4x^5 p^2 + 12x^4 yp + 9 = 0.$ ANS. $x^3(2cy - 1) = c^2$; sing. sol., $x^3 y^2 = 1.$

10. $4y^2 p^3 - 2xp + y = 0.$ ANS. $y^2 = 2c(x - 2c^2)$; sing. sol., $8x^3 = 27y^4.$

11. $p^4 + xp - 3y = 0.$ ANS. $5x = 4p^3 + cp^{1/2}$ and $15y = 9p^4 + cp^{3/2}.$

12. Do exercise 11 by another method.

13. $xp^2 + (k - x - y)p + y = 0$, the equation of exercise 5, page 14.
     ANS. $xc^2 + (k - x - y)c + y = 0$; sing. sol., $(x - y)^2 - 2k(x + y) + k^2 = 0.$

14. $x^2 p^3 - 2xyp^2 + y^2 p + 1 = 0.$
     ANS. $x^2 c^3 - 2xyc^2 + y^2 c + 1 = 0$; sing. sol., $27x = -4y^3.$

15. $16xp^2 + 8yp + y^6 = 0.$ ANS. $y^2(c^2 x + 1) = 2c$; sing. sol., $xy^4 = 1.$

16. $xp^2 - (x^2 + 1)p + x = 0.$ ANS. $x^2 = 2(y - c_1), y = \ln|c_2 x|.$

17. $p^3 - 2xp - y = 0.$ ANS. $8x = 3p^2 + cp^{-2/3}$ and $4y = p^3 - cp^{1/3}.$

18. Do exercise 17 by another method.

19. $9xy^4 p^2 - 3y^5 p - 1 = 0.$ ANS. $cy^3 = c^2 x - 1$; sing. sol., $y^6 = -4x.$

20. $x^2 p^2 - (2xy + 1)p + y^2 + 1 = 0.$
     ANS. $x^2 c^2 - (2xy + 1)c + y^2 + 1 = 0$; sing. sol., $4x^2 - 4xy - 1 = 0.$

21. $x^6 p^2 = 8(2y + xp).$ ANS. $c^2 x^2 = 8(2x^2 y - c)$; sing. sol., $x^4 y = -1.$

22. $x^2 p^2 = (x - y)^2$.   ANS. $x(x - 2y) = c_1$, $y = -x \ln |c_2 x|$.

23. $xp^3 - 2yp^2 + 4x^2 = 0$. See exercise 10 above.

                     ANS. $x^2 = 4c(y - 8c^2)$; sing., sol., $8y^3 = 27x^4$.

24. $(p + 1)^2(y - px) = 1$.

                 ANS. $(c + 1)^2(y - cx) = 1$; sing. sol., $4(x + y)^3 = 27x^2$.

25. $p^3 - p^2 + xp - y = 0$.

             ANS. $y = cx + c^3 - c^2$; sing. sol. with parametric equations,

                             $x = 2\alpha - 3\alpha^2$ and $y = \alpha^2 - 2\alpha^3$.

26. $xp^2 + y(1 - x)p - y^2 = 0$.   ANS. $xy = c_1$, $x = \ln |c_2 y|$.

27. $yp^2 - (x + y)p + y = 0$.

        ANS. $py = c \exp (p^{-1})$ and $px = y(p^2 - p + 1)$; sing. sol., $y = x$.

Index